Evolution and the Mechanisms of Decision Making

Strüngmann Forum Reports

Julia Lupp, series editor

The Ernst Strüngmann Forum is made possible through the generous support of the Ernst Strüngmann Foundation, inaugurated by Dr. Andreas and Dr. Thomas Strüngmann.

This Forum was supported by funds from the Deutsche Forschungsgemeinschaft (German Science Foundation) and the Stiftung Polytechnische Gesellschaft

Evolution and the Mechanisms of Decision Making

Edited by

Peter Hammerstein and Jeffrey R. Stevens

Program Advisory Committee:
Nick Chater, Peter Hammerstein, John M. McNamara,
Reinhard Selten, and Jeffrey R. Stevens

The MIT Press

Cambridge, Massachusetts
London, England

Series Editor: J. Lupp
Assistant Editor: M. Turner
Photographs: U. Dettmar
Lektorat: BerlinScienceWorks

The book was set in TimesNewRoman and Arial.

Library of Congress Cataloging-in-Publication Data

Hammerstein, Peter, 1949–
Evolution and the mechanisms of decision making / edited by Peter Hammerstein and Jeffrey R. Stevens.
p. cm. — (Strüngmann forum reports)
Includes bibliographical references and index.
ISBN 978-0-262-01808-1 (hardcover : alk. paper), 978-0-262-55150-2 (pb)
1. Decision making. 2. Cognition. I. Stevens, Jeffrey R., 1974– II. Title.
BF448.H3496 2012
153.8'3—dc23

2012026122

Contents

The Ernst Strüngmann Forum

Founded on the tenets of scientific independence and the inquisitive nature of the human mind, the Ernst Strüngmann Forum is dedicated to the continual expansion of knowledge. Through its innovative communication process, the Ernst Strüngmann Forum provides a creative environment within which experts scrutinize high-priority issues from multiple vantage points.

This process begins with the identification of themes. By nature, a theme constitutes a problem area that transcends classic disciplinary boundaries. It is of high-priority interest, requiring concentrated, multidisciplinary input to address the issues involved. Proposals are received from leading scientists active in their field and are selected by an independent Scientific Advisory Board. Once approved, a steering committee is convened to refine the scientific parameters of the proposal and select the participants. Approximately one year later, the central meeting, or Forum, is held to which circa forty experts are invited.

Preliminary discussion for this theme began in 2008, when Peter Hammerstein brought the initial idea to our attention. Together with Jeff Stevens, the resulting proposal was approved by the Scientific Advisory Board and from June 23–25, 2010 the steering committee was convened. The committee, comprised of Nick Chater, Peter Hammerstein, John M. McNamara, Reinhard Selten, and Jeffrey R. Stevens, identified the key issues for debate and selected the participants for the Forum, which was held in Frankfurt am Main, from June 19–24, 2011.

The activities and discourse that surround a Forum begin well before participants arrive in Frankfurt and conclude with the publication of a Strüngmann Forum Report. Throughout each stage, focused dialog is the means by which participants examine the issues anew. Often, this requires relinquishing long-established ideas and overcoming disciplinary idiosyncrasies, which have the potential to inhibit joint examination. However, when this is accomplished, a unique synergism results from which new insights emerge.

This volume conveys the synergy that arose from a group of diverse experts, each of whom assumed an active role, and it contains two types of contributions. The first provides background information to key aspects of the overall theme. Originally written in advance of the Forum, these chapters have been extensively reviewed and revised to provide current understanding on these topics. The second (Chapters 7, 12, 15, and 20) summarizes the extensive group discussions that transpired. These chapters should not be viewed as consensus documents nor are they proceedings. Instead, their goal is to transfer the essence of the discussions, expose the open questions that still remain, and highlight areas in need of future enquiry.

An endeavor of this kind creates its own unique group dynamics and puts demands on everyone who participates. Each invitee contributed not only their time and congenial personality, but a willingness to probe beyond that which is evident. For this, I extend my gratitude to all. A special word of thanks goes to the steering committee, the authors of the background papers, the reviewers of the papers, and the moderators of the individual working groups: Nick Chater, John McNamara, Sam Gosling, and Rob Boyd. To draft a report during the week of the Forum and bring it to its final form in the months thereafter is never a simple matter. For their efforts and tenacity, I am especially grateful to Ed Hagen, Kevin Gluck, Sasha Dall, and Thomas Bugynar—the rapporteurs of the discussion groups. Most importantly, I wish to extend my sincere appreciation to Peter Hammerstein and Jeff Stevens. As chairpersons of this 11th Strüngmann Forum, their guidance proved invaluable and ensured a vibrant intellectual gathering.

A communication process of this nature relies on institutional stability and an environment that encourages free thought. The generous support of the Ernst Strüngmann Foundation, established by Dr. Andreas and Dr. Thomas Strüngmann in honor of their father, enables the Ernst Strüngmann Forum to conduct its work in the service of science. The Scientific Advisory Board ensures the scientific independence of the Forum. Supplemental support for this theme was received from the German Science Foundation and the Stiftung Polytechnische Gesellschaft.

Long-held views are never easy to put aside. Yet, when this is achieved, when the edges of the unknown begin to appear and gaps in knowledge are able to be defined, the act of formulating strategies to fill such gaps becomes a most invigorating exercise. We hope that this volume will convey a sense of this lively discourse. Most importantly, we hope that this multidisciplinary examination of the cognitive mechanisms that govern decision making—mechanisms that were shaped over evolutionary time through natural selection—will initiate an alternative to the existing axiom-based theory of decision making.

Julia Lupp, Program Director
Ernst Strüngmann Forum
http://www.esforum.de

List of Contributors

Geoffrey K. Adams Center for Cognitive Neuroscience, Duke University, LSRC Building, Durham, NC 27710, U.S.A.

Benjamin Bossan Institute for Theoretical Biology, Humboldt University Berlin, 10115 Berlin, Germany

Robert Boyd School of Human Evolution and Social Change, Arizona State University, Tempe, AZ 85287-2402, U.S.A., and Santa Fe Institute, Santa Fe, NM 87501, U.S.A.

Henry Brighton Center for Adaptive Behavior and Cognition, Max Planck Institute for Human Development, 14195 Berlin, Germany

Gordon D. A. Brown Department of Psychology, University of Warwick, Coventry CV4 7AL, U.K.

Thomas Bugnyar Department of Cognitive Biology, University of Vienna, A-1090 Wien, Austria

Nick Chater Behavioural Science Group, Warwick Business School, University of Warwick, Coventry CV4 7AL, U.K.

Jan Crusius Universität zu Köln, Humanwissenschaftliche Fakultät, Sozialpsychologie I, 50931 Köln, Germany

Sasha R. X. Dall Centre for Ecology and Conservation, University of Exeter, Campus Biosciences College of Life and Environmental Science, Penryn, Cornwall TR10 9EZ, U.K.

Peter Dayan Gatsby Computational Neuroscience Unit, University College London, Alexandra House, London WC1N 3AR, U.K.

Niels J. Dingemanse Behavioural Ecology, Department Biology II, Ludwig-Maximalians University of Munich, Planegg-Martinsried, and Evolutionary Ecology of Variation Group, Max Planck Institute for Ornithology, Seewiesen, Germany

Ido Erev Faculty of Industrial Engineering and Management, Technion-Israel Institute of Technology, Technion City, Haifa 32000, Israel

Jessica C. Flack Center for Complex Systems Analysis and Collective Computation, Wisconsin Institute for Discovery, Madison, WI 53715, U.S.A., and Collective Social Computation Group, Santa Fe Institute, Santa Fe, NM 87501, U.S.A.

Simon Gächter School of Economics, University of Nottingham, Sir Clive Granger Building, University Park, Nottingham NG7 2RD, U.K.

Charles R. Gallistel Rutgers Center for Cognitive Science, Rutgers University, Piscataway, NJ 08854, U.S.A.

Kevin A. Gluck Air Force Research Laboratory, Human Effectiveness Directorate, Wright-Patterson AFB, OH 45433–7955, U.S.A.

Samuel D. Gosling Department of Psychology A8000, University of Texas, Austin, TX 78712, U.S.A.

Thomas Griffiths Department of Psychology, University of California at Berkeley, MC 1650 Berkeley, CA 94720–1650, U.S.A.

Edward H. Hagen Department of Anthropology, Washington State University, Vancouver, WA 98686-9600, U.S.A.

Peter Hammerstein Institute for Theoretical Biology, Humboldt University Berlin, 10115 Berlin, Germany, and Santa Fe Institute, Santa Fe, NM 87501, U.S.A.

Alasdair Houston School of Biological Sciences, University of Bristol, Bristol, BS8 1UG, U.K.

Keith Jensen Biological and Experimental Psychology, Queen Mary University of London, London E1 4NS, U.K.

Alex Kacelnik Department of Zoology, University of Oxford, Oxford OX1 3PS, U.K.

Tobias Kalenscher Department of Comparative Psychology, Heinrich Heine Universität Düsseldorf, 40225 Düsseldorf, Germany

Yaakov Kareev Center for the Study of Rationality, The Hebrew University of Jerusalem, Edmund Safra Campus, Jerusalem 91904, Israel

Martin Kocher Department of Economics, Ludwig-Maximilians-Universität München, 80539 München, Germany

David C. Krakauer Wisconsin Institute for Discovery, WI 53715, U.S.A., and Collective Social Computation Group, Santa Fe Institute, Santa Fe, NM 87501, U.S.A.

Jens Krause IGB, Abt 4, Biologie und Ökologie der Fische, 12587 Berlin, Germany

Robert Kurzban Department of Psychology, University of Pennsylvania, Philadelphia, PA 19104, U.S.A.

Graham Loomes Department of Economics, University of Warwick, Coventry CV4 7AL, U.K.

John M. McNamara Department of Mathematics, University of Bristol, Bristol BS8 1TW, U.K.

Thomas Mussweiler Department Psychologie, Universität zu Köln, 50931 Köln, Germany

Rosemarie Nagel Department of Economics, Universitat Pompeu Fabra and ICREA, 08005 Barcelona, Spain

Daniel Nettle Centre for Behaviour and Evolution, Institute of Neuroscience, Newcastle University, Newcastle NE2 4HH, U.K.

Danny Oppenheimer Department of Psychology, Princeton University, Green Hall, Princeton, NJ 08540, U.S.A.

John Pearson Center for Cognitive Neuroscience, Duke University, LSRC Building, Durham, NC 27710, U.S.A.

Michael L. Platt Center for Cognitive Neuroscience, Duke University, LSRC Building, Durham, NC 27710, U.S.A.

Alexandra G. Rosati Center for Cognitive Neuroscience, Duke University, Durham, NC 27708, U.S.A.

Laura Schulz Department of Brain and Cognitive Sciences, Massachusetts Institute of Technology, Cambridge, MA 02139-4307, U.S.A.

Reinhard Selten Juridicum, University of Bonn, 53113 Bonn, Germany

David W. Stephens Ecology, Evolution and Behavior, University of Minnesota, St. Paul, MN 55108, U.S.A.

Jeffrey R. Stevens Department of Psychology, University of Nebraska-Lincoln, Lincoln, NE 68588-0308, U.S.A.

Andrew R. Todd Department of Psychology, University of Iowa, Iowa City, IA 52242, U.S.A.

Peter M. Todd Cognitive Science Program, Indiana University, Bloomington, IN 47405, U.S.A.

Bernhard Voelkl Institute for Theoretical Biology, Humboldt University Berlin, 10115 Berlin, Germany

Felix Warneken Department of Psychology, Harvard University, Cambridge, MA 02138, U.S.A.

Karli K. Watson Center for Cognitive Neuroscience, Duke University, LSRC Building, Durham, NC 27710, U.S.A.

Franjo J. Weissing Theoretical Biology, Centre for Ecological and Evolutionary Studies, University of Groningen, 9747 AG Groningen, The Netherlands

William C. Wimsatt Department of Philosophy, University of Chicago, Chicago, IL 60637, U.S.A.

Max Wolf Department of Biology and Ecology of Fishes, Leibniz-Institute of Freshwater Ecology and Inland Fisheries, Berlin, Germany, and Theoretical Biology Group, Centre for Ecological and Evolutionary Studies, University of Groningen, 9751 NN Haren, The Netherlands

Alex M. Wood School of Psychological Sciences, University of Manchester, Manchester M13 9PL, U.K.

1

Six Reasons for Invoking Evolution in Decision Theory

Peter Hammerstein and Jeffrey R. Stevens

> It is in biology and psychology that economists and other social scientists will find the premises needed to fashion more predictive models.—E. O. Wilson (1998:206)

In 2012, a "Mega Millions" lottery in the United States set a world record with a jackpot of $656 million. The estimated odds of winning the jackpot were 1 in over 175,000,000. Despite the terrible odds, this lottery generated a frenzy of ticket purchases: 650 million tickets were sold in three days. Why would so many people gamble on such poor odds?

On a cold winter morning, a young black-capped chickadee (*Poecile atricapillus*) awakens from sleep, very hungry. She faces the choice of flying off to an area with a low but consistent supply of seeds or an area with the possibility of many or no seeds (Barkan 1990). Which should she choose?

In both of these examples, a decision must be made on whether to gamble on a risky option or stick with a safe option. How do organisms make these decisions? The study of games and decisions has long been guided by a philosophical discourse on concepts of rationality and their implications. This discourse has led to a large body of mathematical work and kept generations of researchers busy, but few serious attempts have been made to understand decision making in the real world. Over the last decades, however, decision theory has moved toward the sciences and developed its "taste for the facts." Research is now guided by experimental economics, cognitive psychology, behavioral biology, and—most recently—neuroscience. Despite the increasingly empirical leanings of decision science, the explanatory power of evolutionary theory has been neglected. We proposed this Strüngmann Forum to rectify this oversight, with the goal of initiating an alternative to the existing axiom-based decision theory by developing a theory of decision making founded on evolutionary principles.

Tackling this task requires a broad disciplinary base, and thus we assembled a group of researchers from a diverse range of fields—including evolutionary biology; cognitive, evolutionary, comparative, and developmental psychology;

neuroscience; computer science; economics, and philosophy—to approach this goal.

Toward a Darwinian Decision Theory

In 1654, Blaise Pascal and Pierre Fermat laid the foundation for our current theory of decision making by developing the notion of *expected value* (Daston 1995). Accordingly, a reasonable organism need only estimate the value of an outcome, weight this value by the probability of the outcome occurring, and choose the outcome offering the highest expectation. This view was formalized mathematically in the 20th century, but for the most part it has remained unchanged. Despite the beauty of the theory, its descriptive validity has been plagued by contradictory data. In numerous cases, humans and other animals systematically violate the predictions made by decision theory (Kahneman and Tversky 2000; Gigerenzer et al. 1999; Rosati and Stevens 2009): They seek risks when the theory predicts that they should avoid them. They prefer immediate rewards when they should wait. They place a premium on items in their possession when trading offers a better deal. In short, decision theory all too often fails to predict behavior.

In the rush to catalog violations of decision theory, economists and psychologists failed to propose a viable alternative to the existing theory. When they did propose alternatives, they simply patched up the existing theory (e.g., prospect theory; see Kahneman and Tversky 1979). Whenever a new violation appeared, a new patch was placed over the old theory. Repairing the old theory, however, poses two problems. First, having too many patches greatly complicates the theory and suggests that the foundation of the theory is flawed. Second, the original decision theory and its successors ignore two factors critical to understanding decision making: evolution and cognitive mechanisms.

The current mismatch between decision theory and data highlights the benefits of starting afresh in our study of decision making. For the purposes of this Forum, we anticipated initiating a *theory of decision making that rests on an evolutionary foundation* by building from first principles. Note that this theory was intended to encompass not only human decision making but rather a theory of decision making that transcends a particular species and explores the general principles of how evolution can generate decision-making agents, biological or artificial.

This Strüngmann Forum assembled experts from varied disciplines to integrate a careful understanding of evolution with precise models of cognitive mechanisms. Our aim was to develop a springboard for the construction of a *Darwinian decision theory* from which deeper insights could be gained about the functionality of cognitive design. To this end, we discussed four key components of a Darwinian approach to decision making: (a) understanding the origins of decision mechanisms, (b) exploring why these mechanisms are robust,

(c) accounting for variation between and within individuals, and (d) investigating the pressures of social life on decision making. In this introductory chapter, we explore a few general themes that draw on discussions across these four areas. In particular, we enumerate six of the many reasons why a Darwinian approach could aid our understanding of decision making.

Reasons for Invoking Evolution in Decision Theory

What would a Darwinian decision maker look like? Would she make transitive, consistent, logical decisions? Perhaps. Chater (this volume) refers to this as formal rationality (i.e., an emphasis on consistency in decision making). It seems unlikely, however, that natural selection would favor transitivity, consistency, and logic per se. Selection should favor *successful* rather than *consistent* decisions. While consistency may pay off in some situations, it may in others provide no benefits or even be costly when behavioral variation is favored (Brown et al., this volume). This idea maps onto Chater's notion of substantive rationality, which refers to decisions that require "an external standard, against which the quality of a decision can be measured." A Darwinian approach would provide substantive rationality because evolutionary fitness provides the ultimate external standard that selects for successful decision mechanisms (Stevens 2008; McNamara and Houston 2009; Hagen et al., this volume). Bearing in mind the evolutionary version of substantive rationality, we can begin to develop general principles of Darwinian decision making.

Adaptive Specialization

From a biological point of view, the evolved decision mechanisms of a given species are expected to operate very efficiently in those environments that typically occurred during the evolutionary past of the species. Since a high degree of efficiency cannot be achieved by a general-purpose device, these mechanisms have likely been tuned to the environmental circumstances under which they have been used. This *adaptive specialization* is a key reason for "invoking Darwin" in the study of decision making.[1]

Though not without controversy (see Bolhuis and Macphail 2001), the notion of adaptive specialization in animal cognition and decision making is widespread (Shettleworth 2000). From spatial memory in caching birds (Kamil et al. 1994) to temporal preferences in primates (Rosati et al. 2007), we observe examples of cognitive abilities and decisions that differ among

1 A Darwinian account is not restricted to adaptive explanations of behavior. A comprehensive Darwinian account includes influences of genetic drift, mutation, gene flow, and phylogenetic history. Though we do not consider these alternative mechanisms and constraints on evolution here, phylogenetic history is critical to the comparative approach taken by many decision scientists.

phylogenetically closely related species in ways that match the adaptive problems faced by those species in their natural environments. We, therefore, have evidence of adaptive specialization in animals.

Evolutionary psychologists use the concept of modularity to highlight the adaptive specialization of cognition in humans. Kurzban (this volume) argues that, rather than using Fodor's (1983) very strict and multifaceted definition of a cognitive module, the *evolutionary module* is defined by functional specialization. For instance, one of the classic examples proposed for evolutionary modules is cheater detection. Cosmides et al. (2010) argue that although people have difficulty solving abstract reasoning problems, they succeed when the problem is couched as an evolutionarily relevant situation, namely detecting cheaters. The authors contend that our minds did not evolve to solve content-free, abstract logical puzzles using *P*'s and *Q*'s. We did evolve, however, to avoid playing the sucker. When the logical puzzle is phrased as a social exchange, our "cheater-detection module" kicks in to solve the problem. Kurzban acknowledges the controversy surrounding the notion of an evolved module and provides justification for its utility.

In the evolutionary psychology tradition, Cosmides and Tooby (1994) liken the modular brain to a Swiss army knife. Analogously, in the judgment and decision-making field, Gigerenzer et al. (1999) endorse the "adaptive toolbox" approach of investigating the cognitive mechanisms used in specific decision contexts. Based on insights from Simon's (1956) concept of bounded rationality, the adaptive toolbox approach emphasizes fast and frugal heuristics that organisms use to solve adaptive problems. In the biological tradition, these heuristics are called "rules of thumb" (Hutchinson and Gigerenzer 2005). Though specific for a particular context rather than general purpose, these decision rules are not necessarily limited to a content domain. Adams et al. (this volume) provide behavioral and neural data on decision making in rhesus macaques (*Macaca mulatta*) which suggests that the same decision algorithms can apply across social and nonsocial domains. In particular, they argue that when macaques forage for food or search for information in their social environment, they use a comparable decision rule that simply weighs their return (on food or information) against a fixed threshold.

Work on foraging in humans has demonstrated that people use similar decision rules in very different types of foraging; for instance, physical foraging for food in an artificial fishing task and cognitive foraging for words in memory (Hutchinson et al. 2008; Wilke et al. 2009). Though applicable across content domains, these rules are by no means general purpose. Instead, they are likely tuned to the statistical structure of the environment (Hills et al. 2008). Thus, we might expect similar decision rules to operate in the foraging, social information search, and memory retrieval domains if all of these problems involve a common statistical problem to solve, such as finding objects in a clumped distribution.

Despite the advantages of specialized decision rules, disadvantages exist as well. An important difficulty faced by the toolbox approach is the *strategy selection problem* (Gigerenzer and Gaissmaier 2010; Marewski and Schooler 2011). As discussed by Kacelnik (this volume), Kurzban (this volume), and Mussweiler (this volume), having an adaptive toolbox requires selecting the correct tool for the current circumstance. How does an organism "know" which rule or strategy to use in which situation?

Kacelnik (this volume) criticizes the decision rules approach for ignoring the role of learning. He argues that, rather than selecting among specific decision rules, organisms can and do use reinforcement learning to solve many different problems. The question is: What is learned? Kacelnik assumes that the behaviors are what are learned. There is evidence, however, that strategies are learned as well. Humans, for instance, can learn to use the appropriate strategy for a given decision environment (Rieskamp and Otto 2006). Learning, therefore, leads to selecting the correct tool from the toolbox and thus a solution to the strategy selection problem. Nevertheless, more work is needed to address how organisms choose among the many possible decision strategies that they can implement. A Darwinian account could provide fruitful insight into this question.

A second (and related) critique of the decision rules approach is that we already have good evidence of a general-purpose decision-making mechanism, namely reinforcement learning (Kacelnik, this volume; Dayan, this volume). Though learning can apply across a range of circumstances, it is by no means unbounded. A Darwinian approach reminds us of the restrictions on learning. As an evolved mechanism, learning faces constraints and biases tailored toward evolutionarily relevant problems.

Learning Prepared by Evolution

Since learning plays an extremely important role in human decision making, it is often thought that evolutionary biology cannot contribute much to the study of human behavior. In fact, many scholars in the humanities and social sciences consider decision making as a process governed mainly by experience and the imprints of culture in our minds (e.g., Sahlins 1976). However, experience needs to be acquired and cultural knowledge has to be gained. Evolution certainly has shaped the fundamental learning mechanisms by which this can be achieved. More generally, for any given animal species, the learning mechanisms are likely tailored in such a way to facilitate learning the specific things that matter for survival and reproduction under ecological conditions typical for their evolution. In other words, rather than being an omnipotent tool, learning is prepared to satisfy the particular (historic) needs of a species in an efficient way. The evolutionary preparedness of learning is thus fundamental to understanding mechanisms of decision making. It explains the impact of culture on humans' choices of action as well as the difficulty a rat has in learning to associate nausea with the ring of a bell (see Hammerstein and Boyd, this volume).

The specificity argument has its limitations in that evolution cannot easily respond to emerging needs of a species by developing appropriate mechanisms entirely from scratch. It usually has to modify existing machinery, and the modified machinery will thus be subject to "optimization under phylogenetic constraints." Furthermore, there seems to be little use of loading the same brain with a great variety of different learning procedures, one for each problem. Even if almost unlimited mental resources allowed this to happen, an evolving species could then easily be trapped by "overfitting" its learning machinery to environments that are short-lived on the evolutionary timescale (Gluck et al., this volume). In the evolutionary picture of learning, there is thus space for psychological laws and principles that are valid under a wide range of conditions.

Such principles have been found in experimental psychology, but they percolated little through the disciplinary boundaries between psychology and biology. Kacelnik (this volume) rightly complains that, from its start, behavioral ecology addressed behavioral mechanisms as if experimental psychology never existed. In his view, established psychology—with its search for generality—was replaced by a search for rules of thumb which were generated and abandoned in an ad hoc fashion to interpret experiments within the narrow scope of their setting. Kacelnik considers one of the "flagships" in theoretical behavioral ecology, the marginal value theorem (Charnov 1976), and argues that predictions obtained from "sailing this ship" can be significantly improved and almost independently achieved with the findings of experimental psychology in mind.

Dayan (this volume) demonstrates nicely how the reasoning about general psychological principles can be combined with reasoning about specific adaptations. Looking at Pavlovian control of behavior, he stresses the broad range of problems that can be addressed through this mechanism. Dayan also emphasizes that animals have a rather limited repertoire of specific actions that evolution has "found to be useful" and which are triggered through Pavlovian learning. Pavlovian control thus seems to combine generality with specificity in a way that allows animals to cope quickly with variation in their typical environments.

Social animals are prepared to learn both individually and socially. Evolution tunes the balance between these mechanisms for acquiring information and made the human species uniquely dependent on social learning. Spreading rapidly all over the globe, humans needed quick responses to new environments which included the development of new tools and social arrangements. Hammerstein and Boyd (this volume) describe how learning from each other enables us to accumulate information across generations and acquire the tools, beliefs, and practices that single individuals could never have invented. Almost paradoxically, the accumulation of knowledge about adaptive practices hinges to a large extent on the fact that people often do not understand *why* culturally transmitted behavior is adaptive. The crucial point here is that children learn to do what they are supposed to do without much cognitive interference. Human learning is biased toward conformism, and children receive most of their cultural information from older individuals who have a tendency to discourage questions by the young learner. This has the advantage that we do not waste our time trying to figure out what may be difficult or impossible to understand.

Do we really find it convincing that conformist elements in human learning govern our decisions toward adaptive behavior? Conformism means that we also learn to do things that have no adaptive value and which may even be harmful. Despite this drawback, evolution can favor forms of conformism strongly enough to induce occasional maladaptive "herding effects" in populations. A population then goes through phases where most, if not all, of its members make harmful decisions (Hammerstein and Boyd, this volume). This insight from evolutionary theory is of relevance to financial markets and helps us understand what economists call the "foolishness of the crowd." Understanding the adaptive nature of and constraints on different types of learning can inform how learning is used in decision making.

Mental Adaptations to Sociality

As discussed in the previous section, human learning is evolutionarily prepared to lead us through life in a highly social environment. Without this preparedness, particularly the conformist tendencies, human culture could hardly exist, and no one would have ever built a kayak, airplane, or spaceship. Human sociality is also based on a number of mental mechanisms beyond learning, which will be described below. These are superb cognitive features but not the ones on which economic decision theory is built. Economists typically envision a merchant who has learned arithmetic at school, effectively treating information processing as a "black box" (Bugnyar et al., this volume). In our evolutionary past, however, there was no teaching of mathematics and our Stone Age ancestors did not possess an abacus, slide ruler, or pocket calculator. They had to rely on means other than calculation to make their decisions. For this reason, and guided by empirical research on child development, Warneken and Rosati (this volume) advocate an alternative way of thinking about altruism and cooperation.

In Warneken and Rosati's view, a decision to help or collaborate may be driven more strongly by intention attribution than by explicit calculations of costs and benefits. Children are able to differentiate intentions from behavioral outcomes and can tell, for example, whether a person is unwilling to hand them a toy or is unable to do so. This attribution of intentions makes it possible for the child to direct cooperative behavior preferentially to persons with cooperative intentions. The identification of intentions seems more important to social partner choice than engaging in the kind of payoff calculations around which economic theory is built. Chimpanzees (*Pan troglodytes*) share with us some of the intention-attribution skills, yet we seem to outcompete them in our ability to form *joint intentions*—an important mental adaptation to sociality (Tomasello et al. 2005).

Emotions are another prominent feature of our minds, and sociality crucially depends on some of them. Conventional decision theory is surprisingly devoid of this issue and, as Jensen (this volume) puts it, philosophers have long viewed emotions as the enemy of reason. Jensen gives his picture of how the emotions govern positive and negative concerns for others. Like Warneken and Rosati, he emphasizes our ability to detect others' goals and describes positive social concern as an emotional state that motivates the actor to reduce the suffering of others and to seek their emotional well-being. Jensen disagrees with the idea of "psychological hedonism," in which altruistic behavior is seen as a selfish attempt to obtain internal rewards with no genuine concern for the recipient of the altruistic act. In his view, the primary objective of altruism is the well-being of others and not the internal reward that comes with it.

Not all social concern is positive, as suggested by the terms "moralistic aggression," "punitive sentiments," or "moral outrage." People (including children) do punish others for causing harm even if they have not been harmed themselves. Jensen (this volume) promotes the idea that human properties like spitefulness, schadenfreude, jealousy, and envy may be important enforcers of cooperation that stabilize prosocial behaviors and may be regarded as the backbone of human prosociality. According to Jensen, much less evidence exists for such a backbone in the social life of chimpanzees. Hammerstein and Boyd (this volume) emphasize that conventional decision theory has always made humans look intellectually superior to chimpanzees but it failed to grasp the important emotions to which Jensen discusses in his chapter. For example, game theorists found a flaw in the logic of deterrence and convinced themselves that it would never work in a world of rational decision makers. Their conclusion was that in real life, deterrence can only work because humans are not rational and also do not view their opponents as rational players.

Social psychologists and anthropologists, on the other hand, have actually studied a commitment device used in deterrence: anger (Nelissen and Zeelenberg 2009). The emotion anger prevents in-depth reasoning, causes us to take great risks, and sometimes carries us into absurdly costly forms of retaliation (e.g., road rage). Hammerstein and Boyd view emotions as both promoters and inhibitors of sociality and make the point that the targets of anger are shaped by cultural evolution. The degree of violence in response to transgressions, for example, may differ dramatically between two cultures if people value personal honor more strongly in one than in the other. Anger can, in principle, stabilize any norm defined by culture, not all of which are beneficial to society. Anthropologists have indeed described a variety of norms that are deleterious, such as mortuary cannibalism (Whitfield et al. 2008).

Mussweiler (this volume) argues more generally that the complexity of social information in our everyday life can only be handled through a number of selective steps that determine (a) the information to which we should attend, (b) the information-processing tools to be used, and (c) the set of behavioral options from which one will finally be chosen. These three steps are ignored in conventional decision theory, where decision makers make use of all information they have, possess only one tool for processing it (utility maximization), and take all behavioral options into account. Evolution, however, has fabricated a fundamentally different decision maker who may look less ingenious to mathematicians but works astonishingly well in practice. To understand real decisions, we must understand how natural selection has shaped the selection steps highlighted by Mussweiler. Adams et al. (this volume) agree that social stimuli engage specialized mechanisms for the acquisition and processing of social information, but they assert that the decision rules used may be the same as those used in nonsocial situations. These perspectives highlight the need to explore the specificity of cognitive mechanisms to social situations. Nevertheless, they indicate that a "bottom-up" approach of investigating the cognitive mechanisms is critical to understanding the evolution of decision making in a social world (see also Bugnyar et al., this volume).

Error Management

We have already seen that a Darwinian decision maker may be biased in (a) the contexts in which the decision mechanisms operate adaptively, (b) the types of information that can be learned easily, (c) the way that social information is filtered and processed, and (d) the triggers of emotional responses. Error management theory (Nettle, this volume) predicts that we will see biases in the types of errors made when we make inferences about the state of the world (Haselton and Buss 2000). Not all errors are created equally.

From the evolutionary perspective, the handling of errors in the decision-making machinery must reflect the effect of these errors on an organism's survival and reproduction (fitness). When different types of errors result in different costs, we would expect to see biased decision making. Natural selection will likely favor the avoidance of even small errors if they incur high costs in terms of fitness. In contrast, seemingly large errors (e.g., a male mating with a member of the wrong species) may not face strong selective pressure if they have little impact on fitness. This is referred to as the "smoke detector principle" (Nesse 2001b). A smoke detector's job is not to minimize the numbers of errors it makes. Its job is to detect the presence of smoke in order to save lives. In doing its job, smoke detectors are biased to give an alarm when no smoke is present (a false alarm) rather than fail to give an alarm when smoke is present (a miss). Many inattentive cooks have suffered the indignity of fanning at a smoke detector after burning a meal. Despite the inconvenience and embarrassment of a false alarm, fanning a smoke detector or even evacuating a building is preferable to a smoke detector not triggering in the presence of a real fire. We will pay the minimal costs of false alarms to avoid the devastation of a miss.

Kareev (this volume) argues that this bias toward avoiding misses is a core property of human decision making based on the potential benefits of using our limited short-term memory to detect patterns in our environment. This is an appropriate strategy when the costs of misses outweigh the costs of false alarms, as is seen in the smoke detector example, as well as when avoiding predators, predicting the impact of a looming object, attributing agency to objects in the environment, and detecting signals of sexual interest from potentials mates (Nettle, this volume). When false alarm costs outweigh those of misses, however, we see the opposite pattern with greater sensitivity to false alarms. For instance, females may be biased toward accurately detecting honest signals of male parental investment (Haselton and Buss 2000). Missing an investing male is not as costly as succumbing to the false advertising of a deadbeat.

Considering the costs of errors, therefore, is critical to understanding why we observe biases in decision making. Though signal detection theory and expected utility theory also incorporate the costs of errors, Nettle (this volume) argues that the evolutionary approach via error management theory makes key predictions about the kinds of contexts in which we would expect to see biased decision making; namely, contexts with important implications for evolutionary fitness. Moreover, due to the lag in natural selection's ability to adapt organisms to their environment, error management theory can explain potential biases in situations in which no current cost differential exists, though historically strong evolutionary pressures may have resulted in divergent costs. Thus, while the occurrence of errors as such relates to mechanistic properties of the mental machinery, the management of these errors cannot be understood without exploring the evolutionary question of which errors in decision making are tolerable and which are not.

Robustness and the Mechanisms behind It

From an engineer's point of view, robustness is the ability of a system to maintain its functionality across a wide range of operational conditions. Different conditions arise, for example, from environmental variation, noisy input, sloppiness or breakdown of system components, and subversion by parasites. In the course of evolution, organisms are expected to adapt their behavior to the variety of conditions under which they have to survive and reproduce. Robustness as a concept is, therefore, of great relevance to the Darwinian approach to decision making. Flack et al. (this volume) describe the various ways in which biologists have used this concept at different levels of organization: from molecular systems to individual decision making and animal societies.

Most organisms operate in a highly variable and complex world. How do organisms make decisions when facing large temporal variation and spatial heterogeneity? To meet the enormous challenges, humans and other species must detect regularities in their new physical and social environments. At first glance, our brain seems poorly equipped for this task. Our short-term memory, for example, can hold just a few items at a time. This may seem ridiculously small for a storing device. Kareev (this volume) makes the interesting point, however, that limited short-term memory actually has a number of advantages when individuals are in search of unknown regularities. It forces us to do much of our "mental statistics" on the basis of small samples. Correlations are then likely to appear stronger than they actually are and variance is typically underestimated. This amplification of correlations makes it easier to detect the regularities and, in this sense, improves our mechanisms for exploration. There is a drawback, of course, since occasionally we will find correlations that do not exist at all (i.e., the world will look more regular than it is). But, returning to error management theory, it may be more costly for individuals to overlook important regularities than to "fantasize" a few.

Kareev (this volume) offers further arguments why it can pay to rely only on a very limited number of recent experiences in decision making. Such a self-imposed restriction can make it easier for organisms to follow changes in the environment because it helps them avoid overfitting to conditions that vary in space and time (Gluck et al., this volume). Furthermore, remembering too many past events involves the risk of always lagging far behind in fast-changing environments. Dayan (this volume) emphasizes the fact that information stored in our working memory degrades more or less gracefully as memory is taxed. The evolved working memory's design thus differs considerably from that of a human-designed computer, which stores information with extreme reliability in addressable locations until a voluntary act of deletion occurs.

Gallistel (this volume) has difficulties accepting the idea that the brain is fundamentally different from a computer in that it lacks an addressable read-write memory. He explains why such a memory is taken by computer scientists to be the foundation of any powerful computing machine. He then reviews evidence that behavior is mediated by computational information processing that deals with extensive data structures, as demonstrated, for example, by experiments that reveal the contents of the cache memory of food-caching birds (Clayton et al. 2001a). He asks whether the conceptual chasm between the computer science understanding of the essential role that a symbolic, read-write, addressable memory mechanism plays in any powerful computing machine and the neuroscience conception of memory, which is not symbolic, not read-write, and not addressable, is a problem for computer science or a problem for neuroscience.

Among the obstacles in addressing environments are the inherent uncertainties that need to be handled, which arise from our ignorance as well as from environmental change. Dayan (this volume) discusses how evolution has prepared our brains for the challenges posed by uncertainty. For example, in his view the brain "offers" itself an exploration bonus when assessing the payoffs relating to actions. This bonus supposedly drives our exploratory behavior.

But how do we know about the existence of such a quantity? Dayan interprets some experimental findings from neuroscience as keys to how the bonus manifests itself in the brain. Novel objects, for example, generate temporary activity in the dopamine system that resembles the activity triggered by unpredicted "true" rewards.

Discussing robustness at a higher level of abstraction, Dayan (this volume) sees two sources of noise that pose a major threat to it. There is noise associated with incomplete and inefficient learning in what he calls the model-free system, and there is noise inherent in the complex calculations performed by the brain with its limited potential for computation. The latter limitation becomes particularly visible when our decisions are based on internal models of reality. These models enable us, however, to predict events under changed conditions long before learning could achieve anything.

Kurzban (this volume) takes a perspective on robustness in which the brain is already equipped with a number of tools that allow us to act as if we had indeed modeled some aspects of reality. He argues, for example, that our ancestors typically encountered a spatially and temporally autocorrelated world when searching for water, food, and other important items. The a priori expectation of autocorrelation may thus be one of the innate biases that evolution has implemented in our decision-making machinery. This would explain why human predictions are often based on the implicit assumption that events come in "streaks" and are particularly likely to occur after they just occurred—the "hot-hand" phenomenon (Wilke and Barrett 2009).

A final word must be said about redundancy, an extremely important design principle used by human engineers in their efforts to create robust machinery (Flack et al., this volume). Space ships are, for example, equipped with several computers, each of which performs the same calculations. If one of them makes a calculation error, this computer can be "outvoted" by the other two machines. Dayan (this volume) argues that the robustness of animal and human decision making is also supported by a fundamentally different kind of redundancy. Instead of just having duplicates of subroutines or other mechanisms, we have different computational devices that rely on very different procedures to perform similar tasks. This kind of redundancy protects against both errors in computing and maladaptive properties of the implicit models on which these computations are based. Dayan views decision making as a permanent struggle for the "brain's attention" by different mechanisms that provide the kind of redundancy just described. Evolution seems to have equipped our minds with elements of internal competition to maintain robust decision making.

Biological Roots of Variation

If there is one thing that remains constant in decision making, it is variability. Despite the sincere wish of economists—that a single equation and set of parameters can hold for all decision makers in all situations—this is not to be. For instance, many researchers are interested in the rate of discounting that people employ (i.e., the rate at which future benefits are devalued). Frederick et al. (2002) contend that measures of discounting have not converged on a single discount rate as has happened with the speed of light. In fact, studies measuring discount rates yield values ranging from a negative discount rate (meaning a preference for delayed rewards over immediate rewards) to an infinite discount rate (meaning the strongest possible preference for immediate rewards over delayed rewards). Economists tend to sweep variation like this under the rug or treat it as noise.

Can a Darwinian approach account for variation? Though variation is a key component of natural selection, evolutionary game theory models typically predict a single best solution or small set of best solutions to an adaptive problem. We observe individual differences in behavior across a broad spectrum of species and in a wide range of contexts. In addition, within individuals we see frequent variation in behavior. Both between- and within-individual variation greatly exceeds that expected by many of these models, and understanding the role of this variation will be important in developing a theory of decision making.

To this end, biologists have turned to psychology as a field that takes variation seriously. To minimize anthropomorphism, some biologists refer to individual differences in the behavior of animals as "behavioral syndromes." Bolder researchers use the same term as applied to humans: personalities. Evolutionary models are beginning to explore adaptive accounts of the breadth of variation between individuals in their behavior (Dingemanse and Wolf, this volume; Dall et al. this volume). However, these models tend to focus on behavioral polymorphism and do not account for many interesting aspects observed in individual differences.

A key attribute of personality is not just consistent individual differences but also stability across contexts. For instance, one of the classic personality traits in animals is the bold-shy continuum, in which individuals range from novelty seeking to neophobic. Dingemanse and Wolf (this volume) describe early work on stickleback, in which fish who act aggressively toward conspecifics intruding on their territory are also more likely to approach potential predators. This correlation of boldness appears across a range of species, and Dingemanse and Wolf (this volume) highlight a growing interest in the genetic and physiological mechanisms underlying similar behavior across contexts. For instance, researchers are conducting quantitative genetic analyses which show strong genetic components, molecular genetic analyses that reveal candidate genes, neuroendocrine analyses of coping styles and stress response, and correlations of metabolic rates with behavioral differences.

In addition to mechanisms underlying personality, biologists are now beginning to ask critical questions about possible advantages of both behavioral consistency and correlated traits (Dingemanse and Wolf, this volume). Do correlated traits provide benefits to individuals or are they simply by-products of the genetic and physiological mechanisms underlying behavior? Research on spiders shows that, though bold individuals gain benefits in foraging situations, their aggressiveness imposes costs on fertility when aggressive females cannibalize potential mates before copulation (Johnson and Sih 2005). Thus, the correlated trait results in adaptive trade-offs. Would an individual be better off bold with food and shy with mates? Regardless of the adaptive benefits of correlated traits, they are real and therefore require our attention. Dall et al. (this volume) emphasize that this reality means that we cannot necessarily treat different adaptive problems as independent. As the previous example illustrates, the mating game is not independent from the predator avoidance game, and this has critical implications for how we model behavior. Understanding the adaptive value of and constraints on behavioral consistency and correlations will offer key insight into the evolution of decision making.

Individual differences provide one type of variability, but we also observe variation within individuals. Brown et al. (this volume), suggest that within-individual variation can result from noise, context, mood, life span changes, and prior experience. Noise in behavior can provide benefits when individuals want to exhibit unpredictable behavior. This can occur when trying to avoid predators or to make a credible threat of irrational behavior. Though some decisions appear rather capricious, a careful examination of the situation may highlight important context-specific predictors of behavior. Facebook notwithstanding, we often behave differently among family, friends, and strangers. Similarly, members of other species decide differently with a (potential) mate, rival, or dominant present. Even in the same context, our moods can have critical influences our actions. Brown et al. (this volume) describe how positive moods and pessimism shape decision making in humans and other animals. For instance, inducing a positive mood shifts people's risky decision making away from focusing on probabilities of reward and toward focusing on outcomes (Nygren et al. 1996).

Variation in decision making also occurs over the life span. Early in life, juvenile animals and human children demonstrate changes in decision making as various cognitive capacities come online during development (Jensen, this volume; Warneken and Rosati, this volume). At the other end of the continuum, we see modifications in decision making as adults age. Brown et al. (this volume) review research which suggests that cognitive aging results in increases in risk avoidance, patience for future rewards, weighting of losses compared to gains, and altruistic behavior. They also highlight the role of prior experience in decision making. The decision-by-sampling approach accounts for how sampling from the distribution of prior experience via long-term memory can influence decisions (Stewart et al. 2006). For instance, risky choices for mortality-related decisions correlate with the risk of mortality faced in a given participant's country (Olivola and Sagara 2009), suggesting that risky choices match the distribution of experience in the world. Thus, individuals with different experiences, or more specifically different memories, will exhibit different choices.

The study of variation from a biological approach remains in its infancy. At the moment, psychology has a lot to offer to an evolutionary account of variation in behavior both between and within individuals. Despite the standard tendency to "control for" this variation, the psychology of individual differences will be critical in informing a Darwinian approach to decision making.

Concluding Remarks

After describing several reasons for invoking evolution in decision theory, we now discuss at a more general level to what extent this helps us understand the mechanisms of decision making. Let us first play the devil's advocate and question the importance of marrying evolution with cognitive science. Most of the advancements in behavioral ecology, for example, were made by explicitly ignoring cognition and treating evolution as if natural selection acted directly on behavioral traits (the so-called "behavioral gambit"; Fawcett et al. 2012). Research programs in this field were particularly successful because they used the shortcut of circumventing the nitty-gritty of cognitive machinery. We must admit that using this shortcut has its rationale. Evolutionary theory cannot predict mechanisms as such, since in principle many different mechanisms can serve as a tool for solving the same problem. To use an analogy from engineering, there are many ways to design a clock, but all that a well-engineered clock can tell us with high accuracy is what time of the day it is.

As convincing as this multiplicity argument may sound in defense of classical research programs, it is also quite misleading. We use our wristwatch at different temperatures, for example, and may even leave it on while swimming. The mechanism operating the watch must therefore tolerate changes in temperature and pressure and continue to work adequately while being submersed in water. More generally, the more we know about the conditions under which a mechanism has to operate, the better we are able to reflect necessary specifications for its design. So, despite the fact that evolutionary theory cannot simply predict the entire mechanisms of decision making, it can inform us about fundamental properties that the evolved mechanisms can be expected to possess. As the chapters of this volume document nicely, these properties include (a) biases in learning, error management, and information usage, (b) robust responses to variation in the environment, (c) variation within and between decision makers, and (d) specializations for coping with complex social situations.

It is also important to state properties that are unlikely to exist. For example, one would not expect consistency to be a general property of evolved decision mechanisms (see, however, Chater as well as Hagen et al., both this volume). An evolutionary theory of decision mechanisms, therefore, strongly undermines the approach that dominated decision theory in economics for more than the last hundred years. Research combining evolution and cognition does indeed give us good reasons to knock economic decision theory off its pedestal.

There are two more important aspects of evolutionary analysis which deserve general attention. One is the phylogenetic approach in which ancestral mechanisms are considered as starting points from which decision mechanisms evolved. Studying decision making in phylogenetically closely related species such as chimpanzees (Jensen, this volume; Warneken and Rosati, this volume) can provide unique insights into the human condition. Humans and chimpanzees differ, for example, in how they learn from others. However, chimpanzees do show sophisticated forms of social cognition that offer a foundation for understanding our own social decision making. The second important aspect of evolutionary analysis is the comparative approach; that is the study of how different species cope with similar kinds of problems. The comparative approach offers a glimpse into how the environment shapes decision mechanisms by comparing decisions across species that both face similar environmental pressures and are adapted to different environments. For instance, from honeybees to hummingbirds and locusts to starlings, we see similar effects of context on decision making that violate classical decision theory (Rosati and Stevens 2009; Hagen et al., this volume). The ubiquity of these behaviors across species suggests that natural selection has shaped decision mechanisms to solve a widely applicable problem. On the other hand, differences in environmental pressures can be used to predict differences in decision making, such as when foraging ecology matches temporal preferences in chimpanzees and bonobos (Rosati et al. 2007).

In summary, evolutionary theory is, of course, far from being anything like an omnipotent explanatory device. It can shed light on why all sorts of biases and apparently odd effects exist in human decision making, why *Homo sapiens* is far from being anything like a "relative" of *Homo economicus*, and why we are nevertheless quite successful in addressing our everyday problems. From an evolutionary perspective there seems to be a logic behind decision making in humans and animals, but it is a logic that makes individuals successful in real life without caring about axioms of rationality. We thus see the contours of a new decision theory and wish to merge cognition and evolution further in order to root this theory firmly in empirical grounds and make sense of the facts.

Acknowledgments

On behalf of all of the participants, we wish to thank the Ernst Strüngmann Forum for their generous funding of this Forum and resulting publication. We are also grateful to Gerd Gigerenzer, who encouraged us to pursue this theme, and to the Santa Fe Institute, where some of the planning took place.

We are particularly indebted to the Forum's program director, Julia Lupp, and her staff, Marina Turner and Andrea Schoepski, who managed everything in a highly professional and inspiring way. Their admirable personal engagement enabled us to focus entirely on the scientific subject matter and enjoy the chairing of this Forum without having to endure any of the headaches that often accompany the necessary organization. Julia, Marina, and Andrea impressed us deeply with both their warmhearted and firm way of handling us scientists, who can be difficult at times.

The Forum also benefited greatly from contributions made by Nick Chater, John McNamara, and Reinhard Selten as members of the program advisory committee, as well as from the support received by the moderators of the four groups. Furthermore, this volume would not have come into existence without the brave efforts of the four rapporteurs—Ed Hagen, Kevin Gluck, Sasha Dall, and Thomas Bugnyar—who had to pull many strings together as they created the first drafts of group reports during a single afternoon (and long night) of the Forum while the rest of us were able to enjoy the sites and culinary delights of Frankfurt. Needless to say, we appreciate these efforts and are grateful to the help offered from everyone during the revisions of the chapters for this book.

Biological and Cognitive Prerequisites

2

Putting Mechanisms into Behavioral Ecology

Alex Kacelnik

Abstract

This chapter contrasts two approaches to the study of mechanism and function in decision making, one based on rules of thumbs or heuristics and another based on the contributions of experimental psychology and psychophysics. The former is the most frequently used by behavioral ecologists. It implements a behavioral gambit by which researchers deal with hypothetical decision problems without reference to independently known cognitive processes. Typically, cost-benefit analyses of the problem are carried out to identify adaptive solutions, and then simple rules are envisaged and tested for their level of performance. As a final step, not always followed, the properties of one or more of these rules are sought in behavioral data. The alternative approach shares the interest in the functional consequences of behavior, but shows greater subordination to empirical research on behavioral and cognitive mechanisms. In this case natural selection is seen as acting on processes that tune behavior to the environment across broad domains, such as the need for behavior to respond to causal relations and to process sensory information across exceedingly large ranges in the input. Associative learning and Weber's Law are two putative evolutionary responses to such challenges. In the second approach these independently known traits, rather than ad-hoc rules or heuristics, are considered as candidates for effecting decisions, and this can often lead to asking for the functional problem a posteriori, querying what selective pressures might have led to the presence of the trait. It is argued, based on examples from foraging research, that for a majority of decision problems investigated across vertebrates the second approach is preferable. It is also recognized that dedicated rules are preferable when the relevant information acts across generations and involves little learning, as is the case with life-history adjustments or responding to lethal threats that offer no second choices.

> What changes in evolution is the norm of reaction of the organism to the environment. —T. Dobzhansky (1937:22)

Introduction: The Dichotomy between Function and Mechanism

Behavioral ecology emerged in the 1970s as a dynamic scion of ethology, the discipline defined most clearly by Niko Tinbergen (1963) as the joint study of phylogeny, development, adaptive value, and mechanism of behavior. There is little doubt that this disciplinary branching followed changes in the theoretical Zeitgeist, rather than new techniques or major factual discoveries. The new field was characterized by a lessening of interest in development and mechanism, a shift toward a quantitative handling of both phylogeny and adaption, and greater contact with population ecology, evolutionary theory, and microeconomics. Whatever its causes, the birth of behavioral ecology, consolidated and organized intellectually through the books with that name edited by John Krebs and Nick Davies, caused by itself new theoretical advances, the development of new techniques aimed at dealing with novel questions, and the discovery of a wealth of factual information about the natural world. It was, thus, a successful shift of paradigm.

Some of Tinbergen's own work had preannounced behavioral ecology. For instance, his analysis of the adaptive significance of eggshell removal by black-headed gulls was typical of what behavioral ecology was to become (viz. Tinbergen et al. 1963). Starting with a behavioral observation (that gull parents remove empty eggshells from the nest shortly after hatching), he and his coworkers ran simple field experiments that determined that hatchlings' survival from predation by crows was enhanced by eggshell removal, presumably because the crows used the shells as markers for the presence of vulnerable youngsters. From this they inferred that the adaptive explanation of eggshell removal was to promote crypsis as an antipredator device. They discussed other possible adaptive functions, such as that eggshells could cause injury to the newborns, but did not explore them experimentally (see Arnold 1992). Crucially for present purposes, Tinbergen and coworkers never assumed that the birds understood the link between eggshells and predation risk, or that their previous experience might have given opportunities to shape the response by trial and error followed by differential outcomes. They suggested that simple heritable responses to specific stimuli controlled eggshell removal. Their search for details of the triggering stimuli showed that from late incubation onwards gulls remove many different objects resembling eggshells in some respect, and even noticed that some unnatural colors appeared even more effective than the real one in controlling behavior. The trial and error with differential consequences was done by natural selection, not by individuals. I will come back to this, but the reason for this historical digression is to emphasize that the dual treatment of function and mechanism pre-dated behavioral ecology and has remained a core constituent of it. At present the issue is experiencing renewed attention.

With the development of behavioral ecology, hypotheses about adaptive function became more sophisticated, but although the theoretical duality was preserved, in practice researchers tended to show a neglect of mechanism. For mainstream behavioral ecology, a divorce from psychology was consolidated. This divorce was the equivalent in behavioral research of the "phenotypic gambit" (Grafen 1984), by which the adaptiveness of phenotypic traits is discussed without reference to their genetic underpinnings. At the behavioral level, optimality arguments are used to predict behavior using a functionalist viewpoint that ignores how the behavior might come about, focusing instead on predicting and observing overt behavior. Like the phenotypic one, this is a gambit because it involves a risk: that cognitive mechanisms capable of generating the predicted optimal behavior do not in fact exist, or are impenetrable to our research tools.

I will exemplify the issue using *optimal foraging theory* (OFT), a research program that addresses functional aspects of foraging behavior. Although I use OFT as a main source of examples, I intend for my reflections to be of wider significance, since studies of mate choice, sperm competition, territorial behavior, sex allocation, and most topics in behavioral ecology have similar virtues and vices. I will discuss in some detail the *marginal value theorem* (MVT), a model proposed independently by Parker and Stuart (1976) and Charnov (1976) to identify the optimal switching behavior for consumers exploiting heterogeneous (patchy) environments.

The Marginal Value Theorem

By the time it was applied to foraging theory, the MVT's rationale was well established in economics. It can be summarized as follows: When a consumer accumulates benefits by investing money, time, or other limited resource in discrete sources of benefit that are encountered sequentially, the maximum long-term benefit is achieved by investing the limiting resource in each source for as long as the local rate of benefit exceeds the rate that can be achieved in the whole environment, including transitions costs. For instance, while a bird is foraging in a bush, it can decide to carry on in that bush or to fly away to exploit alternative foraging sites. The benefit rate from staying is directly experienced, but that to be expected from flying away is a joint function of the costs of the upfront investment in traveling and the distribution of food densities in the bushes the bird may expect to find. The rate-maximizing policy results from comparing these two rates. If the rate of gain in each bush drops as a consequence of the bird's foraging, the rate-maximizing policy is to abandon each bush when the local rate falls below the potential benefits expected from departure. Under certain assumptions, this can be computed by differentiation: a rate-maximizing consumer would compare moment to moment the derivatives of gains versus time in its present site against that in the environment, including travel costs. By choosing the higher derivative constantly, every unit of time is invested where the payoff is greater.

Some early criticisms, now fortunately mostly forgotten, argued that the optimality formulation was flawed because the necessary computations to achieve the solution were too hard for foraging animals or that the relevant functions may not be differentiable (e.g., by involving discrete captures distributed stochastically in time). Behavioral ecologists rightly pointed out that it had never been assumed that the procedures used by the scientists to identify optimal strategies had to be followed by the agents to implement these strategies. To expect this would have been equivalent to criticizing Tinbergen's interpretation for the adaptive value of eggshell removal with the argument that the gulls had no chance to learn the relation between predation risk and eggshell presence, or that mosquitoes could not bite you because they don't understand hemodynamics. Animals, argued optimal foraging researchers, could follow psychologically realistic algorithms to converge to the optimal strategy identified by scientists through calculus. I bring this up because although such criticisms are not normally brought up in that form today, real problems of implementation do exist in most examples of functionally oriented behavioral modeling.

These "implementation" issues inspire the present Forum and within it this discussion paper. I intend to defend the view that neglect of mechanism impoverishes functional analysis, and that the time for the functionalist gambit in behavioral ecology (i.e., the assumption that natural selection can always find a way to make animals maximize their own fitness) may be over.

Rules of Thumb

The most popular answer to the function-mechanism dichotomy is that animals could follow simple rules that respond to environmental stimuli so as to generate a behavioral output close to the theoretical optima. In early behavioral ecology these rules were referred to as *rules of thumb* (RoT), a phrase used in colloquial English to describe a procedure that is simple to apply and does not require understanding, but is often efficient in generating the right outcome to a well-defined problem. Krebs and Davies (1993:58) expressed the point thus: "We have been thinking of animals as problem solvers making decisions that maximize an appropriate currency, but of course we do not believe that bees and other animals calculate their solutions in the same way as the behavioral ecologist. Instead the animals are programmed to follow rules of thumb that give more or less the right answer." A very similar notion is called *heuristic* in cognitive psychologists' treatment of human decision makers (Gigerenzer et al. 1999). Todd et al. (2000:727) explain: "We explore fast and frugal heuristics—simple rules in the mind's adaptive toolbox for making decisions with realistic mental resources."

Such rules can indeed be very effective for narrowly defined domains. When members of a species face regular problems generation after generation, natural selection can create patterns of behavior that are channeled rigidly through development to produce appropriate responses to relevant stimuli. For example, if gulls always placed their nests against similar backgrounds and faced similar predators at similar densities, then eggshell removal as a response to the sight of the empty shells would work just fine. If on rare occasions the crows were not a danger to nests, then the removal would not increment survivorship, but would anyway be performed provided that the cost is not too large. As Tinbergen et al. (1963) remarked, these actions may occur for only a few seconds in the life of an animal but could be crucial for breeding performance and hence be under strong selection. However, when circumstances vary within and between each life span, the work of natural selection is harder. What needs to evolve then is a more flexible norm of reaction that generates a successful balance across many circumstances, and most RoTs would flounder. This is very close to the point made earlier by Dobzhansky (1937), who argued that it wasn't specific genotypes that evolved, but norms for building genotypes in the multiple environments faced by the species.

The use, and perhaps abuse, of RoTs can be illustrated with the MVT. As I described earlier, the optimal policy for a bird foraging in depleting bushes is to move on when the expected local yield falls below the average rate of gain in the environment. This is a much more dynamic problem than the regular circumstances hypothesized for nesting gulls. First, local gain rate can be a constantly varying quantity as a function of consumption. Second, expected rate in the background is a changing function of season, time of day, competitors' density, bush distribution, and many other factors. The very idea of a "local" rate in the current patch is to some extent underdefined, since foraging birds often move as they forage so that food density in the field ahead is not predictable from that experienced recently. For instance, when a starling feeds on grubs by probing the ground, it often follows a nonrecursive trajectory, always acting on new ground. As a consequence, the interval between prey encounters does not follow a monotonic increase as assumed in most applications of the MVT. Intercapture interval increases, however, when the birds load captures in their beaks to transport them to feed nestlings. This is just to say that efficient foragers need highly flexible behavioral mechanisms. These difficulties were found very early in experimental tests of the MVT and other optimal foraging predictions, and behavioral ecologists proposed ingenious variations of RoTs that could do a reasonable job under different simplifying assumptions.

Some examples of RoTs for the MVT had descriptive, memorable names, such as *giving-up time* (GUT), *hunting by expectation* (HbE), or *patch residence time* (PRT). Some of these RoTs were supplemented with subrules such as *leave after capture* (LAC); others were more sophisticated, implying, for instance, that the consumer used a Bayesian procedure for updating its leaving policy (see, e.g., Green 1984). Some of these rules are described in Box 2.1. As an early contributor to such literature, I do not think today that this is the most profitable approach and prefer instead to advocate investigating the relevance of broad-domain psychological mechanisms for functionally inspired problems. I will return to the MVT to discuss alternatives to treating the implementation problem with RoTs later, but wish to highlight one particular role in which the RoTs can be important.

Box 2.1 Rules of thumb for the patch departure problem.

GUT: Giving-up time. The consumer moves after searching unsuccessfully for some critical interval. This may work because the reciprocal of intercapture intervals gives an estimate of expected local rate.

HbE: Hunting by expectation. The consumer moves after counting a given number of captures. This works when the distribution of items per patch is very narrow (as when eating seeds from a fruit). In such cases the amount consumed gives an estimate of remaining food density in the present patch.

PRT: Patch residence time. The consumer leaves after a certain exploitation time. As with HbE, this may be a good index of local expected rate when patch density has little variance. PRT is also used as an observable dependent variable, rather than a hypothetical decision rule.

LAC: Leave after capture. A refinement can work with other rules. The rationale is that under some (non-Poisson) capture regimes the probability of capture may increase with previous time searching, as when one waits for buses that leave their garage at regular times and gather limited variability in the traffic. In this case, even if the present patch's estimated average is judged to have declined, it can make sense to wait until a prey is caught and then leave immediately.

Note: The RoT approach has also been used in other areas. For instance, in mate choice papers (e.g., Wiegmann et al. 1996), authors discussed mate choice strategies examining analytically or by simulation the relative advantages of a sequential search rule (where the acceptance threshold is adjusted until a mate above the present threshold is met) versus a "best of *n*" rule (where *n* candidates are considered and the best of that set is taken).

A Role for Rules of Thumbs: Documenting the Plausibility of Optimal Strategies

I believe that hypothetical, dedicated, behavioral rules, such as foraging RoTs or the psychologists' heuristics, have useful roles. To respond to the criticism that optimal strategies are "too hard for agents to implement," it is useful to explore *in silico* whether rules of extreme simplicity can perform close to the predicted optima. This is not direct research on the mechanisms underlying behavior, because whatever the outcome, it doesn't prove that any such rules are in fact used by the organism; it does, however, weaken the anti-optimality criticism. I first came across this function for RoTs when working on the problem of balancing exploration and exploitation in the so-called *two-armed bandit paradigm* (Krebs et al. 1978). To exemplify this role for RoTs, I describe this older work in some detail.

The analysis of the exploration-exploitation balance was inspired by coffee-room comments made by the late Mike Cullen. He made the (with hindsight) obvious comment that as an animal forages it collects both food for immediate consumption and information that might increase its performance later on. Maximizing simultaneously the current rate of gain and rate of information acquisition (which improves performance later) may be incompatible, opening the opportunity for research into how animals solve inter-temporal information trade-offs. This constitutes a fundamental problem in studies of decision under uncertainty, namely when the decision maker does not know the probabilities associated with each available action: as its experience accumulates, the agent transforms the problem by gaining knowledge of the probabilities, and uncertainty migrates toward risk.

Together with John Krebs, we implemented the problem in the laboratory, offering small birds (great tits, *Parus major*) two feeding machines, each with a different reward probability that was stable during a session but which changed from one session to the next. Early in each foraging session, the animal would have collected a few items and done some work in one or both machines, thus possessing some information, however unreliable, to compare the machines. The exclusive allocation of effort to the favored option at an early stage, however, would have a high probability of misidentifying the richer source. This probability declines as more information is gathered in both machines. Thus, the behavioral allocation that maximizes information gain, and hence future performance, is less biased than the one that maximizes immediate gain expectation.

Without going into technicalities here, an ideal technique to find optimal solutions in this kind of problem is *stochastic dynamic programming* (SDP). Using this technique we identified properties of the optimal allocation of effort between the two sources for each amount of experience, as a function of different histories of reinforcement, and compared this predicted allocation to that shown by the birds. The great tits achieved payoffs close to the maximum predicted. The use of SDP is interesting in the present context because it is clearly not intended as an algorithm that could be used by the birds or any other actor: it identifies the optimal strategy from the perspective of an observer, without discussing how the decision maker may achieve it. The technique itself is conceptually simple but computationally heavy, requiring multiple loops and the building of arrays with the results of these calculations. Besides, it proceeds backward—an unlikely computational procedure for any actor, including humans. The good approximation by the great tits to the predictions of the SDP-based optimal strategy raised the issue of whether plausible mechanisms simple enough to approximate the theoretical optimum might exist, and within those, whether the choice mechanisms actually used by the birds could be identified and compared to the hypothetical rules. This is the implementation problem that I mentioned earlier. In my opinion, this sort of situation is when RoTs can show their worth. In the particular study of the two-armed bandit, we examined a few alternative heuristics that according to intuition could respond reasonably to the two-armed bandit problem (Houston et al. 1982). These rules are described in Box 2.2.

The output of these RoTs was examined by Monte Carlo simulations. The outcome of the exercise can be summarized as follows:

1. Both absolute and relative performance of the rules depended on the details of the problem and the choice of parameters. For instance, for certain combinations of the memory parameter and pattern of environmental fluctuation, a rule that uses a moving average of experience and responds to the option that has so far been richer can be close to the optimum, but for other parameters (viz. some combinations of rate

Box 2.2 Rules of thumb for the two-armed bandit problem.

CULLENn: Allocates effort equally and keeps track of δ, the difference in cumulative number of rewards between the alternatives, with *n* as a parameter. When $\delta > n$ the agent switches to allocating all effort to the option with higher current score. Note that (a) the algorithm does not need to remember the cumulative number of rewards or responses, only the reward difference; (b) the parameter *n* implements the trade-off between exploration and exploitation: the larger *n*, the more conservative is the switch to exploitation, with lower probability of error at the expense of a longer "sampling" phase (the rule does not by itself specify how *n* is chosen); (c) the assumption of equal allocation until decision can be replaced by random allocation, provided the optimal *n* is chosen appropriately.

IMMAX: Stands for "immediate maximizing." Keeps a running average of rewards per unit of effort in each machine and allocates the next response to the option with the richer average. IMMAX does not ever commit itself: if the preferred alternative yields no rewards, its score drops and a reversal occurs.

LINOP: Stands for "linear operator" and is the algorithm proposed originally by Bush and Mosteller (1955). In the version used by Houston, Kacelnik, and McNamara (1982), each alternative was assigned a moving value following $E_t^i = \alpha E_{t-1}^i + (1-\alpha) R_{t-1}^i$, where E_t^i is the value of option *i* at time *t*, α is the memory of the system ($0 \leq \alpha \leq 1$), and R_{t-1}^i is the reward obtained at time *t* (*R* can be 0 or 1). Effort at time $t + 1$ is always allocated to the alternative with higher E_t^i. The optimal value of the parameter α is an additional interesting problem.

MATCH: Like IMMAX but here effort is allocated probabilistically among the options, with probability of receiving the next response being equal to the current estimate of each option over the sum of the estimates in both options.

RANDOM: This heuristic was a benchmark that allocated effort at random. This is not trivial because some of the other rules can do worse than random for certain parameter values.

of depletion and the size of the averaging time window) the same rule can be at phase opposition to the patches' relative quality, inducing the subject to choose systematically the worse alternative present. It is easy to see why this can happen: since the rule uses memory to choose, in a continuously changing environment memory values will always lag behind present parameters, and if the forager faces cycles of renewals, this lag can place preferences precisely at the opposite of the optimal solution. In such cases, the moving average rule does considerably worse than random choice.

2. If parameters are chosen appropriately for the current problem, many rules can do remarkably well. This proves that it is not necessary to follow the unrealistic SDP process to achieve a satisfactory outcome, but highlights that parameter choosing becomes the central difficulty.
3. When tested for descriptive power, namely how each rule's output resembled the dynamics of the distribution of effort by the birds, the rules that were easiest to reject were deterministic ones: adding probabilistic noise to any rule protects any algorithm from "sticking its neck out." This is, however, a Pyrrhic sort of triumph, as gaining resistance to refutation does not contribute any further insight.
4. None of the rules were a particularly good platform to generalize the findings to other (nonbandit) problems, because they had not sprung from previous knowledge about animal learning, but had instead been proposed as potential solutions to a preconceived, idealized problem. This is a definite weakness of the RoT and heuristics approaches, as we will examine below.

The Problem of Narrow Task Domain

The RoT approach treats all tasks as Tinbergen and collaborators treated eggshell removal. Given a well-specified functional problem, simple algorithms that are efficient without demanding understanding or unlikely computations are proposed and tested. The implicit assumption is that either natural selection or previous individual experience may somehow have shaped such rules and subsidiary procedures to adjust their critical parameters. This is appropriate for many real-life situations. For instance, among parasitoid wasps (insects that lay eggs in the body of other invertebrates so that their offspring nourish from the tissues of their host), some species face a narrow distribution of number of hosts per patch whereas others face highly variable host numbers. For the former it makes sense to use variants of hunting by expectation to decide when to abandon host patches, adjusting the estimated potential of each patch by "counting down" after each oviposition, as this reflects the depletion of opportunities from a fixed starting number caused by using up each host. For the latter, instead, as patches may vary greatly, each oviposition may be an index of patch quality so that an optimally behaving wasp would up its estimate of the local patch's host richness after each capture ("if I found a host this patch is probably better than I thought"). Indeed, species of wasps do vary in their response to hosts, some increasing and others decreasing their tendency to leave patches after each oviposition. The ecological consistency between the rule genetically preprogrammed in each species and their different ecologies is not yet fully mapped (van Alphen et al. 2003), but the existence of both diverse rules and diverse ecologies is highly promising and highlights the importance of attending to the ecology of the species under study.

The diversity in relative performance of different rules according to variations of the problems is an argument against the wholesale application of the RoT or heuristics method. This is especially true for generalist consumers with life spans much longer than the foraging events being studied, as is the case with most vertebrates. In nature, an individual bird (such as a great tit) forages on dozens if not hundreds of food sources, some patchy and some not, faces multiple configurations of intra- and extra-specific competition, and does not normally face straight simultaneous choices between just two sources of reward with fixed probabilities. The species inhabits large-scale heterogeneous habitats, so that circumstances differ between generations. This means that to apply a suitable RoT with the correct parameters, a bird would need a large library of heuristics as well as a large number of subsidiary rules to select the optimal parameters and then rank the performance of each rule. The problem is not just that the library of rules or heuristics would be large, since many relations can indeed be coded in vertebrate brains. The real problem is that this is compounded with how to choose appropriately which rule to use, given that the decision maker does not have an a priori description of the problem it faces. To use a suitable heuristic, the agent has to characterize the problem, consider a family of potentially suitable rules for such problem, identify the optimal parameter for each rule, and then apply the best performing rule with its best parameter values. Rather than supplying fast and frugal shortcuts, by the time this process has been completed, the problem faced by the agent is likely to have changed. Further, even a competent statistician would not be able to choose a rule and its best parameters in a reasonable time unless they were informed first of the nature of the problem and the stability of the environment.

The Problem of Psychological Realism

Although the RoT or heuristics approach is aimed at addressing decision mechanisms, it neglects the psychological literature on learning and choice, at least in the animal applications with which I am most familiar. I believe it is fair to say that behavioral ecology has from the start addressed behavioral mechanisms as if experimental psychology had not been invented. Established psychology was replaced by newly conceived putative decision rules aimed at solving problems identified as having ecological and evolutionary interest, such as the already described patch or mate choice, patch residence time, cooperation, etc. There are interesting sociological reasons for this cross-disciplinary neglect, but my point here is that redressing it can lead to a more successful program and that future work should not maintain the weaknesses developed in the past.

Addressing ecological problems in terms of behavioral mechanisms leads to more testable predictions. Just as important, it often reframes functional questions: mechanistic and functional questions are separable, but they are not independent. Just as neglecting issues of adaptation may lead psychological research into blind alleys, neglecting the psychology of real organisms may lead the behavioral ecology program to wasteful debates about the relative performance of rules that have no biological relevance.

At this point it is pertinent to ask for methodological alternatives, since it is easier to criticize than to offer alternatives. A proper answer requires some choices about the level of analysis, because what is seen as a mechanism at one level of analysis is always an effect when viewed at a more fundamental level. For instance, we may treat learning as a mechanism underlying choice, but learning is a consequence of neuronal mechanisms, which are themselves effects of membrane phenomena; while what qualifies to be called a mechanism is a semantic problem, it makes no sense to try to explain issues of behavioral ecology, such as foraging behavior, at the level of membrane physiology. For the purpose of this chapter, which addresses approaches to research into behavioral mechanisms in behavioral ecology, I shall stick to this high level of analysis.

With the preceding caveats, there are many known phenomena that can be treated as mechanisms by behavioral ecologists, and which differ from RoTs in that they have not been proposed as solutions to specific ecological problems but have been constructed through independent experimental research, often with entirely different protocols. Examples of such mechanisms that we have found useful in dealing with issues of function include scalar timing (Kacelnik and Brunner 2002), Weber's Law (Kacelnik and Brito e Abreu 1998), state-dependent valuation learning (Pompilio et al. 2006), and choice by latency cross-censorship (Kacelnik et al. 2011). All these mechanisms are testable, offer clear avenues for further experimentation, and can turn around how one models decision making from a functional perspective. For reasons of space and deadlines, I will limit the following discussion to one such example, perhaps the oldest and best known: the many variations of Thorndike's Law of Effect, which I will loosely call reinforcement learning.

Reinforcement Learning

I use this label here in its broadest sense, meaning roughly "do more of what works and less of what doesn't." *Reinforcement learning* (RL) has been around for over a century and applies across a wide range of well-established learning phenomena. Before putting it forward as a good candidate, it is best to raise some caveats. RL does not apply to all forms of learning, as shown for instance by the omission paradigm (Williams and Williams 1969), in which a signal is followed by reward only if the subject refrains from acting toward it; in this case the force of classical (Pavlovian) associations is such that many animals are driven by the pairing of the signal and reward to sustain a high level of responding, and lose rewards as a consequence. Even when it works, as a theoretical construct RL is not free of difficulties. To apply it, the agent needs at least implicit memory for recent events (its own actions and events in its environment) and criteria to identify outcomes as having positive (or negative) value. Nonetheless, these problems are better defined, extensively investigated by behavioral analysts, and more tractable than those posed by RoTs. To illustrate my point, I return to a foraging problem mentioned earlier, the MVT, and contrast insights derived from RL with the RoTs handling of the same issue.

The Marginal Value Theorem: Reinforcement, Travel Time, and Patch Exploitation

As discussed earlier, one of the classic predictions of the MVT is that intensity of exploitation of each patch should decrease when background opportunities in the environment increase, and vice versa. Everything else being equal, overall gain rate in any environment declines when travel time between sources of reward increase. Thus, exploitation of each patch visited should be more exhaustive when travel time is longer. This prediction is probably the best-documented empirical finding in foraging theory. To my knowledge it has been supported whenever it was tested. Should this widespread result be expected from the action of nondedicated mechanisms known with independence of the MVT? In particular, what would a simple and widespread process such as RL predict for this paradigm?

To conceptualize the MVT problem from an RL perspective, we first identify a meaningful pairing of a critical action and its consequence. The obvious choices are, respectively, the departure from each patch (when the consumer takes the decision biologists try to predict) and the encounter with a new patch (a meaningful consequence of the previous decision). In the associative learning literature, the temporal proximity between any action and its consequence (known as contiguity) has been known for a century to facilitate conditioning (Dickinson et al. 1992). In our example, patch-leaving will be reinforced by patch-finding more strongly when travels are short because shortening travel means greater contiguity between the action of departure and the outcome of patch finding. Because patch-leaving is reinforced more strongly in short-travel environments, patch-leaving occurs earlier for shorter travel times in accordance to both the predicted optimal behavior and the observed results. A classic, tested process, postdicts the results of experiments aimed at testing an optimality model. So far so good, one may say, but not all predictions of the MVT and RL are consistent. What happens when they differ? Let's consider a few examples.

Reinforcement, Travel Time Variance, and Patch Exploitation

I first consider environments that differ only in the variance (not the mean) of travel times. According to the MVT, the optimal level of exploitation of each patch is independent of travel variance, because the theorem refers only to average background opportunity. This is because the model was developed for infinite time horizons. In contrast, RL processes are highly sensitive to variability in the delay between action and reward. The effect of contiguity on acquisition is a declining, convex function (i.e., negative first and positive second derivative) of the temporal association of action and reward (Dickinson et al. 1992, Fig. 6). As a direct consequence of this nonlinearity, we should expect conditioning strength to be a positive function of delay variance (Kacelnik and Bateson 1997). It follows that the reinforcing effect of patch-finding on patch-leaving should increase with travel variance. Everything else being equal, consumers will be more "trigger happy," leaving patches earlier when travel times are more variable even if mean travel time is unaltered. This has been corroborated experimentally (Kacelnik and Todd 1992). So, in this case there is no need to invent a new RoT, because a well-established, wide-domain principle of learning suffices to account for the data.

Reinforcement, Patch Depression, Travel Time, and Patch Exploitation

The MVT patch-leaving predictions are particularly relevant for cases in which capture rate in each patch drops as the consumer exploits it. Because background opportunities are independent of the exploitation of the present patch, if the reward rate in the consumer's current patch remained constant (i.e., no depletion), from the perspective of rate maximizing there would never be a reason to leave, because the ideal agent compares two rates that are both invariable in time. A patch should either be left on arrival or never, if it yields lower or higher rates than the background, respectively. Of course, consumers will in practice move, due to factors other than rate maximization (such as satiation leading to pursuing other goals), but if patches do not deplete, the MVT per se predicts that travel time should have no systematic effect on departure times.

The conditioning account differs. As mentioned earlier, the act of patch-leaving is reinforced with a declining function of travel time whatever the local rate of returns. Everything else being equal, patch-leaving decisions are reinforced more strongly in environments with shorter travels. When a consumer is in a patch, the pull of the background environment is stronger when travel is shorter, and this should cause earlier departures for shorter travels even if patches do not deplete.

The different predictions between the MVT and the RL account can be examined in light of an experimental study that combined field and laboratory tests (Cuthill and Kacelnik 1990). Cuthill and Kacelnik first examined starlings that were feeding young by central place foraging in the field. They reported that even when the birds were exploiting nondepleting artificial patches, they increased their average load size to the brood as a function of travel time. This appeared to support the RL account, violating the basic MVT analysis. However, they also noticed that carrying a bigger load took somewhat longer, and that when the birds were tested with a nondepleting patch in the lab, where they did not have to transport the load to a central place, the effect disappeared, and, as predicted by the MVT for this situation, they did not respond to travel manipulations. In this case, then, the RL account is clearly insufficient, as the MVT seems to add predictive power when enriched with other components of the foraging task. While RL gave a better account than the MVT with respect to travel time variance, it does not seem to do the same with respect to mean travel costs.

Since this is not a review of optimal foraging theory but a discussion of why mechanisms matter in functional analyses, I will not belabor this by adding further examples, but instead advance some conclusions. MVT is a functional tool to predict foraging behavior, and RL is a mechanism of behavioral change as a function of experience. We have seen that some predictions of MVT and RL coincide. When they do not coincide, in some cases RL is a more robust predictor, but in others it fails to capture some issues highlighted by the optimality analysis. It seems to me that an obvious conclusion is that RL—a simple and highly general principle—has the properties demanded by McNamara and Houston (2009) in their discussion of this issue. McNamara and Houston call for the search for simple rules that perform well in a variety of problems (rather than complex rules for simple environments). RL does precisely what they predicate, and since RL most likely did not emerge as a rule specifically to solve the patch problem, investigating the adaptive role of this psychological mechanism across a wide diversity of circumstances is more productive for functionally oriented research than inventing specific heuristics for the MVT.

Why would RL have evolved if it sometimes causes loss of potential yield? I believe that the conditions under which RL does a good job are so widespread, including domains such as social interactions, antipredator strategies, keeping warm, cooling down, avoiding toxins, as well as foraging, that the evolutionary pressure and economy of the mechanism is maintained or discovered again if for some rare twist of evolutionary history a species has lost it. The ecological frequency of situations in which RL causes a loss of intake is likely not to be sufficiently high to counteract its benefits. Rather, in specific cases when the rule causes some notable loss, some bells and whistles are added to correct for this. In the example of travel time impact on nondepleting patches, the correction may simply be that reinforcement is not just patch arrival, but some measure of experienced rate of returns. This suggests that the functional analysis should be reframed: One may ask why and when is RL adaptive, and use the answer to generate hypotheses about the organism's ecology. Asking what are the acts being reinforced and the events acting as reinforcers uses RL as a guide for understanding function, and may be more effective than boxing oneself inside idealized optimality models and derive putative dedicated heuristics to solve it. Such reframing can be hard because we tend to believe (often without justification) that we know from the start what the relevant issues were in the design history of the species with which we work.

Risk Sensitivity

I have focused on the MVT to show different approaches to the study of mechanisms, but this was an arbitrary choice on my part. Similar arguments comparing the use of bespoke rules for each problem against the use of broad-domain psychological processes already described in animals can be applied to other ecologically meaningful cases. I shall restrain from piling up examples, except for a further illustration of the mechanism discussed so far, RL, in the context of animals choosing between outcomes of different variability. Here, instead of considering a patchy environment and the decision of when to leave each patch (as in the MVT), the subjects' choices are between simultaneously present sources of reward.

The dominant approach to the problem of reward variance in behavioral ecology is *risk sensitivity theory* (RST) (for reviews, see Houston and McNamara 1999; Stephens and Krebs 1986). When an agent allocates behavior between variable food sources, an interesting issue to explore is the relative influence of mean, variance, and skew of the outcomes of each source on the agent's preference. It is trivial that everything else being equal, agents will always be predicted to prefer sources with higher average yield, but everything is not always equal, and the case in point is when two or more alternatives differ in outcome variance. The question is whether any component of subjects' preferences explained by variance will add or subtract from preferences due to other factors such as average yield; in other words, whether subjects will be risk prone or risk averse.

Being a functional treatment of the problem, RST does not address the mechanisms by which organisms might implement preferences. Its foundation (and that of models of risk sensitivity in economics) is that variance adds to an option's value when the relation of fitness (or utility) versus reward is convex (i.e., has positive second derivative) and lowers value when it is concave (i.e., has negative second derivative). Textbook versions of RST base predictions on the budget of the agent, namely the relation between expected gains and needs. The *budget rule* (reviewed and fully explained by Stephens and Krebs 1986, McNamara and Houston 1999), for instance, states that the organism should be risk prone when expected gains in the least variable option are below needs, and the opposite otherwise, because needs thresholds are likely to be inflection points and this means a change in the curvature of the fitness-gains function from convex to concave. This theoretical statement of what a specific optimality discussion predicts is sometimes cited as an empirical fact, but this is unjustified by available evidence (see review in Kacelnik and Bateson 1996). The empirical test of such predictions are much harder than can be envisaged at first because the function linking fitness to gains (including the needs threshold) is rarely, if ever, known, and because the timescale over which decision makers assess rate of gain and its variability is also typically unknown to the researchers.

The mechanism already discussed as an example of a simple, robust rule for other problems, namely RL, also makes predictions for this case. To discuss it, I focus on variance in two dimensions of rate of gain: the size of rewards and the delay to rewards.

Consider first the effect of variance in the delay between an action and its outcome. As stated above reinforcement effects obey a declining, convex function of delay to reward. Because of this convexity, variance in delay increases conditioning power. In the MVT example, the convexity of the effect of contiguity on conditioning results in an increase in reinforcement of the departing action and consequent shortening of patch residence times as a function of travel variance, but in the present problem, simply because what is reinforced is a choice, and not a departure, variance in delay to experience the outcome of a choice adds to the option's subjective value (for a graphical explanation of this phenomenon, see Kacelnik and Bateson 1997). The more variable in delay is an option, the higher the conditioning effect of the outcome on the action to choose it. This leads to what behaviorally can be described as "risk proneness" even if there is no functional implication that such added value is beneficial. Now consider variance in size, rather than delay of the reward. Fewer experiments have tested variation in conditioning power as a function of reward size, but it is a strong bet that this relation will be concave, because larger outcomes produce saturation. This concavity implies that the likely effect of variance in reward size will be to reduce the reinforcement of a choice action as a function of variance in the size of the action's outcome, leading to risk aversion for reward amount.

Whereas RST predicts changes in risk appetite as a function of the state of the subject, with no reference to mechanism, RL predicts risk proneness for delay and risk aversion for amounts, independently of the subject's state. Most experimental studies do find a strong bias toward risk proneness for delay, with inconclusive effects of variance for amounts in the direction of risk aversion. RST's predicted reversals in the sign of preference with state are sometimes reported but tend to be elusive, and firm overarching conclusions are hampered because studies with negative results are rarely published. However, the point of this example is that the single, broad-domain mechanism known as reinforcement learning, for which properties such as the nonlinear effect of action-outcome contiguity are well established elsewhere, produces robust predictions for decisions of when to leave a patch as a function of mean and variance of travel time, and for decisions on which of two food sources to choose when both are presented simultaneously. No special RoT or dedicated heuristic needs to be derived to deal with these three different problems.

Overall Conclusions

I have contrasted two approaches to the use of optimality models in behavioral ecology. Both share the use of optimality as a technique to identify and model ecologically meaningful scenarios, and both recognize that functional and mechanistic analyses are complementary and worth pursuing. The differences between the two approaches refer to the preferred approach to suggest mechanisms of choice for empirical tests.

After modeling optimal solutions to each problem, one approach proposes hypothetical rules, or heuristics, whose outcome, if applied by the organism, would approximate these optimal solutions. These rules can be invented and explored in a computer. The other approach, which I admit to prefer, replaces ad hoc RoT or heuristics with established processes of learning, choice, and perception culled from experimental psychology, adding new algorithms when the experimental evidence suggests them.

There are practical and theoretical differences between these approaches and although I focus here on how they may conflict for the sake of the discussion, I am convinced that both have useful roles. From a practical point of view, one difference is the response to discrepancies between predicted optima and observed behavior. In the case of rules devised ad hoc for each problem, a frequent response is to invent new rules or suggest remedial modifications to the original one. This response is more constrained in the second, more empirically driven approach, because the rules explored tend to be phenomena that have been well established in other paradigms and it makes no sense to propose rules that are known to conflict with preestablished evidence. If a prediction based on operant conditioning fails to describe behavior in a foraging task, one could hardly propose that operant conditioning does not exist, but could instead reinterpret the organism's problem and test what actions are reinforced in the situation being tested.

On the theoretical side, if an established and ubiquitous behavioral process such as reinforcement learning causes suboptimal decisions in a given experimental protocol, a reasonable guess is that problems of the class represented by the experimental protocol have not had major significance in the adaptive history of the species. Thus, mechanistic research leads to the reformulation of the functional enquiry.

In summary, the interest for functional problems and optimality theorizing is a fundamental and progressive feature of behavioral ecology, but I surmise that research into mechanisms of behavior should be closely associated to the constructs and methods of experimental psychology, rather than by the formulation of hypothetical rules of thumb or heuristics. It would be a mistake to read this contribution as proselytism for reinforcement learning. I only used reinforcement learning as one illustration of a behavioral mechanism that has major relevance to behavioral ecology but originated elsewhere. Other equally suitable examples could have been chosen. My goal has been to defend the notion that injecting mechanism in functionally oriented disciplines like behavioral ecology or economics is better supported by cross-fertilization with experimental psychology in its various styles than by the continuous proposal of newly invented rules of thumb or heuristics.

3

Machinery of Cognition

Charles R. Gallistel

Abstract

A Darwinian approach to decision-making mechanisms must focus on the representation of the options between which the animal decides. For example, in matching behavior, is the animal deciding between the different locations in which to forage or simply whether to leave its current location? A neurobiologically informed approach must be concerned with the mechanism of representation itself. In the computational theory of mind, options are represented by the symbols that carry information about them forward in time. In conventional computing machines, symbols reside in an addressable read-write memory. Current theorizing about the neurobiological mechanism of memory rejects this form of memory in favor of an associative memory. The problem is that the associative bond—and its neurobiological embodiment, the plastic synapse—is not suited to the function of carrying acquired information forward in time in a computationally accessible form. It is argued that this function is indispensable. Therefore, there must be such a mechanism in neural tissue, most probably realized at the molecular level.

Introduction

The computational theory of mind—that brains perceive the world and control behavior through computation—is the central doctrine of cognitive science. It is now widely though not universally embraced by neuroscientists; however, there is, no consensus about its neuroscientific implications. Does it imply that the brain has the architecture of a computer? Or does it imply that the brain has its own architecture, which is in some way superior to that of a computer? These questions are relevant to an evolutionary approach to decision making, because the brain is the organ that makes decisions, and those decisions depend on a representation of the options between which the decision is to be taken; what is not represented cannot be decided on. Moreover, if the animal's brain represents the decision to be made in a way other than the way in which the experimentalist and theorist conceive of it, the choices made may seem paradoxical and nonoptimal.

Before considering the machinery that makes decision-relevant representation possible—the machinery of memory—I review contrasting models of matching behavior. I do this to illustrate the importance in an evolutionary approach to mechanisms of decision making of two considerations that figured prominently in the discussions at the Forum: the correct identification of (a) the animal's representation of the options and (b) the function of the behavior.

When animals forage at more than one nearby location, moving back and forth between them, the durations of their visits are exponentially distributed, with expectations whose ratios approximately match the ratios of their current incomes from those locations (Heyman 1982; Gibbon 1995; Gallistel et al. 2001). The distinction between income and return is important: The income from a location is the amount of food obtained there per unit overall foraging time; its computation is independent of the amount of time spent there. The return is the amount of food obtained per unit of time spent at there; that is, per unit of time invested there. Matching equates the returns, not the incomes. It is often taken as more or less self-evident that (a) the function of the behavior is to maximize overall income, (b) the options between which the animal decides are the two locations, and (c) the decision-making strategy is based on the relative returns. When the return from one option is perceived to be greater than the return from the other, the animal adjusts its strategy so as to prolong its visits to the location with the higher return and shorten its visits to the location with the lower return (Herrnstein and Vaughan 1980; Vaughan 1981; Herrnstein and Prelec 1991), a strategy called melioration.

On the basis of data from my lab, we have suggested that all of these assumptions are wrong (Gallistel et al. 2001, 2007). First, the function of the behavior is to gather information about each location (what kinds of food are currently to be found there in what amounts and at what intervals). On this view, the harvesting of found food is ancillary; the animal simply eats what it finds (or not, depending on its motivational state). Second, the options between which the animal decides during a visit are whether to leave that location or not. Third, the decision to leave depends only on the incomes, not the returns.

Reviewing the evidence for our arguments and the model of matching that they lead to would digress too long from my main theme, which is the machinery that makes the representations possible. In this evolutionary context, however, I note that one of our arguments is that matching is not optimal from an income-maximizing perspective (Heyman and Luce 1979), but it is optimal from an information-gathering perspective (Gallistel et al. 2007).

The Machinery of Memory

In the computational theory of mind, options and decision variables are represented by symbols in the brain (Gallistel 1990; Gallistel and King 2009). For example, in most contemporary reinforcement learning models, decisions depend at a minimum on the values of different options and, at a maximum, on a model of the relevant aspects of experienced environment (see Dayan, this volume). In a computing machine, symbols reside in memory. Their function is to carry acquired information forward in time (Gallistel and King 2009). The function of a symbol in memory is analogous to that of an action potential in an axon. However, the function of the action potential is to carry information from one location in the brain to another, whereas the function of the symbol is to carry information from an earlier location in time to a later location—in a computationally accessible form. In the world of the computer scientist, this function is implemented by means of an addressable read-write memory.

The problem is that most contemporary neuroscientists would not regard an addressable read-write memory as the mechanism by which options and evidence are represented in nervous systems. In most neuroscientific theorizing, altered synaptic conductances redirect the flow of signals between the input and the output. In the computer scientist's conception of memory, its most basic property is its ability to encode acquired information (write to memory) and redeliver it upon request (read). In at least the strongest version of the neuroscientist's conception of memory, the encoding of information is not on the list of the properties that the memory mechanism is thought to possess. In this latter view, the rewiring of a plastic brain by experience explains the effect of past experience on future behavior. The rewiring is mediated by changes in synaptic conductances. This theory is the extension to neuroscience of the associative theory of learning. The alterations in synaptic conductances are the material realization of the associative bond. This theory explicitly rejects the assumption that an addressable read-write memory is central to the machinery of cognition.

There are two schools of thought about whether the rewiring of the brain by experience implies that the brain has acquired the information about the experienced world that is implicit in the altered behavior. One school of thought holds that the brain does not store information in the explicit form in which a computer does; the brain is "sub-symbolic" (Smolensky 1986). This is consonant with the anti-representational stance that has always been a salient feature of associative and behaviorist theorizing in a materialist framework. In associative theorizing, the only elements of the brain that undergo enduring structural change are the associative bonds. The associative bond, however, was never intended to carry information forward in time. Associative theories of learning do not attempt to explain how associative bonds encode the information extracted from experience by sensory/perceptual processing, precisely because the associative bond is not conceived of as a carrier of information. Neither it nor its material realization, the plastic synapse, can readily be adapted to this function (Gallistel and King 2009).

Another school of thought holds that changes in synaptic conductances do, in some sense, encode information. This information, however, is distributed across the synapses of a complex multielement circuit in such a way that one cannot readily specify how or where it is encoded. It is, so to speak, lost in a cloud. Martin and Morris (2000:650) explain this cloud view as follows:

> [Long-term potentiation] LTP may serve a universal function in the encoding and storage of memory traces, but what gets encoded and how is an emergent property of the network in which this plasticity is embedded, rather than of the mechanisms operating at the synapse in isolation.

Similarly, Neves, Cooke, and Bliss (2008) write:

> Two facts about the hippocampus have been common currency among neuroscientists for several decades. First, lesions of the hippocampus in humans prevent the acquisition of new episodic memories; second, activity-dependent synaptic plasticity is a prominent feature of hippocampal synapses. Given this background, the hypothesis that hippocampus-dependent memory is mediated, at least in part, by hippocampal synaptic plasticity has seemed as cogent in theory as it has been difficult to prove in practice. Here we argue that the recent development of transgenic molecular devices will encourage a shift from mechanistic investigations of synaptic plasticity in single neurons toward an analysis of how networks of neurons encode and represent memory, and we suggest ways in which this might be achieved. In the process, the hypothesis that synaptic plasticity is necessary and sufficient for information storage in the brain may finally be validated.

While Koch (1997) puts it this way:

> And what of memory? It is everywhere (but can't be randomly accessed). It resides in the concentration of free calcium in dendrites and the cell body; in the presynaptic terminal; in the density and exact voltage-dependency of the various ionic conductances; and in the density and configuration of specific proteins in the postsynaptic terminals.

In considering whether an addressable read-write memory is central to the functioning of a computational brain, one must understand the role it plays in computation. Computation is the composition of functions. A logical constraint of fundamental importance in shaping the architecture of a computing machine is that functions of arbitrarily many arguments may be realized through the composition of functions of two arguments, but not through the composition of functions of one argument. Examples of two-argument functions are the basic logical operations, AND and OR, the basic arithmetic operations ≥, +, –, *, ÷, and the CAT (concatenation). Examples of one-argument functions are NOT, log, sine, and abs. This logical constraint explains why an architecture consisting of one or more central processors coupled to a read-write (aka fetch/store) memory is a universal feature of engineered computing machines (Figure 3.1). The results of computations performed on earlier inputs to the machine are stored in memory (written). When they are needed in subsequent computations, they are retrieved from memory (read). The read-write memory frees composition from the constraints of space and time. It allows information acquired at different times to come together to inform current behavior.

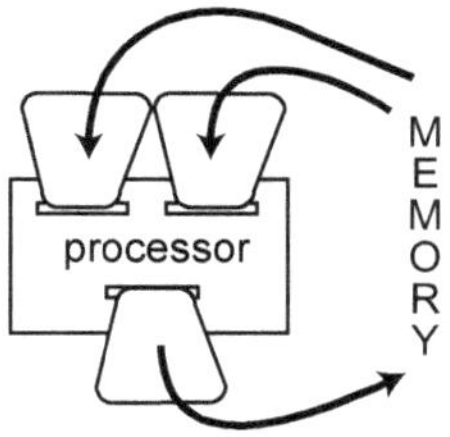

Figure 3.1 The essential architecture of a computing machine. The physical realization of the symbols that may become the arguments (inputs) to a two-argument function reside in memory. They are brought from memory to the machinery that effects the function (the processor). The result (processor output) is returned to memory, where it is carried forward in time so that it may become an argument of a further function. Fetching the inputs from memory is the read operation. Returning the result to memory is the write operation. The symbols to be fetched are located by means of their addresses. It is this architecture that makes possible the composition of functions in all known computing machines.

Although some computer scientists have suggested dispensing with this architecture (e.g., Backus 1978), it has not been discarded. There is currently little thought within computer science that it will or can be. Even highly parallel architectures, like the NUMA architecture in multiprocessor supercomputers, do not dispense with read-write memory. Nor do discussions of quantum computing envisage its absence if such machines are ever built. If any company that manufactures computing machines contemplates building one without a read-write memory, they are keeping it a closely guarded secret.

The constraint that leads to this architecture is physical and geometric: a large number of different symbols may serve as inputs to the combinatorial machinery (the processor). The number of the pairwise combinations grows as the square of the number of individuals that might be paired. This is why, in the design of computing machines going back to Babbage, the symbols to be processed have been stored in a readable memory, whence they are fetched by the processing machinery. It is possible to configure solids in three dimensions so that arbitrarily many are adjacent (Wilson 2002). Neurons, which can intertwine in complex ways, might be thought to be well suited to this manyfold spatial adjacency. There are, however, additional constraints beyond the basic constraint that the participants in a physical interaction must be at the same location at the same time (the spatiotemporal adjacency constraint). The constituent structures must encode information; that is, they must be physically realized symbols, like the coding sequences in DNA, and some of them must be capable of the basic combinatorial operations that underlie neural computation, whatever those basic operations may be. In short, to dispense with a read-write memory in a computing machine, it would be necessary to replace it with an architecture in which coding structures and processing structures were intertwined in such a way that there was universal adjacency of coding structures (every coding structure was adjacent to every other) and every pair was adjacent to every elementary processing structure. To my knowledge, no one knows whether such architecture is possible.

The passages quoted earlier from the works of leading computational neuroscientists would seem to suggest that neuroscientists believe not only that such an architecture is possible, but also that it is realized in the brain. These passages make clear, however, that neuroscientists are frustratingly vague about what this alternative architecture is—too vague to offer guidance, for example, to the engineers at Intel. They are particularly vague about how the different bits of information gleaned from experience at different times are encoded in such a way that they may be retrieved on demand to serve as the inputs to the computations that inform the daily behavior of animals. Thus, there exists at this time a conceptual chasm between what seems to be required by the behavioral facts, some of which I review below, and what neuroscientists are inclined to believe about how memory works in the brain. Several leading cognitive scientists recently put the problem this way (Griffiths et al. 2010:363):

> In our view, the single biggest challenge for theoretical neuroscience is not to understand how the brain implements probabilistic inference, but how it represents the structured knowledge over which such inference is defined.

The problem that Griffiths et al. refer to arises because there is no plausible read-write addressable memory mechanism known to contemporary neuroscience. That is why theoretical neuroscientists generally try to find a way of doing computations without such a mechanism. They sometimes find this so nearly impossible that they end up positing, for example, "context units" (Elman 1990)—a form of read-write memory. Hidden units in a net write their current activity state to the context units, which then supply copies of that activity state to all the other hidden units in the next cycle of operation. In making this assumption, however, as in making the back-propagation assumption, neural net modelers surrender all claim to neural plausibility. The widespread use of context units in neural modeling work over the last two decades is a testimony to the seemingly indispensible role that a read-write memory plays in most computations—a role that was already understood by McCulloch and Pitts (1943), who were perhaps the first to conjecture that reverberating loops could serve this function. Whether or not reverberating loops are a plausible mechanism for carrying information forward over short intervals may be argued, but few would argue that they are a plausible mechanism for carrying information forward over hours, days, months, and years. Thus, contemporary neuroscience does not provide us with a mechanism capable of encoding information and carrying it forward over intervals of indefinite duration in a computationally accessible form.

The question arises: Whose problem is this? Is this a problem for cognitive science and computer science? Should they worry that neuroscience does not recognize the existence in neural tissue of an addressable read-write memory mechanism capable, as Griffiths et al. (2010) posit, of "represent[ing] the structured knowledge over which...[behaviorally consequential] inference is defined." Or is this a problem for neuroscience? Should neuroscientists be looking for the addressable read-write memory mechanism that most cognitive scientists take for granted in their theorizing? Put another way, which science should we look to in pondering what may bridge this conceptual chasm in contemporary thinking about the machinery of cognition?

Building on the seminal work of Alan Turing (1936), computer scientists have developed a mathematically rigorous analysis of what is required in a powerful computing machine of any kind. In this analysis, a read-write memory is essential. In addition, as already noted, in all practical computing machines, the contents of memory must also be addressable, for theoretically well-understood reasons. Importantly, this is not pure theory; we are surrounded by working computing machines, all of which rely on addressable read-write memory.

According to most neuroscientists, however, there is no addressable read-write memory mechanism in the brain. Thus, we must ask: Which science provides a more secure foundation for reasoning about how the brain computes the symbols on which experientially informed decision making presumably depends? Relevant to this question is the following list of questions that I believe most theoretical neuroscientists would admit are both foundational and without consensus answers at this time:

- How is information encoded in spike trains (rate, interspike intervals)?
- What are the primitive combinatorial operations in the brain's computational architecture (its instruction set)?
- How does the brain implement the basic arithmetic operations?
- How does it implement variable binding?
- How does it implement data structures?
- How can changes in synaptic conductance encode facts gleaned from experience (e.g., distances, durations, directions)?

It would seem that a science that has yet to answer questions this basic is unable to provide a good foundation for theorizing about the machinery of cognition.

Behavioral Manifestations of the Composition of Functions

Many animals, perhaps most, have a home base from which they venture forth in search of food and reproductive opportunities and to which they must then return (Figure 3.2). This involves navigation, and navigation relies on the storage of information acquired from experience.

Experiments show that if the ant whose outward foraging path and homeward track is shown in Figure 3.2a were captured and displaced into unfamiliar territory, it would nonetheless run the same compass course it ran in returning to its nest (dashed line in Figure 3.2) for approximately the same distance and

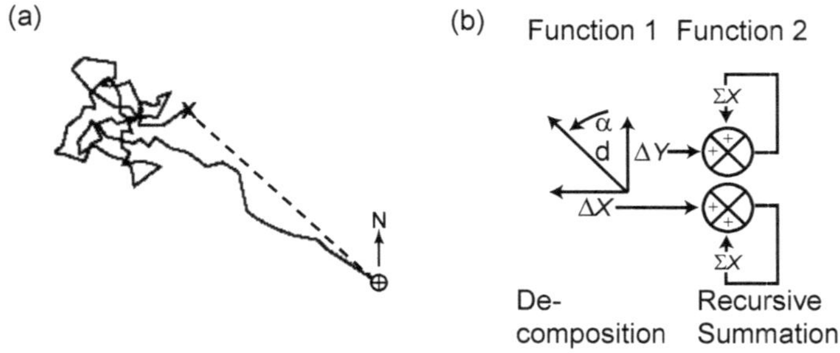

Figure 3.2 (a) Outbound (solid complexly twined line) and home bound (straight dashed line) track of a foraging ant (*Cataglyphis bicolor*) that found food at X (Harkness and Maroudas 1985). (b) The dead-reckoning computation that mediates the behavior seen in (a) requires Function 1—the decomposition of successive displacements into orthogonal components (ΔX and ΔY), which then serve as the inputs to Function 2—the recursive summation of successive orthogonal displacements. This second function requires carrying the sums forward in time via memory in a computationally accessible form, because the results of the current displacement are continually added to the sum of the previous displacements.

then begin a search for its nest (Wehner and Srinivasan 1981). This ability to run a prescribed course for a prescribed distance through unfamiliar territory implies dead reckoning. In the large literature on animal navigation, there is a broad consensus that dead reckoning plays a fundamental role. It also provides a clear example of the composition of functions, combining information gleaned from experiences spread out over time, as shown in Figure 3.2b. Dead reckoning is the integration of the velocity vector with respect to time to obtain the position vector as a function of time. In discrete terms, it requires the summation of successive displacement vectors, with the current displacement vector being continually added to the sum of the previous displacement vectors (Function 2 in Figure 3.2b). This recursive summation is itself an instance of the composition of functions; the output of a previous application of the function (the previous sum) serves as an input to the current application of the function. The other input is the current displacement vector. This recursive composition of a simple function (addition) requires a memory mechanism capable of carrying forward in time the information acquired from earlier experience (earlier displacements) in a form that permits that information to be integrated with subsequently acquired information (later displacements).

Animals, including ants and bees, use the Sun to maintain their compass orientation while dead reckoning. To do this, they learn the current solar ephemeris; that is, the compass direction of the Sun as a function of the time of day. The solar ephemeris varies with the season and the latitude of the navigator. Learning it involves the creation of a data structure, combining an encoding of the panorama around the hive or nest, the time of day as indicated by the brain's circadian clock, and the parameters of a universal ephemeris function, a function that specifies what is universally true about the solar ephemeris

(Lindauer 1957, 1959; Dyer and Dickinson 1994; Dickinson and Dyer 1996; Budzynski et al. 2000). The information encoded in the solar ephemeris may then be combined with information obtained either from its own foraging expeditions or from attending the dance of another forager to compute the current solar bearing of a food source; that is, the angle to be flown with respect to the Sun (Figure 3.3). Again, we see that information about different aspects of the experienced environment—in this case the solar ephemeris and the location of a food source—gleaned from very different sources by different computations at different times in the animal's past enter into the computation that determines the animal's present behavior. Insofar as we currently understand computational machines, and assuming that the brain is one, then we must conclude that the brain possesses the kind of memory that makes this possible, which is to say, an addressable read-write memory.

Setting courses between known locations is another simple example of the composition of functions with inputs derived from different experiences at different times in the past. Menzel et al. (2011), using radar tracking of individual bee foragers, have recently published the following ingenious experiment (Figure 3.4a): A foraging bee is shaped to visit one feeding station (F1 in Figure 3.4). Another forager is shaped to visit a different station (F2 in Figure 3.4). The first forager is seen to follow the dance of the second forager within the hive, thereby learning from the dance the location of F2. On a subsequent visit to its feeding station (F1), this first forager finds the beaker there empty. It then flies directly from F1 to F2. However, it does so only when the angle between the hive and the two stations is less than some critical value.

As a supplement to the dead reckoning of their location, ants and bees also use compass-oriented snapshots to wend their way through complex environments (Collett 2009). This implies data structures that are more elaborate than location vectors. A data structure is a structured vector or array of symbols

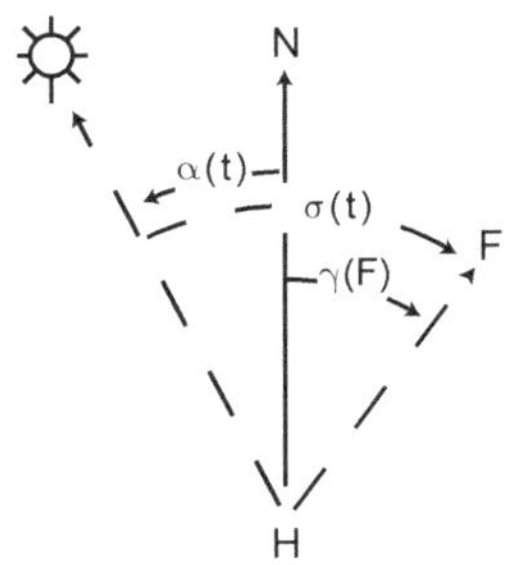

Figure 3.3 Setting a course from the hive to a food source whose location has been stored in memory, either during previous visits to it or from having attended the dance of another forager returning from it, involves the composition of the function σ(t), which gives the compass direction of the Sun as a function of the time indicated by the brain's circadian clock, with the function γ(F), which gives the compass bearing of F from the hive (H), to obtain α(t), the function that gives the angle which must be flown relative to the Sun to get to F from H.

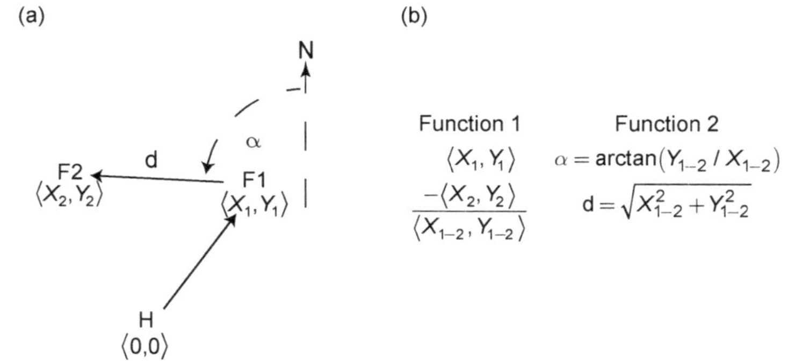

$$\begin{array}{r} \langle X_1, Y_1\rangle \\ -\langle X_2, Y_2\rangle \\ \hline \langle X_{1-2}, Y_{1-2}\rangle \end{array} \qquad \alpha = \arctan(Y_{1-2} / X_{1-2}) \quad d = \sqrt{X_{1-2}^2 + Y_{1-2}^2}$$

Figure 3.4 Schematic of the Menzel et al. (2011) experiment. (a) The forager learns the location of one feeding station (F1) by direct experience and travels repeatedly back and forth between it and the hive (H), bringing artificial nectar from the beaker at F1 to the hive. While in the hive, it observes the dance of a forager returning from F2, from which it learns the location of F2. On a subsequent visit to F1, it finds the beaker there empty. It then flies directly to F2, but only if the angle ∠F1HF2 is less than a critical value (k). (b) To set its course from F1 to F2, the bee must first compute the difference vector (Function 1), then transform that difference vector into its polar form (Function 2) to get the range (d) and bearing (α) of F1 from F2. In deciding whether to pursue this new course, it checks that the angle ∠F1HF2 is less than a critical value (Function 3). The inputs to Function 1 (the location vectors, $\langle X_{1,} Y_1\rangle$ and $\langle X_{2,} Y_2\rangle$ for the two feeding sources) come from different kinds of experience (direct vs. communicated) at different times. The location information, which is itself a simple data structure (because vectors are ordered number lists), must have been carried forward in time by a memory mechanism to make it accessible to these functions. This same memory mechanism enables the composition of these functions to inform the observed behavior of the bee.

such that the locations of the symbols within the array encode the relations between the symbols. For example, locations on a two-dimensional surface are encoded by ordered pairs of symbols for quantities: either two distances, if the vectors are Cartesian, or a distance and an angle, if the vector is polar. When we say that these pairs are ordered, we mean that the reference of a symbol depends on the order of the symbols. For example, by convention, latitude is given first, then longitude, or range, then bearing. In the case of a compass-oriented snapshot, the machine must store the snapshot—that is, a vector (string of symbols) that encodes the appearance of the landmark—together with a vector that encodes the compass direction in which the ant was looking when it took that snapshot. Yachtsmen will be familiar with the compass-oriented views and their function in navigation, because they are an important feature of pilot manuals. These manuals contain drawings or photographs of harbor entrances as seen from various approach directions. Recent research shows that for each compass-oriented snapshot, ants remember a sequence of bearing angles (Collett 2010). This implies a multifaceted data structure whose components are the snapshot, the compass orientation of the ant's visual system when the snapshot was made, the panorama behind the snapshot, and the sequence of bearing angles.

As explained below, the addresses in an addressable read-write memory make variable binding possible, and with it, the encoding of arbitrarily complex data structures. Because computational neuroscientists assume that there is no addressable read-write memory mechanism in the brain, variable binding and the creation of complex data structures are unsolved problems in contemporary computational neuroscience (Smolensky 1990; Sun 1992; Lòpez-Moliner and Ma Sopena 1993; Browne and Pilkington 1994; Sougné 1998; Browne and Sun 2001; Frasconia et al. 2002; Gualtiero 2008). When memory is everywhere and nowhere and not addressable, it is hard to see how to use it to bind values to variables.

As a final example, consider the implications of the brilliant series of experiments on food-caching scrub jays conducted by Clayton, Dickinson, and their collaborators in recent years (Clayton and Dickinson 1998, 1999a, b; Clayton et al. 2001b, 2003, 2009; Emery and Clayton 2001a, b; Emory et al. 2004; Dally et al. 2005, 2006). In the wild, these jays make tens of thousands of caches every fall, spread over many square kilometers in their high mountain habitat. They live for months by retrieving the contents of their caches. The experiments from the Clayton and Dickinson laboratory show that jays remember not just *where* they made each cache, but also *what* sort of food they put into it, *when* they made it, and *who* was watching (Figure 3.5). In choosing a cache to visit, they compare how long it has been since they made that cache (subtracting from the current date and time the date and time at which they made the cache) to what they have subsequently learned about how rapidly that particular content (*the what*) takes to rot, the current state of their preference

Figure 3.5 The food-caching and retrieving behavior of the scrub jay implies that each time it makes a cache—here only three are shown—it encodes the location of the cache (where), the date and time at which it was made (when), the kind of food cached (what), and which other jays, if any, were watching (who). These separate encodings of different aspects of a single episode must be stored in memory in such a way that the jay can recover the where, the when, the what, and the who for each different cache. This implies a sophisticated data structure. We know how to physically realize such structures—and their utilization—in computing machines, but this is possible *only* in machines with addressable read-write memory.

hierarchy for the different kinds of food they have buried, whether they have already emptied that cache, and which potential thieves were watching when they made it. Again, these behavior results imply that the record of each of many thousands of caches is a complex data structure, encoding in recoverable form disparate potentially useful information.

In reviewing their results, Clayton et al. (2006), contrast what they call "mechanistic" accounts of the causation of the jay's behavior (by which they mean associative accounts) with what they call "intentional" or "rational" accounts. They conclude, ruefully, that the intentional account better predicts their experimental findings. I say "ruefully" because Dickinson, at least, is a confirmed associationist. The implication of their contrastive language is that the intentional accounts are not mechanistic accounts. If one assumes that the brain does not have an addressable read-write memory—if one assumes that its memory is purely associative, which is the working assumption in most contemporary theoretical neuroscience—then, indeed, it is hard to see how one is ever going to be able to suggest a plausible reductionist account of these results (an account that specifies the underlying neurobiological mechanisms and the architecture that integrates their functioning). However, if one adopts the computational theory of mind and its materialist implications—namely, that the brain has an addressable read-write memory mechanism because such a mechanism is a sine qua non for an effective computing machine—then to contrast "mechanistic" with "intentional" (read "computational/representational") is bizarre. We know the kind of machine that can implement their intentional account. Their experimental results do not pose any profound computational mysteries. We can build autonomous robots that behave like these scrub jays, but only if we are allowed to put an addressable read-write memory in the robot's onboard computer. If there is a way to build an autonomous robot that mimics the scrub jay's behavior but does not possess an addressable read-write memory, no one knows what it is.

Addressability, Variable Binding, and the Creation of Data Structures

Implementing Variable Binding

The memory elements of an addressable read-write memory are bipartite: one part encodes the information to be carried forward; the other part is the address. This second part enables the machine to find the information conveyed by the first part when it needs it. The bipartite structure of the memory element makes it possible to bind a value to a variable, because the coding part and the address part use the same symbolic currency. The symbols in the coding part of a memory element take the form of bit patterns, strings of 1s and 0s, or, more physically speaking, sequences of binary voltage levels, or directions of magnetization, or some other physical realization of a binary number. Regardless of what the bit pattern encodes, it can be regarded as a number. This is a profoundly important point in thinking about physical (neurobiological) structures that could encode information: if a structure can encode a number, it can encode any kind of information whatsoever. Thus, in thinking about encoding, it suffices to think only about how to encode numbers. The address is also a bit pattern (hence, a number). Thus, the address of one piece of information can be stored in the encoding part of another memory element. When so stored, the address becomes the symbol for the variable. This makes the variable accessible to computation on the same terms as its value. It also implements variable binding. The machine finds the value of a variable by going to the address of the variable, where it finds the address of the value. This is called *indirect addressing*.

In short, giving addresses to memories not only makes them findable, it creates and physically implements the distinction between a variable (e.g., cache location or cache content) and its value (e.g., the vector that specifies the location of a particular cache or its content). This distinction is fundamental to computation. Without it, functions have no generality. Physically implemented widely useful functions, such as addition and multiplication, generate the values of output variables when given the values of input variables.

Building Data Structures

Addressable memory makes possible the creation of the arbitrarily complex data structures that reside in the memories of contemporary computers. Data structures are physically realized by array variables. An array variable is a symbol specifying the first address in a sequence of addresses where the data in a data structure are stored. The machine gets to those addresses by way of a computation performed on the symbol for that address. The computation makes use of the fact that the other addresses in the array are specified by where they are in relation to the first. The linear order of the addresses is used to encode which variable is which. That they occur within the sequence following the array address encodes the fact that these variables all relate to the entity symbolized by the array address. Thus, for example, the symbols at the addresses of the latitude and longitude variables give the addresses of where the values of latitude and longitude can be found for the cache symbolized by the array address.

The Universality of This Manner of Encoding Complex Structure

One might suppose that this way of physically realizing an encoding of complex structure is peculiar to modern computing machines were it not for the fact that complex organic structure is encoded in the structure of DNA in the same way. The gene is the memory element that carries heritable information forward in time, and it, too, has a bipartite structure. The coding part specifies the sequence of amino acids in a protein. The system gains access to the information in the coding part by means of the other part, the promoter. As in computer memory, the information in the coding part of a gene is more often than not the address of another gene. It is the recipe for constructing a protein called a transcription factor. A transcription factor binds to the promoter of a gene to initiate the transcription of that gene; that is, the decoding of the information that it contains. It has the same function as the address probe in random access computer memory (RAM). As in computer memory, a promoter may give access to a sequence of genes. If a foreign gene is inserted into the sequence, it will be activated when the sequence is activated by a transcription factor that binds to the promoter of that sequence.

The indirect addressing of the information conveyed in the coding parts of genes enables the hierarchical structure in the genetic code. For example, it makes possible a gene for an eye, so that whenever this gene is activated, an eye develops (Halder et al. 1995). The coding part of this gene does not specify the amino acid sequence for any protein in the eye that develops. Instead, it specifies the amino acid sequence of a transcription factor. This transcription factor binds to the promoters of several other genes, which themselves encode transcription factors, and so on, until one gets to genes that encode the proteins that make up the structure of the eye. The analogy to the manner in which data structures are encoded in random access memory is a close one.

If there were a large number of different architectures for the physical realization of the encoding of complex structures—structures with many different components relating to one another in many different ways—then the close analogy between the architecture of information conveyance in DNA and architecture of information conveyance in RAM would be a remarkable coincidence. If, on the other hand, there were only one or a small number of architectures that are physically realizable and effective, then this coincidence is no more remarkable than the striking similarity between the eyes of gastropods, arachnids, and vertebrates. These architectures have evolved independently, but their structure has been strongly constrained by the laws of optics; that is, by the function that they serve.

In contrast to the structure of a synapse, the structure of a DNA molecule is transparently suited to the carrying of information forward in time. It is no mystery how to encode a number in the nucleotides sequence of DNA; hence, it is no mystery how to encode any information whatsoever into the structure of this molecule. Moreover, the structure of DNA is so thermodynamically stable that it endures for years after the animal has died—longer than many computer disks endure. From an engineering standpoint, DNA is a perfect marvel of a universal information-conveying medium. Because DNA and the complex molecular machinery for reading the information it contains already provide most of what is required in an addressable read-write memory—everything but the machinery for writing to it—it is tempting to conjecture that the neurobiological memory mechanism has co-opted either DNA or its close cousin RNA to perform this indispensable function in the machinery of cognition.

4

Building Blocks of Human Decision Making

Nick Chater

Abstract

This chapter considers the types of building blocks out of which the mechanisms for decision making in humans may be constructed. Two distinct types of mechanisms are distinguished: mechanisms involving *substantive* rationality, which embody specific patterns of behavior, values, or actions; and mechanisms that embody *formal* rationality, which seek to make an agent's beliefs, values, and actions as coherent with each other as possible. It is suggested that both types of mechanisms are likely to play an important and complementary role in human decision making, although different theoretical positions place very different weight on the relative importance of each.

Introduction

Many apparently unity biological functions are carried out by a wide variety of processes. For example, digestion involves a wide variety of organs and physical and chemical processes to carry out what might at first appear to be a unitary task. Similarly, locomotion, for humans, involves not merely the legs, but complex motions of the whole body; by contrast, an aquatic bird utilizes very different systems to walk, swim, and fly. From a Darwinian perspective, there is no reason to suppose that a single function should be achieved by a single system—indeed, the reverse seems to be the norm.

A Darwinian view of decision making, then, raises the possibility that decision making is not the result of the operation of a single, seamless process. Instead, decisions arise from a complex interplay of different cognitive mechanisms: from the integration and interpretation of perceptual information; to drives, motives, and preferences; to processes of inference, planning, and problem solving; and the application of relevant memories of past experiences, inferences, and decisions. While it is possible, from an abstract standpoint, to outline a uniform decision-making system based purely on beliefs and utilities, from a Darwinian point of view, we might expect that a much more elaborate and heterogeneous analysis is required to understand the cognitive and neural mechanisms that underpin decisions. This more heterogeneous approach to decision making is the focus of this chapter.

Thus, although an abstract perspective on decision making is often applied in models of animal (including human) decision making, in the form of various types of optimality analysis (Anderson 1990; Houston and McNamara 1999), any such approach must be complemented by an analysis of the computational building blocks from which decisions are constructed. In this chapter, my primary focus is on human decision making; it is likely that comparative analysis of decision making across species, which is beyond the scope of this chapter, might cast new light on the human case.

I begin by setting the scene, discussing the variety of criteria by which decisions may be assessed, distinguishing substantive and formal criteria for the evaluation of decisions and of rationality more generally, and considering how these relate to different styles of theoretical explanation of decision making. A key argument will be that formal and substantive rationality are complementary: both are required to construct a theory of decision making. Nonetheless, the balance between the two approaches is a major source of legitimate theoretical debate. Thereafter I consider possible building blocks for decision making which relate to substantive aspects of decision making and address the question of whether there are also building blocks which relate to formal aspects of decision making. A brief discussion considers wider implications of these considerations for theories of decision making in humans and nonhuman animals.

Two Ways of Evaluating Decisions

There are two types of criterion against which human or nonhuman animal decision making may be evaluated: substantive and formal. This distinction is closely related to Hammond's (2007) important distinction between correspondence and coherence theories of rationality.

A *substantive* evaluation of a decision requires an external standard against which the quality of decision can be measured. Standards can be quite varied. For example, in ethics, one possible line is that the goodness of a decision might be evaluated in terms of the "goodness" of its consequences (e.g., Broome 1991). Similarly, decisions concerning matters of art and design, broadly construed, are evaluated according to (rather loosely defined and frequently contested) aesthetic standards. In early economic theory, it was often assumed that each individual agent (perhaps animal, as well as human) possesses some internal measure of hedonic state, and that decisions could be evaluated in terms of how far they lead to positive hedonic states and avoid negative hedonic states (for discussion, see Cooter and Rappoport 1984). Evolutionary biology has suggested a further substantive standard; namely, that decisions can be evaluated by their impact on inclusive fitness.

Formal criteria for the evaluation of decisions focus not on the impacts of any individual decision on some external criterion of interest, but rather on consistency relationships between decisions. From such a formal standpoint, there is "no accounting for tastes." Any set of substantive criteria—whether ethical, aesthetic, hedonic, or concerned with fitness—are presumed to be required to adhere to standards of consistency. A paradigm example is *transitivity*: if A is preferred to B, and B is preferred to C, then, prima facie, A should be preferred to C; if this is right, it holds in virtue of the structural relationships between these preferences, and independently of their substantive basis (i.e., it does not matter what A, B, or C actually are). Formal criteria for the evaluation of decisions have been the focus in economics and philosophy.

The relationship between substantive and formal criteria is contested. One possible viewpoint is that substantive criteria are all that is required. For example, it might be argued, from the point of view of natural selection, that all that matters about a certain decision is whether it increases or decreases inclusive fitness. Whether collections of decisions are consistent with each other may appear to be a secondary question of little or no evolutionary import; in any case, to the extent that a person or nonhuman animal single-mindedly pursues any standard, such as maximizing inclusive fitness, consistency will then follow automatically. Advocates of "ecological rationality" (Gigerenzer et al. 1999) can be viewed as adopting this type of standpoint.

Conversely, formal criteria may be seen as primary. For example, if formal consistency constraints, such as transitivity, are assumed to hold, then it is possible to "reveal" the utility that an agent (perhaps implicitly) associates with possible states of affairs, either in ordinal terms (i.e., providing a rank ordering of preference across state of affairs) or even, with slightly stronger assumptions, in cardinal terms (so that each possible state of affairs can be associated with a real number, and decisions whose outcomes are uncertain can be made by evaluating the expected utility of the decision; i.e., the expectation of this real valued quantity). From this standpoint, as long as decision makers are consistent, they can be characterized by a set of utilities over different states of affairs (and, perhaps, a set of subjective probabilities of different states of affairs coming about): these utilities are "revealed" by the preferences the decision maker exhibits. This viewpoint suggests the possibility that the theory of decision making can be cleanly separated from mere questions of taste; that is, the different utilities associated with each possible outcome are not really part of the theory of decision making proper.

I will argue that substantive and formal criteria for rationality must interact closely to provide a credible analysis of decision making: neither is sufficient to provide a theory of decision making, and the balance between the two is of critical theoretical importance. First, however, let us consider, in turn, putative cognitive building blocks that appear directly relevant to substantive and formal factors in decision making.

Cognitive Building Blocks 1: Substantive Factors in Decision Making

Decision making is easy if the agent possesses a single goal and if it is always clear which action will serve to best achieve that goal. Of course, in real-world decision-making environments, this is not the case. Human decision makers typically face a complex mix of conflicting goals. Even if the goal is fixed (e.g., maximizing fitness, eating as many calories as possible, maximizing happiness) the agent typically has no way of directly assessing the impact of a particular action on that goal: principles of formal rationality alone are not sufficient to establish which choices will best serve the overall goal. Two reasons account for this: (a) the agent has *insufficient* information about the enormously complex real-world environment to determine the fitness consequences of its actions and (b) even if this information were to be available, the *computation* required to determine the fitness consequences of each action would be intractable.

Simon (e.g., 1955) and many later theorists (e.g., Gigerenzer and Selten 2001; Payne et al. 1993) have argued that some drastic shortcuts are required. One simplification is to have subgoals, which generally help to achieve the goal. So, for example, drives for food, water, and sex as well as the fear of strangers, heights, or snakes may reasonably be viewed as providing useful subgoals for achieving the larger goal of greater fitness. If these subgoals are built into the organism (perhaps being genetically specified), they can provide substantive constraints on rationality: they specify how the organism should deal with specific states of affairs, rather than merely specifying consistency relationships between the beliefs or actions of the organism.

Yet achieving even these subgoals may be challenging: optimizing food intake by hunting or foraging in a natural environment, for example, is an enormously difficult problem on both informational and computational grounds. Further subgoals (i.e., further aspects of substantive rationality) might be internalized (e.g., preferences for certain foods, built-in constraints concerning how long a particular foraging episode should last). The general approach of decomposing complex goals into simpler subgoals is likely to be important (Newell and Simon 1972). It is not clear, however, under which conditions the subgoals are really significantly easier, in informational or computational terms, than the original goal, and when or whether the process of decomposition terminates.

Another complementary type of shortcut involves achieving a goal not by breaking it into subgoals, the solution of which can guide decision making, but by deploying prepackaged behavior. Much animal behavior seems to employ such shortcuts: while an underlying drive or subgoal may orient an animal to the importance of food or sex, the actual process of catching, killing, and eating a prey animal, or performing rituals of courtship and mating, may primarily be prepackaged. Such prepackaging, which presumably derives from natural selection, allows decision-making challenges to be addressed by trial and error over evolutionary time, rather than having to be addressed on the fly by an individual agent. Of course, even prepackaged behaviors must be somewhat flexible. The routine for catching and eating a prey animal must vary on each occasion, depending on the environment as well as on the responses and characteristics of the prey under attack. Thus the problem of decision making is recast, perhaps in a more tractable form, rather than eliminated. To the degree that the agent needs to be able to generate new ways of addressing novel situations, it needs to *relate* knowledge of old cases to handle the new cases. This type of relation between knowledge of old and new (discussed further below) is naturally viewed as a type of *inference*, underpinned by principles of formal rationality. However, to the extent that behavior is prepackaged and inflexible, the question of coherence with other behaviors need not arise; it may be specified directly, as an aspect of substantive rationality.

A further type of shortcut to present decisions is simply to base them on past decisions or experiences: either one's own or those of others. Regarding learning from one's own experiences, there is an enormous literature on animal and human learning which may inform how present decisions can be influenced by past decisions and experiences. As in the discussion above, two types of information may be learned. First, an agent can learn the "value" of particular states of the world in relation, for example, to whether they are close to achieving a subgoal or, conversely, close to a feared outcome. Such values can, of course, be learned from past experience of what followed from the same, or similar, state on a previous occasion. Second, an agent may simply learn what to *do* in a particular state, irrespective of whether the state is good or bad. Both types of information can also be learned by observing the behavior of others. Such learning is usually discussed under the heading of *imitation*, although the range of forms of copying behavior is complex and varied (Hurley and Chater 2005). For example, the reaction of others may indicate that a particular state is good or bad; thus, a parents' overt fear of dogs might be transmitted to their child. Particular behaviors, and the decisions they embody, may be transmitted directly from person to person through emulation, irrespective of whether evaluative or goal-related information is conveyed.

So far I have stressed how certain evaluations, subgoals, and individual behaviors may be either preprogrammed through natural selection or learned by experience, thus simplifying the online decision problem faced by the agent. According to theoretical accounts in which there are several separate mechanisms for learning (e.g., Dayan et al. 2006), these various subsystems will be among the key building blocks for decision making, and competition between them for control of the agents decision making will be of great importance.

Many theoretical accounts also distinguish between relatively basic learning mechanisms and mechanisms that draw directly on memories for past learning episodes. Theories of memory that distinguish between episodic memory and so-called semantic memory (for general knowledge, including, e.g., explicitly taught rules of conduct) will be likely to see both sorts of memory mechanism as building blocks for decision making, at least in humans. More broadly, given the centrality of decision making in cognition, it is not surprising that the building blocks of decision making extended to encompass whatever we believe the building blocks of thought to be.

Mechanisms of learning and memory may have quite a direct effect on decision-making performance, in a range of ways, thus providing further substantive constraints on rationality. For example, Breland and Breland (1961) describe an interesting type of behavioral clash in which animals retrieve and manipulate objects (e.g., a pig is trained to drop objects into a "piggy bank"), and successful performance is rewarded by food. If the animal is extensively trained, it learns the relationships between the object and the food; this learning leads to the emergence of instinctive (perhaps Pavlovian) food behaviors from the mere presence of, and interaction with, the objects. Racoons exhibit "washing behavior" with the objects associated with food, apparently typically preparatory to eating; chickens hammer and peck; pigs root. This behavior sometimes becomes dominant, partially replacing the behavior that is being rewarded, and thus substantially reducing the amount of food that the animal receives. Thus, the "decision" to engage in such apparently task-irrelevant behavior, which seems baffling from an analysis of the structure of the problem the animal faces, can be explained by reference to mechanisms (here, presumably built-in) which wrest control of the animal's behavior. This type of phenomena may result from the interaction of distinct decision-making systems (e.g., Dayan et al. 2006) or it may be viewed as resulting from distinct uses of the same information—the contingency between the trained behavior and food, and the resulting expectation of food (e.g., Chater 2009). In any case, the existence of "substantive" and probably built-in patterns, such as links between food and food-related behaviors, appears to play a crucial role in this type of animal decision making. Other domains in which such factors may be particularly relevant include addiction, time-discounting, and clinical disorders of decision making (e.g., Loewenstein and Elster 1992).

In contexts in which an agent must interact with other agents, building blocks underpinning social cognition may become important. At a broad level, it may be critical to be able to conceive of other agents as possessing minds. This potentially involves seeing others as decision makers like oneself, but perhaps with different goals or beliefs. Such a theory of mind (Wimmer and Perner 1983), if it can rightly be described as a separate cognitive component, is likely to be crucial to reasoning in contexts in which agents compete or collaborate. While perhaps not uniquely human, nonhuman animals appear to have much less elaborate notions of other's minds than humans do. For example, chimpanzees appear to be able to attribute intentions and perceptual abilities to others but may not be able to entertain the possibility that others may be guided by false beliefs (e.g., Call and Tomasello 2008). Failures of mind-reading may explain a number of persistent decision-making biases, such as those exhibiting what Heath and Heath (2006) call the *curse of knowledge*: decision makers tend to assume that other agents know what they do and react accordingly, even when this assumption is plainly unjustified.

At a somewhat more general level, many theorists have argued that limited computational resources provide a crucial constraint on decision making (Simon 1955). In particular, rather than integrating a rich set of information from current perception or from memory, some theorists have proposed that the decision maker may, instead, build cognitive mechanisms for decision making by recruiting, on the fly, relatively simple rules to govern decision making (e.g., Gigerenzer et al. 1999; Payne et al. 1993). According to one instantiation of this view (Gigerenzer and Selten 2001), the decision-making system draws upon an "adaptive toolbox" that contains a range of simple rule-like processes which can be combined to solve the problem in hand. A critical question for this point of view is: How it is possible to assemble the appropriate components from the adaptive toolbox to solve the problem at hand? This might appear to require general-purpose reasoning of considerable complexity. However, one attraction in this approach is that if the appropriate set of "fast and frugal" rules can be found by chance, then such rules may be surprisingly effective and robust, compared to more complex computational methods (e.g., Gigerenzer and Goldstein 1996).

Representational constraints may provide an equally important substantive restriction on theories of decision making. Many theoretical proposals in decision making assume that agents can represent the absolute values of decision-related quantities, such as money, utility, probability, quality, or time. However, it is arguable whether the basis of human perception and judgment is, by contrast, *local* binary comparison. Comparisons are local in two senses: the perceived loudness of a sound, painfulness of a pain, or saltiness of a food is determined by comparison with similar, and temporally close, alternatives (Laming 1997; Stewart et al. 2005). So, for example, the amount people will pay to avoid a small electric shock will depend on the severity of a shock in relation to other recent shocks and the size of the payment in relation to other recent payments (Vlaev et al. 2009). People will therefore pay very different amounts to avoid the same shock, in different contexts, just as people are willing to pay very different amounts for roughly the same coffee in their home or at a cafe. Moreover, and as a consequence, comparison is also local in the sense that people can only compare one *dimension* of a product or experience at a time. Thus, for example, one chocolate bar may be larger, the other may taste better. The cognitive system must use qualitative reasoning to determine which should be chosen: no quantitative integration of the distinct dimensions is available, because values of each dimension are represented only in terms of local comparisons; there is no stable numerical value that can be compared between dimensions (see Stewart et al. 2006).

One implication of this viewpoint is that choice processes are attribute based, rather than object based. That is, rather than making an overall evaluation of each chocolate bar based on its various features and choosing the best, the decision maker must qualitatively decide whether *more* chocolate has greater or lesser importance than *better* chocolate. A powerful line of argument that people do indeed use attribute-based decisions is that people will easily and consistently decide that a chocolate bar that is both slightly bigger and slightly better quality should be preferred, while noisily and slowly deciding between items where the dimensions conflict (e.g., Loomes 2010; Scholten and Read 2010). Nonetheless, where people can freely choose how to sample information about objects and their properties, they often choose to sample object by object, presumably obtaining an overall evaluation of each object; this suggests that people may integrate attributes to some degree (e.g., Payne et al. 1993).

These putative representational limitations provide an alternative explanation to that of computational limitations for the lack of integration between different attributes in decision making: if attributes cannot be represented on an absolute scale, then, a fortiori, they cannot be mapped into a stable, common currency against which they can be compared. It is not that the computational problem of integration is too difficult; rather, the problem is ill-defined.

Much of the discussion so far has assumed implicitly that the space of decision problems is relatively small, so that, in particular, the decision maker may encounter the same decision more than once (and past learning can be brought to bear). In most real-world environments, however, this is not the case. Previous experience must be brought to bear on a current decision problem in a flexible way. In light of accounts of learning and memory, it is common to assume that what is required is some mechanism of generalization, based on the similarity of past and present decisions (e.g., Shepard 1987). A prime mechanism for establishing such generalization is *similarity*.

From this perspective, it may be assumed that similarity is a key substantive building block of decision making (and perhaps one that is cognitively primitive), and some theories of decision making assume this (e.g., Klein 1998). It is equally plausible that similarity should not be viewed as basic, but that similarity must itself be defined in terms of the decision problem: two situations are relevantly similar if they can be solved in the same way (cf. Goodman 1970). Arguably, one lesson from the study of case-based reasoning in artificial intelligence (e.g., Kolodner 1993) and the study of analogical thinking in psychology and philosophy (e.g., Gick and Holyoak 1980) is that a problem-independent notion of similarity cannot easily be defined (for one proposal, see Hahn et al. 2003). A more general, alternative perspective on the application of past knowledge to present circumstances is that this requires some process of inference: this involves bringing beliefs about past experiences and beliefs about present experience into a consistent alignment. Such matters of consistency lead naturally into the domain of formal, rather than substantive, rationality. It is to formal rationality that we now turn.

Cognitive Building Blocks 2: Formal Factors in Decision Making

If the range of decisions that humans make were limited, then formal rationality, which considers structural relationships between different decisions, might be of relatively little interest. Mechanisms of natural selection on learning could separately ensure good (or good enough) decisions, if a particular type of decision is of sufficient importance, and occurs with sufficient frequency, for the successful choices on that type of decision to have a substantial effect on fitness. Thus, natural selection might operate directly on key decisions involved in, say, food choice, foraging, evading predators, courtship and mating, and so on. In these cases, we might conjecture that substantive rationality (i.e., making choices that lead to "good" outcomes) could be achieved directly through the operation of special-purpose "cognitive building blocks" which would carry out each type of decision effectively. Whether such decisions would be fully coherent with each other would be of secondary interest. Moreover, the issue of coherence would arise only in situations in which special-purpose cognitive building blocks to deal with specific types of decision are in competition (e.g., when the best patch to forage is also that with the highest risk of predation).

Yet, as has been hinted at already, in human decision making, and arguably in many areas of nonhuman animal decision making as well, there is a vast space of possible actions, or sequences of actions, from which a choice must be selected. It is therefore important for the agent to be able to transfer information about past experiences and decisions, and any other relevant background knowledge (including, in the case of human decision makers, knowledge that has been learned by verbal instruction), to *infer* the best course of action for the case at hand. Such inferences can naturally be viewed as maintaining consistency between different beliefs, values, or actions: new decisions are made so that they are consistent with previous behavior and experience. Normative theories of how beliefs, values, and so on *should* be updated, by a rational agent, to provide a standard of consistency against which such inferences can be judged. Such normative accounts include logic (which can, among other things, be viewed as a theory of consistency of *beliefs*), probability (which, on the subjective interpretation of probability, provides consistency conditions on *degrees of belief*), decision theory (which provides consistency conditions on *degrees of beliefs, values, and choices*), game theory (providing consistency conditions in strategic interactions between agents), and so on.

Yet how, if at all, might such normative accounts of consistency relate to the cognitive building blocks of decision making? Opinions differ. One viewpoint, already mentioned, is that what matters for cognition is the efficacy of specific decisions, or local decision-making heuristics in particular environments, and therefore that questions of consistency are of marginal importance. As mentioned, this viewpoint is strongest where the space of possible decision types is small, so that the agent can map present decisions directly to past decisions (and hence the effectiveness of a strategy in the past may be a guide for efficacy in the present). By contrast, many cognitive theories assume that the brain embodies machinery whose role is to maintain such consistency (at least, to the extent possible), so that it is possible to generalize flexibly from prior experience to present problems. Such accounts embody the idea that the brain has access to building blocks embodying principles of *formal* rationality, as well as those embodying particular substantive constraints.

Consider, for example, theories of reinforcement learning. Suppose an agent likes state A (eating chocolate) and dislikes state B (feeling sick), and the agent learns that state B often occurs after state A. Consistency conditions may then operate to reduce the agent's liking for state A and consequently to reduce the agent's willingness to perform actions to bring about state A. This style of reasoning (in a much more general and sophisticated form) lies at the heart of modern machine learning–reinforcement learning algorithms, such as Q-learning (Watkins 1989). Different types of information may (or may not) successfully be brought into consistency. Suppose that a previously valued state (eating chocolate) normally leads to no ill effects and any decisions which lead to chocolate eating are eagerly pursued. If the agent now experiences sickness from chocolate (in the case of experiments with rats, sickness might actually be induced by radiation; e.g., Garcia et al. 1955), then how does the agent respond? Aside from the consistency conditions mentioned earlier, which may now devalue chocolate (and may be observed if the agent now refuses to eat chocolate, when it is presented), there is a separate consistency question: Will the agent now eliminate previously learned *behaviors* that would have led it to eat chocolate? With rats, the empirical picture is mixed (for recent data, see Balleine et al. 2005). If, during the learning phase, the agent has learned only that an action is good (but has not stored the reason why, i.e., because the action leads to chocolate), then it is at least possible that the animal cannot recognize the inconsistency between pursuing the action now that chocolate is devalued. The animal may then, for example, continue to work for food that it does not subsequently eat. It seems likely that human and nonhuman animal behavior embodies a combination of information about which *subgoals* are good and which *actions* are good. In part, this has led some theorists to postulate two, perhaps neurally, distinct systems sensitive to different kinds of information (e.g., Dayan et al. 2006). It may also be possible that a single system may learn the values of states (e.g., what is worth pursuing as a [sub]goal) and the value of actions (which may not necessarily be tied to the goal they achieve). In such a unified system, goal devaluation may lead to a "clash of reasons" (Chater 2009) within a single system. That is, the action is no longer justified because it achieves a previously appealing goal which has now been devalued;

however, the cognitive system still has a record that that action is good (even though this record does not specify why), and hence has a prima facie reason for repeating it. For the present argument, the key point is that mechanisms of reinforcement learning involve establishing *consistency*, rather than promoting any particular belief or choice: such mechanisms aim to achieve *formal*, rather than substantive, rationality.

Let us now consider a very different theoretical framework that is even more transparently focused on formal rationality. Recently, in the fields of cognitive science and artificial intelligence, there has been considerable interest in the idea that many aspects of cognition (including perception, motor control, language processing, and high-level thought) can, at least partly, be modeled as Bayesian *inference* (e.g., Chater et al. 2006). Bayesian inference can be viewed as a process of maintaining consistency across an agent's "degrees of belief." The core idea is that degrees of belief across relevant aspects of an agent's knowledge can be represented as probabilities, and these degrees of belief will be consistent only if they adhere to the actions of probability (e.g., Savage 1954). The canonical Bayesian inference occurs when new information is learned; that is, some propositions about which the agent had some uncertainty have their probability set to 0 or 1. Notice that, necessarily, the probabilities (0 or 1) associated with the new information are inconsistent with the previous set of degrees of belief. A process known as Bayesian conditionalization provides a mechanism for restoring consistency, in the light of this new information, and this may involve substantial revision of past assumptions, as when an ambiguous perceptual stimulus is resolved by additional cues or a sentence must be re-parsed when a surprising continuation is encountered (as in the oft-cited example "the horse raced past the barn fell").

The Bayesian approach to cognition may be useful purely as a description of the information-processing problem that the brain faces: it may be provide a "rational analysis" (Anderson 1990) or "computational-level" (Marr 1982) explanation. However, as I have mentioned, it is also possible that there are neural and cognitive mechanisms specifically designed to carry out (approximate) Bayesian inference. Indeed, there are a number of proposals which explore how rapid approximate Bayesian inference might be carried out by networks of (somewhat) neuron-like processors (e.g., Dayan and Abbott 2001). This perspective is particularly well-developed in vision (e.g., Geman and Geman 1984; Yuille and Kersten 2006). Similarly, in more traditional artificial intelligence, in which logical inference, rather than probabilistic inference, is the focus, cognitive scientists have suggested that the brain might have dedicated inference mechanisms (e.g., Rips 1994; Johnson-Laird 1983; although see Oaksford and Chater 2007).

Notice, of course, that Bayesian models do not promote the importance of formal rationality to the exclusion of substantive rationality. Given only an image, and the admonition to stick to the formal canons of Bayesian inference, the visual system would be in a state of inferential paralysis. Rather, a specific

Bayesian model embodies a wealth of substantive knowledge, for example concerning optics, geometry, and visual regularities (Marr 1982). Bayesian principles are used to find the conjecture about the structure of the environment which is most coherent with the observed sensory data and this "prior" knowledge.

Finally, theories of learning typically involve mechanisms for sophisticated inference; hence these two can be viewed as embodying building blocks of formal rationality. As noted above, a behaviorist psychology eschews inferential mechanisms; each individual behavior is assumed to be reinforced, or stamped out, independently based on its reinforcement history. As soon as a reinforcement-based perspective is taken to sequences of actions (i.e., an animal must learn a pattern of responses rather than merely to emit a single response), the picture changes dramatically. Now the value of a present action depends not merely on its immediate consequences, but on what further actions will be taken, and *their* consequences. Thus, for example, the action of caching food may now have high value, if an animal is likely to retrieve that food later, but it would have no value if the food will not be retrieved (e.g., because the agent cannot recall its location). More broadly, if actions are part of a *plan*, then the value of each component action may be dependent on the rest of the plan being carried out successfully. So the values of actions (and, indeed, states) are interdependent and hence cannot be specified arbitrarily: they must obey consistency conditions. Normative decision theory provides a general framework for specifying such consistency conditions. For example, theories of reinforcement learning, such as Q-learning (Watkins 1989), give a specific set of methods to align the values of actions and states in certain sequences of actions.

More broadly, theories of consistency, in whatever domain, provide the basis for *inference*; that is, given a new state or action, such theories can potentially be used to infer the value of the state or action, if sufficient collateral information is available. So, for example, a novel action can be valued, other things being equal, on the basis of the values of the states to which it leads (weighted by their probabilities). Thus, cognitive mechanisms of inference, planning, learning, and decision making can all be viewed as providing building blocks which embody formal rationality.

There are, of course, potentially many further possible domains for which the mind might potentially have special-purpose formal machinery. For humans, these include reasoning about causality (e.g., Pearl 2000), reasoning about other minds (e.g., Wimmer and Perner 1983), and perhaps even reasoning about aspects of mathematics (Dehaene 1997). The degree to which nonhuman animals should be attributed such reasoning abilities, the degree to which such reasoning involves specialized constraints, rather than merely be an extension of general-purpose reasoning, and the question of the appropriate formal characterization of such inferences are all open issues at present.

General Discussion

I have argued that there are two very different types of mechanisms that may potentially serve as building blocks for decision making. One set relates to substantive rationality: these mechanisms help specify the specific goals of the cognitive system or they specify behaviors that directly achieve such goals. A very different set of mechanisms relates to formal rationality. Methods as distinct as reinforcement learning, putative "mental logics," Bayesian networks, or mechanisms for predicting the behavior of others are concerned with bringing different beliefs, values, or choices into a consistent balance (or some appropriation of this, given that detecting inconsistency, let alone fully repairing it, is typically computationally intractable).

Individual research traditions focus differentially on the different types of mechanisms. At one end of the spectrum are approaches that give a limited, or perhaps null, role for formal rationality. From this perspective, a paradigm of decision making might be, for example, a frog reflexively jumping when it sees a dark looming shape or reflexively snapping at moving dark, convex blobs (Lettvin et al. 1959). Special-purpose circuits underpinning such responses (including circuitry in the retina itself) are perhaps substantively rational, in relation to wider goals such as avoiding aerial predators and eating flies, which might themselves be justified in relation, from an evolutionary standpoint, by contributing to fitness. No such "justifications" need be available to the mind of the frog, of course; the reflex is present purely because it generally achieves the frog's goals (and, indeed, the reflex will still operate, mostly likely relentlessly, even when the frog's goals are frustrated, e.g., if the convex blobs are experimental stimuli, not flies).

Clearly, much animal and human decision making is more complex, and more flexible, than the reflexes of the frog. However, the basic explanatory principle may be the same: particular choices are made not because they are justified by consistency with other choices, values, and beliefs (i.e., not in virtue of formal rationality) but because they *work* (i.e., they help achieve the organism's goals, and perhaps ultimately fitness).

At the other end of the theoretical spectrum stand accounts which see formal rationality as key. Here, the paradigm of decision making is not the frog's reflex action but the careful deliberations of the scientist or economist. According to this viewpoint, the decision makers' beliefs form an intricate and seamless web (Quine 1951), such that disturbance in one belief may, by appropriate chains of reasoning, link to arbitrarily distant parts of the web. In addition, the decisions themselves might (though they need not) be assumed to obey consistency constraints from utility theory, decision theory, and game theory to ensure, for example, that the value of an action is an appropriate function of the values of its consequences, weighted by the subjective degrees of belief that each of these consequences might occur (one might think of this as decision making by cost-benefit analysis). This general style of explanation is employed in traditional artificial intelligence, standard psychological accounts of reasoning, cognitive development (Gopnik et al. 1999), Bayesian cognitive science (e.g., Chater et al. 2006), and the majority of psychological and economic models of individual decision making (e.g., Kahneman et al. 1982).

The debate between theoretical approaches based on substantive versus formal rationality is of critical importance for understanding the building blocks of decision making. In cases where cognitive mechanisms can be fully explained using substantive rationality (i.e., without recourse to consistency across cognition), we may expect decision making in each domain to be relatively independent, and perhaps even subserved by dedicated psychological or neural machinery (just as the frog has dedicated circuits underlying its reflexes). Of course, different mechanisms might share similarities or draw on a common "toolbox" (Gigerenzer and Selten 2001). However, each decision must be considered separately: by considering its typical consequences in the typical environments of the decision maker. At the other extreme, if formal rationality is central, one might suspect that there is one (or perhaps more than one) general-purpose decision-making system.

By contrast, to the degree that a particular behavior is generated by formal rationality, it is then dependent on *coherence* with other aspects of the organism's thoughts and behavior. This follows because formal rationality depends on such coherence constraints. It is possible that such coherence constraints operate only within a fairly restrictive domain; a reinforcement learning mechanism, for example, such as Q-learning (Watkins 1989), mentioned above, may act to ensure that the organisms valuation of, say, eating chocolate takes into account any consequences of eating chocolate (such as being sick). Such a reinforcement learning system may, however, be oblivious to (and hence not coherent with) other world knowledge that the organism possesses. So, in humans, aversion to a particular food when followed by sickness might arise from some kind of general-purpose reinforcement learning mechanism (although it might, of course, be generated by a special-purpose, substantively rational mechanism; e.g., Garcia et al. 1955). Nonetheless, this putative reinforcement learning mechanism may be impervious to a person's explicit knowledge that sickness was actually caused by influenza. In the extreme case, in which the organism applies formal principles of rationality in an attempt to maximize the global coherence of its beliefs, desires, and actions, we should expect behavior to be "cognitively penetrable" (e.g., Pylyshyn 1984). That is, behavior should be open to potential influence from arbitrary pieces of information. Behaviors generated directly from substantive rationality will typically not be so readily influenced, if at all, because such behaviors are, by assumption, "hard-wired" by natural selection, rather than generated on the fly in the light of specific circumstances and knowledge. To take an extreme example, the human blink reflex occurs even if experience or verbal reassurance indicates that the looming object will not actually contact the eye—the reflex cannot be overridden. Similarly, a wax apple, for example, may *look* appetizing even if it is known to be made out of wax, yet the *action* of biting the apple can clearly be overridden by such knowledge. Degree of cognitive penetrability provides a powerful criterion for assessing how far a choice or other behavior depends on formal rationality. Practical application of this, however, is not always straightforward., For example, it is not easy to distinguish between actions that are generated by mechanisms which are directly cognitively penetrable (i.e., which embody formal rationality, to some degree) and those which are generated directly by substantive principles (i.e., "primitive" drives or rigidly applied heuristics), which may, nonetheless, be on occasion overridden by central (and cognitively penetrable) cognitive processes. The problem is reduced if we assume that there is a single decision-making system (Chater 2009), but it is amplified if multiple systems are envisaged (Dayan et al. 2006).

It seems natural to suppose that our choices are frequently determined by some combination of mechanisms concerning substantive and formal rationality. Hume (1739/2007) suggested that reason should be slave to passions: in our terms, formal rationality should help achieve the goals set by substantive rationality. For example, formal rationality, such as Bayesian visual processing or reinforcement learning, might help an animal achieve objectives or engage in behaviors that are directly mandated by substantive rationality (e.g., eating, caring for children, sex, and so on). But is this necessarily correct?

One argument might be that reasons should shape, and not merely be slave to, passions. From an evolutionary standpoint, the "drives" and fixed patterns of behavior delivered by mechanisms of substantive rationality are proxies which may, on average, lead to improved fitness. Perhaps one role of formal rationality is precisely to moderate, direct, and control the impulses generated by mechanisms for substantive rationality, so that behavior is more precisely tuned to the moment; reasoning, planning, and self-control all seem to be high-level cognitive phenomena that appear to shape and give coherence to, rather than being blindly guided by, the "passions" of substantive rationality.

A very different challenge to Hume's viewpoint comes from approaches which largely eschew formal rationality, such as behavior-based robotics (Brooks 1991) or an emphasis on "simple heuristics" (e.g., Gigerenzer et al. 1999). From this perspective, formal rationality, which seeks to establish consistency across different beliefs and actions, has no useful cognitive role. Putting the point tendentiously, from this standpoint, reason, as traditionally conceived, can, from a psychological point of view, be dispensed with entirely. More broadly, the division between substantive and formal mechanisms that underpin rationality, and the degree of importance for each, constitutes one of the central issues in the study of human and nonhuman animal decision making.

5

Error Management

Daniel Nettle

Abstract

This chapter briefly introduces *error management theory*, an evolutionary framework for understanding how natural selection should be expected to shape decision-making mechanisms. Selection minimizes not the overall rate of error in decision making, but rather the expected fitness burden of error. This means that where the fitness impact of errors is asymmetric (e.g., when failing to run from a predator that is present is more costly than running from a predator which is not in fact there), evolution will favor mechanisms that choose the cheap error much more often than the costly one. This principle can be applied to decision making in many different domains. This chapter discusses the relationships between error management theory and expected utility theory, and the extent to which error management theory can be invoked to explain the prevalence of biased beliefs.

Introduction

Over the past twenty years, cognitive psychologists have increasingly turned to evolutionary history to understand why human decision-making processes work the way that they do. In particular, they have investigated two related sets of ideas. The first is that decision making is not an undifferentiated mass. Rather than there being some general capacity to think, which works in the same way whether the decisions are about what to eat or choice of friends, different domains of everyday decisions might have different underlying rules. These rules, in turn, would have been sculpted by natural selection to deliver viable behavior reliably under the recurrent features of that behavioral domain across the kinds of environments our lineage has inhabited.

The second set of ideas is that many of the apparent oddities of human cognition in the laboratory—the fact that people so often get things wrong, and look rather foolish, when you contrive tests for them—might not indicate that human decision-making processes are actually flawed. Rather, these oddities might simply be telling us that producing answers to lab tasks which satisfy normative conceptions of rationality might not represent what human decision making was intended to do. By this I mean something fairly precise: the set of decision-making strategies embodied in cognitive performance has been selected to maximize fitness in *real-world* environments. There is no particular reason to expect such strategies to satisfy normative standards of correctness on artificial tasks or abstract domains, except insofar as those tasks and domains are related to things that people do in their everyday lives which actually impinge on their survival and reproduction.

Error management theory (EMT) is a framework for predicting how decision-making mechanisms should be expected to function when there are recurrent fitness consequences to decisions, and those decisions are made under uncertainty. Thus, any cognitive architecture which has been subject to natural selection should embody error management principles. The work which has been published under the EMT framework embodies aspects of both sets of ideas discussed above. That is, it argues that decisions in a particular domain will be affected by parameters specific to that domain, and that apparently poor performance in cognitive tasks might sometimes reflect strategies which are actually adaptive. EMT was primarily developed by Martie Haselton (Haselton and Buss 2000; Haselton and Nettle 2006; Haselton et al. 2009), although key elements of the idea can be found in various guises in other work (e.g., Stich 1990; Guthrie 2001; Nesse 2001b, 2005). Through a paper coauthored with Haselton (Haselton and Nettle 2006), and a related one of my own (Nettle 2004), I have played a small part in its development. In this chapter, I briefly outline the main idea of EMT and review some of the phenomena which it purports to explain. I then consider a number of interpretive issues, including its relationships with expected utility theory, its implications for cognitive mechanisms, and its limitations as an evolutionary model. I have kept the presentation simple and lightly referenced, outlining what seem to me to be the issues, without any particular new claim or conclusion.

Summary of Error Management Theory

The basic idea of EMT is very easy to state and can be built up using the framework of signal detection theory (Green and Swets 1966). Assume some cognitive mechanism whose function is to mobilize a particular adaptive response; for example, a threat-detection mechanism which will initiate flight in response to the presence of a predator. When a predator is present, a quantity of sensory evidence is generated (e.g., rustling). However, a certain amount of sensory evidence is also generated when no predator is present (e.g., by the action of the wind). Although much more rustling is made on average by predators than wind, the loudest wind is louder than the quietest predator. This is important: If the predator and the wind produce nonoverlapping distributions of rustling, the subject is able to know perfectly whether a predator is there or not, and the correct decision is trivial. If, however, the distributions are overlapping, the subject must determine at which level of perceived rustling to flee.

There are four possible outcomes in this situation. The subject could flee when there is a predator (the true positive, or a "hit"); flee when there is in fact no predator (the false positive or "false alarm"); stay when there is no predator (the true negative or "correct rejection"); or stay when there is in fact a predator (the false negative or "miss"). Each of the four scenarios is likely to have a different impact on fitness (Table 5.1). A false positive means expending a few calories and sometimes running; however, a false negative may very well mean death.

What decision rule should we expect natural selection to favor in such scenarios? If a subject flees when faced with a level of rustling evidence, e, expected payoff will be:

$$p(s|e)V_{TP} - p(\neg s|e)V_{FP}, \tag{5.1}$$

where s represents the relevant state of the world (i.e., a predator being present) and the values, V, are as in Table 5.1. If the subject does not flee, expected payoff will be:

$$p(\neg s|e)V_{TN} - p(s|e)V_{FN}. \tag{5.2}$$

Using standard signal-detection mathematics, the optimal strategy is to flee where:

$$\frac{p(e|s)}{p(e|\neg s)} > \frac{p(\neg s)}{p(s)} \cdot \frac{(V_{TN} + V_{FP})}{(V_{TP} + V_{FN})}. \tag{5.3}$$

Since e refers to the strength of information (i.e., rustling) that reaches the subject, the left-hand side of Equation 5.3 represents the likelihood ratio of that amount of evidence being generated when there is and is not a predator present. The overall probability of a predator being present and absent in the environment is represented by $p(s)$ and $p(\neg s)$, respectively. Where the fitness payoffs of all four situations are equal and predators are as likely to occur as not, Equation 5.3 suggests that the subject should flee when the amount of rustling reaching the subject is more likely to have been made by a predator than by the wind (i.e., the likelihood ratio equals 1)—an intuitive result. When predators are more prevalent than wind, the subject should flee even when the

Table 5.1 The four possible payoffs in a signal detection situation.

	Predator present	Predator absent
Subject flees	V_{TP}	$-V_{FP}$
Subject stays	$-V_{FN}$	V_{TN}

level of rustling is drastically below the level that a predator would make, and only average for the wind to make.

Importantly, for EMT purposes, Equation 5.3 shows that the subject's optimal decision rule depends not just on the likelihood of the presence of predators and the amount of rustling they make relative to the wind when present, but also on the fitness impacts of the four possible outcomes. In particular, as the fitness cost of the false negative goes up, the threshold set by the subject should go down, since as V_{FN} becomes large, the right-hand fraction in the Equation 5.3 becomes small. Even a modest asymmetry in the costs of the two errors, with everything else equal, results in a threshold that is much lower than 1 (see Figure 5.1). Such a threshold necessarily entails that the optimally behaving subject will flee much more often when there is in fact no predator than she will stay when there is in fact a predator. Moreover, most flights will be false alarms.

This leads us to state succinctly the main result upon which EMT is based: Where a decision-making mechanism produces a binary response, and the possible outcomes have had different average effects on fitness over evolutionary history, natural selection should produce mechanisms that may make cheap errors, but avoid costly ones. For mechanisms which are defenses, such as the fight-or-flight response, immune reaction, or cough, it is generally plausible that false negatives tend to have a large negative impact on fitness, whereas false positives, though irksome, are less costly. Thus, for these classes of mechanisms, we might expect a general pattern of over-responsiveness, as

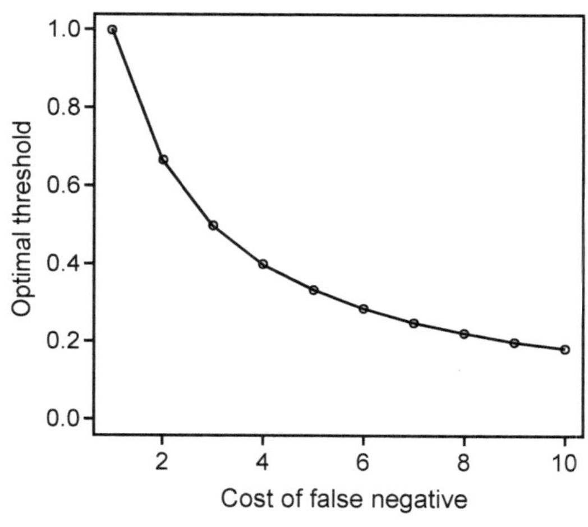

Figure 5.1 The optimal threshold for responding (defined in terms of the likelihood ratio of s and $\neg s$, the event happening and not happening, given the evidence received) for a signal-detection problem, plotted against the cost of the false negative. The background probabilities of s and $\neg s$ are equal, and the fitness values of the other three outcomes are set at 1.

seems to be the case for mechanisms of pain, immune response, vigilance, etc., which produce many false positives (Nesse 2001b, 2005). Conversely, mechanisms designed to obtain a fitness benefit might optimally be expressed much more often than that benefit is actually available. For example, MacFarlane et al. (2010) argue that in species of birds where males contribute little to parental care, male fitness depends almost entirely on how many matings can be achieved. Under such circumstances, missing a mating opportunity is much more costly than undertaking a mating attempt which is inappropriate. MacFarlane et al. explain the prevalence of homosexual behavior in such birds as a by-product of error management: since V_{TP} is recurrently much greater than V_{FP}, the birds evolve sexual motivational mechanisms that are easily triggered, and thus are often triggered by same-sex targets. Thus, the apparently maladaptive behavior of same-sex mating effort, though not adaptive in itself, is a consequence of an optimal policy to ensure that real mating opportunities are seldom missed.

Phenomena Explained Using the EMT Framework

EMT has been used to explain a wide variety of instances in human experimental psychology where mechanisms often seem to give what appear to be wrong or inaccurate answers. Here, I review a few examples (for further discussion, see Haselton and Nettle 2006; Haselton et al. 2009). Neuhoff (2001) has shown that when people are asked to estimate the time of impact of the source of an approaching sound, they systematically err toward estimating a time which is too soon. This apparent bias makes perfect adaptive sense if one considers the respective costs of the two errors. Preparing slightly too *early* for the impact of a projectile has almost no cost, save a second or two of time. By contrast, preparing slightly too *late* for the impact of a projectile could be very costly. Thus, it makes sense for selection to favor mechanisms whose design has the consequence of making the least costly error.

Another class of examples relates to male/female differences in the perception of sexual interest (for full reviews, see Haselton and Buss 2000; Haselton et al. 2009). Psychologists have demonstrated that under many circumstances, men tend to overestimate the degree of sexual interest which women display in a social interaction compared to the estimates made by women themselves or neutral female observers. This bias appears only when men rate women, and only for those women who are potentially mating-relevant targets (e.g., men do not display this when rating their sisters). This bias can obviously lead to problematic social interactions, and worse. At first glance, it seems odd that an ability to assess something so obviously relevant to fitness as a mating opportunity would not be better honed. However, from the EMT perspective, the pattern is explicable. The expected fitness value of a true positive is great, and the expected fitness cost of a false negative is, though potentially embarrassing, presumably less. Thus, it should not surprise us that selection favors mechanisms that respond to only very low levels of evidence that a female is sexually interested, even if those mechanisms often make false positive errors.

As a final example, it has long been established in psychology that only minimal cues are necessary to make people infer a purposive human activity behind some pattern of events (Guthrie 2001). This may account for widespread personal and culturally sustained beliefs in the actions of spirits, ancestors, unseen agents, and so forth. Humans are a highly social species, and one of the largest impacts on fitness will come from the purposive activities of conspecifics, either from within the social group or from neighboring groups. In general terms, it is very important to track evidence that other individuals have particular sets of goals or are undertaking particular types of activities; missing signs of such goals and activities can have a very negative impact on fitness. On the other hand, tracking a random pattern of events and inferring its possible significance in terms of the surreptitious activity of human agents probably does not cost terribly much; perhaps some time and energy. Thus, we might expect "agency detection" mechanisms to be calibrated in such a way that the cues of actual human involvement required to engage them are very slight (Barrett et al. 2005). Consequently, they will often produce errors of inferring agency where there is none, even if they are functioning optimally.

Interpretive Issues

Is EMT Just Expected Utility Theory?

One question that arises is whether EMT is just expected utility theory under another name (McKay and Efferson 2010). Expected utility has long been established in economics and psychology, and can be summed up as the decision rule "for each choice, sum the utility of each possible outcome weighted by the probability of that outcome, and choose the one which gives the highest value." A moment's thought reveals that Equations 5.1 and 5.2 are the expected utilities of taking the action and not taking it, respectively, and that Equation 5.3 can therefore be equivalently stated as the criterion of choosing to act whenever doing so gives a higher utility than not doing so, namely:

$$p(s|e)V_{TP} - p(\neg s|e)V_{FP} > p(\neg s|e)V_{TN} - p(s|e)V_{FN}. \qquad (5.4)$$

Thus, EMT is mathematically isomorphic to expected utility theory. All of the EMT examples can be restated in terms of expected utility. We might prepare relatively early for an incoming object because although the probability of it arriving early is small, the disutility of an unanticipated early arrival is large, and so the expected utility of preparing somewhat earlier than the most likely arrival time is higher than the expected utility of preparing at the most likely arrival time. Men might approach women whose likelihood of sexual receptivity is in fact small, because the utility of their turning out to be receptive is large, and so the expected utility of approaching them is still high. Thus, it seems that EMT merely restates the generally accepted ideas of expected utility.

While acknowledging the mathematical equivalence, I would like to point out some important differences between EMT and expected utility theory as it is usually discussed. First, the maximand in EMT is biological fitness, not some short-term currency such as pleasure or economic gain. Second, and more importantly, expected utility theory is often discussed as if it were both an ultimate and a proximate explanation of behavior. That is, expected utility is the standard by which an adaptively behaving person should behave if they were maximizing (ultimate explanation); however there also seems to be an expectation that the mechanism by which they would do this is to represent internally to themselves the utility and probability of each outcome, do the multiplication, and then choose the one which gives the highest total (proximate explanation). The information on which the decision is based (about the utilities and probabilities) is assumed to have come from the individual's own experience, and the process is thought of as a *domain-general* one which will work the same way whether the decision is to fall in love or avoid a foodstuff.

EMT is somewhat different in this regard. EMT is a theory of why natural selection would favor *domain-specific* decision mechanisms with certain behavioral consequences over other ones. Evolved decision mechanisms should be expected to embody information arising from the evolutionary history of the lineage about the likelihoods of certain states, and recurrent fitness consequences of different actions, in particular domains. There is no reason to expect those mechanisms to be cognitively penetrable or furnish consciously available intermediate representations of the "probability" or "utility" of any given event.

Consider the following example, and my apologies if it is somewhat extended. I offer you a decision—hold up your left hand, and I will give you one hundred dollars—and the probability of anything bad happening to you is close to zero. Hold up your right hand, and I will give you nothing. Naturally, you choose to hold up your left hand, which gives a higher expected utility. Now, I offer you one hundred dollars if you put your hand in the mouth of my defanged python, and nothing if you do not. You know that the python cannot hurt you, because I have shown you a veterinary certificate that he has had an operation rendering him harmless. You do the sums: if you put your hand in the python's mouth, there is a probability of close to 1 that you will come out fine, times a payoff of one hundred dollars; if you do not do so, there is the same probability that you will come out fine, but the payoff is zero. The logical structure of the problem is identical to the case where you just had to hold up your hand. As a consistent expected utility maximizer, then, you should decide to put your hand in the python's mouth. However, as soon as I get Gnasher out his box and he fixes you with his unfathomable reptilian eyes, you begin to experience panic, and the last thing you are going to do is to approach him and insert your hand into his mouth. Your aversion stems from millennia of selection, which favored mechanisms for responding to snakes and similar creatures on the basis of their recurrent fitness impacts. Just because you have some conscious information that things are different on this particular occasion does not mean you will be able to override entirely that phylogenetic legacy to maximize some expected payoff. Under expected utility theory, you look irrational, because you have been offered the same set of expected utilities in two different domains, and you have not been consistent in your behavior (i.e., you held up your left hand in the first task, but "chose" not to insert your hand in the second). By contrast, under EMT your behavior is explicable, because the two mechanisms governing the two domains (putting up a hand, responding to a snake) embody different information from your evolutionary history.

Thus, EMT is really a theory about the design features of particular classes of domain-specific mechanisms (an "evo-mecho" theory; McNamara and Houston 2009), rather than a statement of what to do when faced with a particular life choice. When someone throws a ball at you, you don't stop to ask what the respective utilities of responding early and late in this instance are. That decision has been "built in" to the neural architecture of your response without you having to do anything about it. Thus, EMT will only make the same predictions as expected utility theory if we read expected utility as an "as-if" theory—people will behave *as if* maximizing utility—and understand that the utilities and probabilities involved are not those pertaining to the current instance of the situation. Rather, utilities are average fitness impacts over phylogenetic time and the probabilities stem from some combination of phylogenetically given priors with information received over the individual's developmental history. EMT predicts that logically isomorphic choices will often be approached differently depending on the domain in which they are framed: Is this about snakes? About disease? About social risk? About mating? Expected utility theory, by contrast, demands that a rational actor is consistent in his decisions regardless of how the dilemma was framed. EMT also predicts that in laboratory tasks, the legacy of *ancestral* fitness costs and benefits will often dominate the actual costs and benefits that are obtained in the current experimental trial.

Cognitive Biases, Behavioral Biases, and the Reliability of Belief

EMT has been invoked to explain many cases where there are cognitive biases documented in psychological studies. For example, people overestimate the likelihood of becoming a victim of crime (Box et al. 1988). From an EMT perspective, one could argue that this might in fact be an adaptive bias, as it motivates people to take countermeasures against something which would be extremely detrimental if it did occur. The idea that such a biased belief could be adaptive has captured attention for a number of reasons. First, unrealistic levels of belief about all kinds of things do seem to be quite widespread.

Second, there is a long-standing tradition of arguing that evolution will always favor having a realistic conception of how the world is, and that it is, in effect, natural selection which guarantees the approximate veracity of human belief systems (e.g., Campbell 1974; Millikan 1984). Thus, it is a striking claim that selection might sometimes *favor* distorted beliefs.

However, as McKay and Efferson (2010) have recently pointed out, EMT does not require people to have false beliefs. EMT merely requires people's decision mechanisms to be attuned to the expected fitness impacts of possible events, not just their likelihood. It is agnostic about *how* selection might produce such attunement. For example, falsely believing that crime is likely might be one mechanism for motivating people to (adaptively) lock their houses, but correctly believing that crime is rare while understanding that it would be really bad if it happened would also produce the required action just as effectively, and would not involve any cognitive biases. This illustrates the general point that understanding biological function underdetermines the kinds of mechanisms that might evolve to fulfill that function. As McKay and Efferson go on to argue, for EMT to explain why human cognitive mechanisms produce biased representations of the world requires some extra argumentation. For example, one would have to show that a mechanism based on a cognitive bias was more robust, neurally cheaper, or worked across a greater variety of environments than one which accurately computed probabilities and fitness impacts.

While McKay and Efferson are undoubtedly correct, I would frame the question slightly differently. One of EMT's implicit claims is that things like consciously statable beliefs, or explicitly reportable intuitions about the probabilities of certain events, are probably not the kinds of things that have been targeted by selection. (Nor are they very important for our everyday lives now, unless we are academics or actuaries). Rather, selection works on individuals' norms of action and reaction, given particular environmental contingencies, and these are probably delivered via simple evolved rules of thumb. Thus, when we ask a person to give us a numerical probability that they will be burglarized, or that a particular woman is receptive to a man's advances, we are asking cognitive mechanisms to do something for which they were not designed. The mechanisms were designed to quickly give resource loss and mating the behavioral priority that their large fitness impact merits. When we use them secondarily to assign a numerical probability, or some other ecologically unusual task, the mechanisms produce answers which might look quite dumb. (In most people's minds, the distinction between the probability of crime happening to them and the disutility of it happening to them is probably hazy; being a victim of crime is just a horrible prospect). The mechanisms could be, however, very good at solving the problems they were actually designed to solve; namely efficiently delivering appropriate behavior in real, noisy environments (for extended discussion of related ideas, see Gigerenzer 2000; Haselton et al. 2009; Wilke and Todd 2010). So although evolution does not *require* agents to form biased representations of the world, it is probably indifferent to whether they do or not, as long as the rules of thumb which do evolve lead to adaptive decisions often enough and quickly enough.

In general, then, we should echo Berg and Gigerenzer's (2010) call to assess decision rules not by their conformity to abstract or decontextualized standards of rationality or veracity in tasks, which have little relevance to everyday life, but rather by their impact on people's real-life behavior, health, and well-being.

EMT and Biological Evolution

Let us consider the relationship between EMT and some of the biological work on the evolution of decision-making mechanisms that has appeared recently. A strong point about EMT is that it recasts the "utility" familiar from conventional economic and psychological approaches to decisions into biological fitness, thus replacing a short-term measure with a longer-term one. However, "fitness" in EMT is usually taken to mean something like lifetime reproductive success, and, from a biological standpoint, this may not be long-term enough. Fitness is in fact a measure of the long-term growth rate of a lineage, and the relationship between this and the mean lifetime reproductive success of lineage members is not guaranteed to be straightforward. Perhaps most obviously, EMT assumes that we should take the arithmetic mean of the expected effects of the four possible outcomes on reproductive success as the quantity to be maximized. However, in many instances, fitness is better thought of as a geometric function of lifetime reproductive success rather than an arithmetic one (Metz et al. 1992; Houston et al. 2007a). For example, in a lineage which produces ten descendants in one generation and none in the next, the arithmetic mean (5) is not a good measure of fitness, but the geometric mean (0) comes a lot closer. It is not enough to do well in the average year; you have to not die out in *all* the years there are (see Gluck et al., this volume).

The geometric nature of fitness means that under many circumstances, the occasional consequences of very bad reproductive success can disproportionately affect what evolves. Thus, it is a priority to reconcile the ideas of EMT with models using a more explicit and better motivated measure of fitness (see, e.g., McNamara et al. 2011). EMT already argues that very large possible fitness consequences in one direction or another should weigh heavily in the design features of decision mechanisms, but it could be that true weighing over evolutionary time, particularly for the avoidance of fatal hazards, is even greater. Incorporating better measures of fitness would mean that EMT was no longer mathematically isomorphic to expected utility theory, and thus perhaps establish it as more genuinely novel alternative way of thinking about what we should expect of decision-making mechanisms.

Conclusions

In this chapter, I have outlined what EMT is and provided examples of the kinds of psychological phenomena which it has been invoked to explain. Interpretative issues have been considered—such as whether EMT can be reduced to expected utility theory, and whether it can help explain biased cognition—and a fruitful direction of development has been proposed to integrate it with more explicit modeling of biological fitness. Many of the questions around EMT are not yet resolved, and I am not making any particular novel claim here. However, I do think that evolutionary-based theories such as EMT can importantly help us see that, as a foundation for decision theory, we need to consider the fitness impact of the outputs of different decision rules in real-world, adaptively relevant situations, rather than their ability to produce conscious intuitions which satisfy abstract principles of rationality or accuracy in essentially contrived laboratory tasks.

6

Neuroethology of Decision Making

Geoffrey K. Adams, Karli K. Watson, John Pearson, and Michael L. Platt

Abstract

A neuroethological approach to decision making posits that neural circuits mediating choice evolved through natural selection to link sensory systems flexibly to motor output in a way that enhances the fit between behavior and the local environment. This chapter discusses basic prerequisites for a variety of decision systems from this viewpoint, with a focus on two of the best studied and most widely represented decision problems. The first is patch leaving, a prototype of environmentally based switching between action patterns. The second is social information seeking, a behavior that, while functionally distinct from foraging, can be addressed in a similar framework. It is argued that while the specific neural solutions to these problems sometimes differ across species, both the problems themselves and the algorithms instantiated by biological hardware are repeated widely throughout nature. The behavioral and mathematical study of ubiquitous decision processes like patch leaving and information seeking thus provides a powerful new approach to uncovering the fundamental design structure of nervous systems.

Introduction

In the wetlands of North America, male red-winged blackbirds (*Agelaius phoeniceus*) compete for territorial control over patchy resources, with some males managing to dominate rich territories while others are stuck with low-quality leavings. When the mating season begins, female red-winged blackbirds select a primary male with which to mate and subsequently nest in his territory. Because territory quality is so variable, some highly successful males may attract a harem of up to about fifteen females, while many unsuccessful males will fail to attract a mate at all. Although each female may produce multiple broods during the breeding season, there is a significant chance that she will fail to bring any offspring to fledging. With the stakes so high, her selection of a mate is a critical decision that will have direct consequences for her fitness.

Many factors can impact her success and may, in principle, play a role in her decision: these include male health and parasite load, food density, the presence of acceptable nesting sites, the number of females already present in the harem, the likelihood of receiving paternal care for her brood, and the presence of healthy male neighbors who may provide opportunities for extra-pair copulations. Despite decades of excellent studies of the red-winged blackbird, the extent to which females assess these factors in choosing a mate remains an active area of research (Beletsky 1996).

Mate choice in female red-winged blackbirds is an illustrative example of a complex decision-making problem posed by the particular details of an animal's habitat and social structure. Similar examples are common across a broad diversity of animal taxa: primates select long-term social partners with consequences for their health and reproductive success (Silk et al. 2003; Schulke et al. 2010; Silk et al. 2010); cleaner fish in coral reefs decide between providing good service or "cheating" their clients, and adjust their level of service for each individual client (Bshary and Noe 2003); African buffalo decide the best direction for the herd to move by implementing a vector average of the orientation favored by each cow weighted by her social status (Prins 1996); tens of thousands of honeybees in a swarm select a new nest site by comparing their own assessment of sites based on location, temperature, and humidity with the evaluations made and communicated by other scouts (Vonfrisch and Lindauer 1956; Seeley and Visscher 2004). In each case, the animal's immediate context may reasonably permit multiple possible behaviors, but only one or a few will optimize fitness.

Evolutionary theory and behavioral ecology identify the decision-making problems that animals face in their natural environments, as well as the costs, benefits, and constraints associated with pursuing specific behavioral strategies. In the case of mate choice in female red-winged blackbirds, one of the more successful proposals is the *polygyny threshold hypothesis*, which suggests that territorial resources are the primary factor influencing female choice (Verner and Willson 1966; Orians 1969). According to this model, females choose a mate based on the amount of resources his territory will offer her, accounting for the resources that will be consumed by the male's existing harem. Although other factors may also influence female choice, behavioral studies of red-winged blackbirds in their natural habitat suggest that the polygyny threshold hypothesis provides a good approximation of the true mate choice strategy that females pursue (Beletsky 1996).

Decision neuroethology is concerned with understanding the physiological mechanisms that evolution has produced for solving the complex decision-making problems posed by animals' environments. Although the study of animal behavior has revealed a remarkable diversity of such problems and their solutions, mathematical analysis has demonstrated that dissimilar-seeming problems may be solved with similar strategies; for example, patch and prey foraging (Stephens and Krebs 1986:32). When the fitness impact of a decision is large, there are strong selective pressures on the corresponding decision-making mechanism to behave according to the mathematically optimal strategy. Such a constraint means that the neurobiological mechanisms mediating decisions may tend to be highly conserved or convergent across taxa for a general class of decision-making problems. To understand how decision-making behaviors evolve, the concepts of conservation, convergence, and disparity of mechanisms must be considered.

To illustrate the basic approach of decision neuroethology, we will describe a computational framework for understanding the selective pressures on decision-making mechanisms, and how to approach comparison between species. We will then discuss in greater detail two examples of decision problems in behavioral ecology and their neural implementations. First, we examine patch leaving, a prototype of environmentally based switching between action patterns. Second, we examine social information-seeking behaviors in nonhuman primates. We conclude that this work portends a more general understanding of complex decision problems and, ultimately, endorses the unification of theoretical and experimental work in behavioral ecology and neuroscience.

A Computational, Comparative Approach

In his seminal work on computational vision, Marr (1982) introduced a three-level framework for considering vision as a computational problem and argued that visual neuroscience must consider all three levels. Such a framework is also valuable in considering decision-making problems that must be solved by the nervous system (Figure 6.1). Furthermore, as we will see, this framework permits both a comparative analysis across species and across decision types within a single species.

According to Marr, the *computational level* of analysis is the level at which we consider the actual *problem* a biological system is attempting to solve. From an evolutionary standpoint, this is the level at which selective pressures exist to shape a decision process. Put another way, this is the ecological problem posed by an animal's niche. Of course, in an ultimate sense, this generally reduces to "maximize inclusive fitness." However, given that inclusive fitness may be practically impossible to measure in most contexts, we will often limit analysis to more specific goals such as "maximize resources for offspring" or "maximize long-term caloric intake." In the case of mate choice in the female red-winged blackbird, this level of analysis corresponds to the identification of factors associated with selecting a male that will influence seasonal reproductive success. In some well-known cases, problems posed at the computational level possess optimal solutions, such as the ones obtained by *marginal value theorem* (Charnov 1976; see also patch-leaving decisions discussed below). In this sense, an optimal solution is a relationship between environmental and behavioral parameters guaranteed to produce the best outcome under the posed

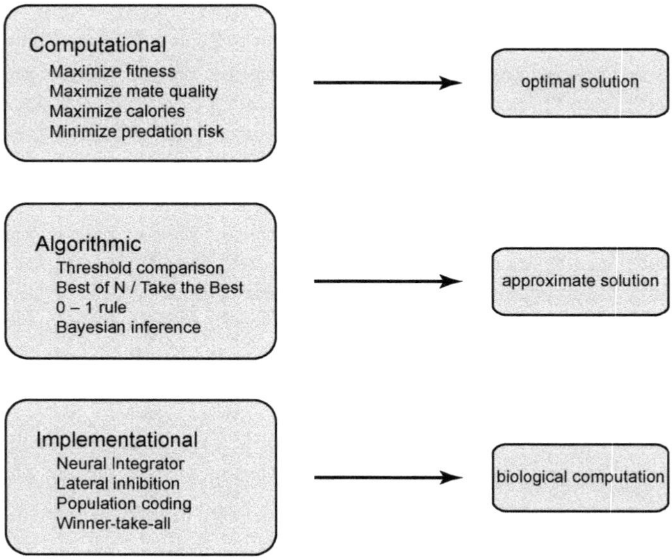

Figure 6.1 Decision-making problems can be analyzed at three levels. The computational level identifies the decision-making problem to be solved; the algorithmic level identifies strategies by which the problem may be solved; and the implementational level identifies the exact biological mechanism by which an organism solves the problem.

problem. There are many examples of organisms producing optimal or near-optimal behavior for solving an identified problem.

At the *algorithmic* and *representational level* of analysis, we are concerned with identifying *representations* (i.e., formal systems for describing features of the environment relevant to the problem) and *algorithms* (i.e., the series of computational steps whereby animals solve the problems posed). A popular choice of representation in neuroeconomic models of decision making is utility, but other representations are possible. The polygyny threshold hypothesis suggests a possible representation and algorithm for female red-winged blackbird mate choice. Representationally, one or more features of a territory are represented together as its quality, and the number of females already in a harem is represented separately. The algorithm consists of dividing quality by harem size and selecting the territory with the best ratio. In principle, there are many possible algorithms corresponding to a single solution; successfully identifying a problem and its optimal solution at the computational level does not yield a unique solution at the algorithmic level. However, some algorithms are more parsimonious than others, and a relatively simple decision-making algorithm should generally be preferred over a needlessly complex algorithm that produces the same result. Furthermore, there is no guarantee that the algorithm an animal uses to solve a decision-making problem need actually produce an optimal solution; such algorithms may be satisficing (i.e., guaranteed to meet some minimal requirements) rather than optimizing (i.e., guaranteed to produce the best possible outcome). These algorithms may, if probed correctly, reveal systematic "bugs"—the equivalent of optical illusions—indicative of the underlying computations involved. In the case of foraging, such algorithms involve decision heuristics like threshold comparison for patch leaving and the 0 – 1 rule for prey selection, as well as more complex rules like *take-the-best* or *elimination by attributes* (Gigerenzer and Goldstein 1996). In some environments (presumably those typical for the species that apply them), these algorithms may perform equivalently to truly optimal algorithms, though there is no guarantee that they will do so in other contexts.

Finally, at the *implementation level* of analysis, we are concerned with the question of which *proximate mechanisms* actually *implement* algorithms like threshold comparison and take-the-best. In principle, there is a many-to-one relationship between implementation and algorithm. The true implementation may involve neuronal or genetic circuits, neuromodulators, hormones, or a complex interplay of these. In fact, a genetic circuit and a neural network may implement the same decision rule—equivalence at the algorithmic level—though their biological details remain entirely distinct. In practice, once we have identified the proximate mechanism by which a decision is made, we can often specify the algorithm, as well.

Decision neuroethology proceeds by considering all three levels of analysis. The computational level identifies the decision-making problem to be solved. The algorithmic level proposes one or more means for computing approximate solutions to the problem, and the implementation level identifies the actual proximate mechanisms by which the algorithm is performed. In practice, the study of algorithms may suggest implementations or vice versa. It is also interesting to consider that implementations themselves may have costs in terms of caloric consumption or ontogenetic complexity, which may be analyzed at the computational level to understand why one implementation is favored over another, or even why some algorithms are favored if they happen to be readily implemented by low-cost mechanisms.

In what follows, we will explore this threefold description as a means of exploring what appear to be striking similarities across taxa in behavioral ecology—similarities, we suggest, that result from convergent evolution at the algorithmic level to common problems posed at the computational level. We examine these ideas through the lenses of two of the most ubiquitous problems in behavioral ecology: patch foraging and social information seeking.

Patch-Leaving Decisions

A well-studied example of an ecological decision is the patch-leaving problem, mathematically analyzed by Charnov (1976) and first tested in a series of experiments in birds by Krebs et al. (1974). This problem considers an animal foraging in an environment with food items distributed in sparse patches. As the animal forages in a patch, local resources are depleted, and the time required to find a new food item increases, thus reducing the rate of food intake. As a result, animals must balance the benefits of diminishing returns against the costs of searching for new patches. The optimal solution to this problem, the *marginal value theorem* (MVT; see Kacelnik, this volume)—that foragers should abandon patches when the local rate of caloric return falls below the average for the environment as a whole—has been demonstrated to hold in a breathtakingly wide array of species, including worms, insects, fish, rodents, birds, nonhuman primates, and humans (Stephens and Krebs 1986; Stephens et al. 2007).

On the computational and algorithmic levels, this result is unsurprising: most animals must forage, and nutrients are often sparse, so many species face computationally equivalent problems. Still, the replication of a simple decision rule across species with such diverse neuroanatomical organization need not imply conserved proximate mechanisms. Rather, algorithms useful for solving ubiquitous problems like patch leaving and prey selection (see below) are more likely to be products of convergent evolution, primarily because they are robust, require only simple components, and do not require a centralized architecture. Such algorithms are thus more likely to be repeated across taxa.

In a recent experiment investigating the neural basis of patch-leaving decisions, Hayden, Pearson, and Platt (2011) designed a laboratory version of the problem in which monkeys chose between "stay" and "leave" options represented by visual stimuli on a computer monitor. Monkeys received juice rewards of diminishing value for selecting the stay option, simulating the effect of remaining in a depleting food patch, whereas selecting the leave option resulted in a cued delay, simulating the travel time between patches. As in previous studies in other species, the authors found that monkeys readily optimized their patch residence times (Figure 6.2). Monkeys' patch-leaving decisions also depended systematically on the "travel time" to the next patch, consistent with the MVT. Notably, monkeys slightly, but systematically, remained in patches for longer than predicted by the MVT. As noted elsewhere (Cuthill et al. 1990), however, this finding is precisely the result expected if monkeys' estimates of reward rate were based on short-run returns, a quantity calculable by simple linear filtering (Anderson and Moore 1979; Bateson and Kacelnik 1996). As a result, this deviation from optimal behavior itself carries information about the underlying decision algorithm, utilizing an average of recent returns that can be linearly updated each trial instead of the full nonlinear computation of reward rate across the entire foraging history.

Simultaneously, Hayden and colleagues recorded the firing patterns of single neurons in the dorsal anterior cingulate cortex (dACC), an area of the macaque and human brain linked to reward monitoring, error signaling, learning, and behavioral control. Neuronal firing rates revealed a strikingly simple implementation of a thresholded decision circuit (Figure 6.3). Single neurons

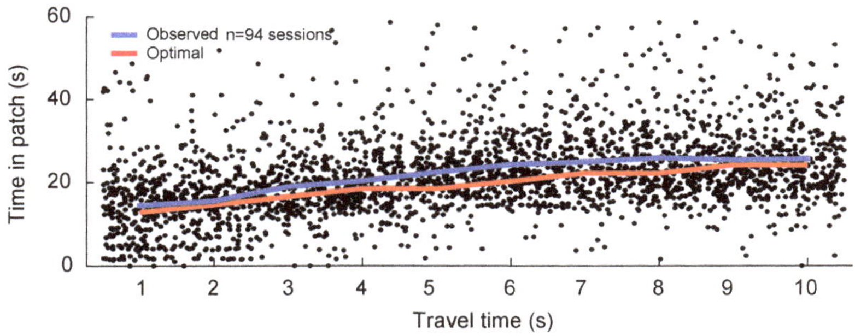

Figure 6.2 Rhesus macaques forage nearly optimally in a computerized patch-leaving task. Monkeys remain in the patch longer as travel time rises, as predicted by the marginal value theorem (MVT). Each dot indicates a single patch-leaving decision (n = 2,834 patch-leaving events) sampled from two individuals. The time at which the monkey chose to leave the patch (Y axis) was defined relative to the beginning of foraging in that patch. Travel time was kept constant in a patch (X axis). Data from both monkeys is shown. Behavior (average is traced by the blue line) closely followed the rate-maximizing leaving time (red line), albeit delayed by 0–2 s.

responded phasically (i.e., with a transient, time-locked change in firing rate) to each decision. Further, phasic neuronal activity increased with each successive decision to stay in a given patch, across multiple actions unfolding over tens of seconds. Finally, firing rates peaked with the decision to abandon a patch and move on to the next. In fact, peak firing not only predicted when monkeys decided to leave the patch, it also differentiated between premature and postponed leave decisions. Furthermore, with increasing "travel time" between patches, the rate of increase in peak firing diminished, in keeping with the longer dwelling times observed behaviorally.

All of these observations are consistent with a firing rate threshold for the leave decision; when neuronal activity in dACC reaches this fixed value, monkeys opt to leave the patch. More specifically, firing rates on patch-leaving trials did not differ statistically across premature or postponed leaving decisions within a given delay condition, further strengthening the conclusion that the threshold is constant for a given neuron. Nonetheless, leaving thresholds for firing rates did increase with delay to the next patch, indicating a flexible control process necessary to reset the threshold to generate optimal behavior for the current environment.

The rise-to-threshold process evident in the decision-related responses in ACC mirrors similar rise-to-threshold processes evident in the activity of neurons in parietal and prefrontal association cortex in monkeys rendering perceptual judgments (Gold and Shadlen 2007). Further, such integrate-to-threshold processes have been theorized to serve as implementations of the sequential probability ratio test (SPRT)—the most efficient solution to this type of binary decision problem (Wald and Wolfowitz 1948; Ratcliff 1978; Ratcliff and McKoon 2008). The observation that a similar process appears to govern both

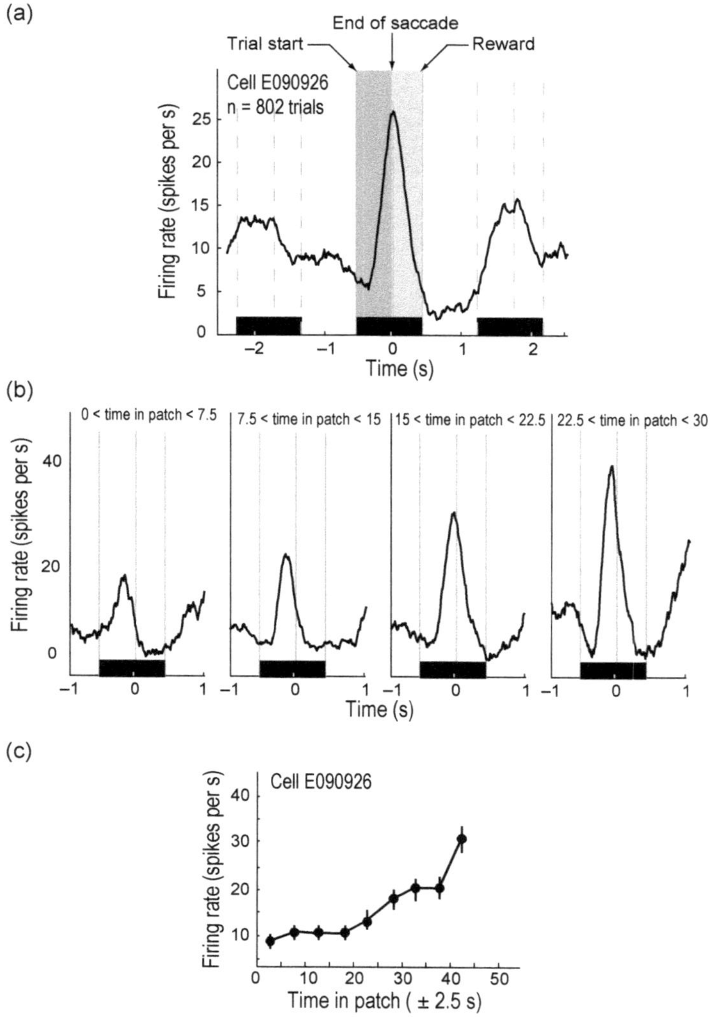

Figure 6.3 The rising value of leaving a patch is represented by single neurons in the macaque frontal cortex. (a) Average reward-aligned peri-stimulus time histograms (PSTHs) are shown for example cell in the anterior cingulate cortex (ACC). Neuronal responses were briefly enhanced around the time of saccades and then fell to a baseline level between trials. Time zero indicates end of saccade, indicating choice. Dark gray box indicates the pre-saccadic epoch; light gray box shows the post-saccadic epoch; and the black rectangle indicates the average duration of the trial. (b) The firing rate during the peri-saccadic period rose with time in patch. Each panel indicates responses selected from one range of patch residence times. (c) Average responses of example neuron occurring in a series of 5-second analysis epochs (gray box in (a)). Firing rates increased as time in patch increased. Error bars represent standard error of the mean.

perceptual decisions and patch-leaving decisions endorses the idea that the brain uses a small suite of common mechanisms to solve diverse problems in multiple domains (Hayden et al. 2011).

Recent evidence also indicates that genes coding for neuromodulatory chemicals may mediate the impact of local environmental conditions on patch-leaving decisions, perhaps by controlling the threshold for patch-leaving decisions. Bendesky and colleagues (2011) compared patch-leaving decisions by a nematode (*Caenorhabditis elegans*) strain from Hawaii with one developed in the laboratory. Hawaiian worms abandoned algae lawns at ten times the rate that the laboratory strains did. The authors also found that differences in patch-leaving threshold among distinct strains resulted from polymorphisms in promoter regions of the tyramine receptor gene (tyra 3b), which controls the expression of a G-protein-coupled receptor analogue of vertebrate catecholamine receptors. Thus worms' patch-leaving thresholds are regulated genetically, as opposed to monkeys' use of flexible thresholding, though both obey the MVT. In other words, completely different proximal mechanisms are capable of instantiating the same algorithm to solve a computationally equivalent biological problem.

Intriguingly, the invertebrate catecholamines tyramine and octopamine, which bind to the tyra3 receptor, are closely related structurally to the vertebrate neuromodulator norepinephrine, which has been hypothesized to regulate exploration/exploitation trade-offs in primates and humans (Aston-Jones and Cohen 2005; Cohen et al. 2007). Here again, a similar design for regulating the patch-leaving threshold is implemented by two separate mechanisms—one genetic, the other cortical—with behaviorally similar results. It is tempting to speculate that individual differences in exploration behavior, an innate tendency to abandon the current behavioral strategy for another potentially more profitable one, might also be mediated in humans through genetic influences on catecholaminergic neuromodulatory systems (Frank et al. 2009).

Social Information Seeking

The ability to select, inhibit, and shift behavior rapidly is particularly important in social species, given the highly dynamic nature of social environments. For example, the act of consuming a food resource is often mutually exclusive with vigilance behavior. This behavioral trade-off is analogous to the explore/exploit decisions made between two foraging sites. Thus the primacy of information in guiding decisions is abundantly evident in the social behavior of primates. Primates have frontally oriented, mobile eyes with a central fovea composed of a high density of cone photoreceptors, and thus are not capable of sampling all regions of the visual field simultaneously. This set of adaptations has led to the evolution of mechanisms which orient the visual system to objects with high information value via overt and covert attention (Moore et al.

2003). In the context of foraging, information is given value as a consequence of the value of the nutriment it may yield, as when an animal evaluates a tree for the presence of ripe fruit. Animals living in complex and dynamic societies can, however, use the same attentional strategies to gather information about others (Klein et al. 2008), including rank (Bovet and Washburn 2003), identity (Parr et al. 2000), group membership (Mahajan et al. 2011), direction of gaze (Lorincz et al. 1999; Ferrari et al. 2000; Deaner and Platt 2003), and emotional state (Sackett 1966). This information must be perceived and evaluated to guide adaptive behaviors, such as abandoning a food resource in the presence of a threatening dominant individual.

Whereas locomotion imposes high energetic costs during foraging, the metabolic costs of information seeking are by contrast quite low. However, information seeking can impose time costs, requiring animals to forego activities such as sleeping, drinking, or eating, which demand postures or behavioral states incompatible with attentive orienting (Figure 6.4). Other costs are social: in the case of rhesus macaques, visual fixation on the face of another individual invites aggression (van Hooff 1967). Furthermore, inappropriate information seeking (e.g., directing attention to low-value information) can result in missed opportunities to gather more useful information elsewhere. These constraints can be considered as part of the computational problem social animals must solve in deciding how to allocate their time and attention in seeking social information.

Thus, as in foraging, the net gains that accrue from information seeking can often outweigh the potential costs, and adaptive decision making depends on the assessment and comparison of these costs and benefits. In an experimental measure of the relative value of different classes of social rewards, Deaner, Khera, and Platt (2005a) demonstrated that thirsty male rhesus macaques will forego a small amount of juice to acquire specific types of social information,

Figure 6.4 Social information is a valuable resource for macaque monkeys. A rhesus macaque on Cayo Santiago assumes a vulnerable posture (left) to drink from a puddle, but periodically interrupts this posture to visually scan the surrounding region for potential threats (right). There are no predators on the island, but aggressive social interactions are commonplace.

such as reproductive signals (i.e., female perinea) or the faces of dominant males, but will not do so for other types of social information, such as the faces of subordinates (Figure 6.5). An alternative measure of value—the duration of time that the monkey chooses to look at the image once it is displayed—shows that monkeys look longest at reproductive signals but quickly avert gaze from both dominant and subordinate faces. Taken together, these two results invite the hypothesis that although both sexual signals and status-related signals contain high information value, there is a high cost associated with an extended period of direct eye contact. In support of this interpretation, a genetic polymorphism in the serotonin system associated with heightened anxiety in humans elicits reduced attention to the faces of other monkeys, greater pupil dilation (a somatic index of elevated autonomic arousal) in response to faces of dominant males, and reduced reward value for viewing the faces of dominant males (Watson et al. 2009) in rhesus macaques.

Patterns of neuronal responses also support the notion that information gathering has value for making decisions. Dopamine neurons, which respond to primary reinforcers (such as nutritive rewards) when unpredicted and to cues that predict them, also encode monkeys' preferences for advance information about impending choices between primary rewards (Bromberg-Martin and Hikosaka 2009). Moreover, the firing of neurons in the lateral intraparietal area (LIP)—a region of visuomotor cortex thought to encode a salience map of the visual world (Goldberg et al. 2006; Bisley and Goldberg 2010)—varies with the value of social images displayed in the neurons' receptive fields. Neurons in LIP not only encode the value of the juice reward they will gain for orienting to a particular location (Platt and Glimcher 1999), but also the value of the social information they receive for orienting to the same location (Klein et al. 2008). Importantly, social and gustatory value are encoded independently using the same coding scheme, suggesting that LIP plays a role in assigning value to a particular location in space.

In another experiment, a subpopulation of LIP neurons showed increased activity when a centrally positioned monkey face oriented its gaze toward the neuron's receptive field. This result is consistent with the well-known *Posner effect*, in which reaction times toward an eccentric target are reduced when attending a target and increased when attending away from the target (Posner et al. 1980). Both of these studies bear obvious relevance to the phenomenon of *joint attention* (Scaife and Bruner 1975), in which gaze is reflexively drawn in the direction of another's line of sight, an example of socially facilitated information gathering that is severely disrupted in neuropsychiatric disorders such as autism and schizophrenia.

Regions in the visual perceptual cortex in the temporal and occipital lobes are specialized for processing social information in humans and macaques, suggesting adaptation for the rapid assessment of visual social information. Humans and macaques both possess multiple brain regions, identified by functional imaging and confirmed by recordings from single neurons, exquisitely

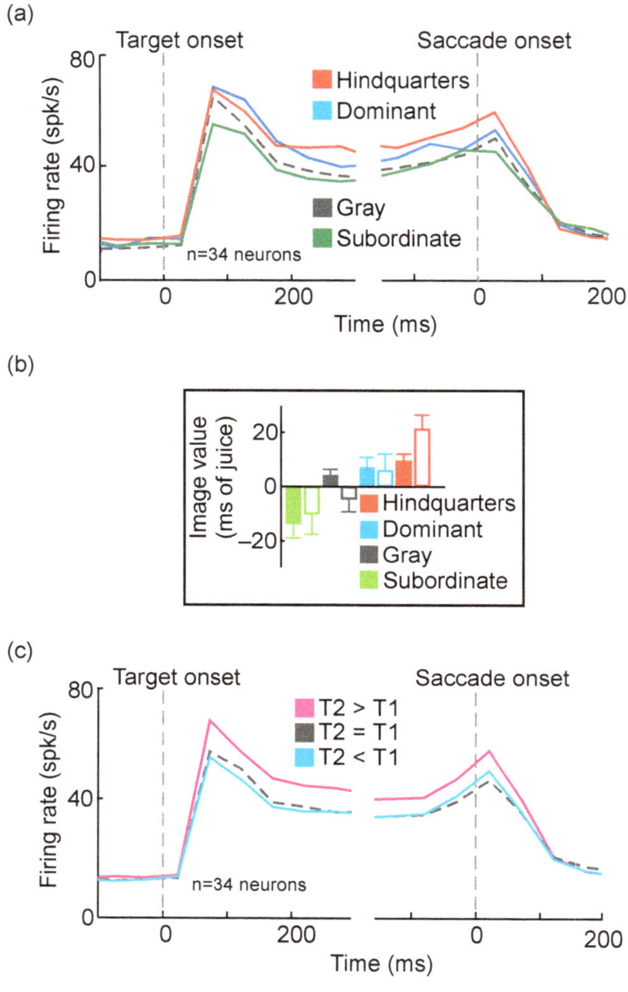

Figure 6.5 The value of social information is signaled by neurons in the macaque visual-orienting system. (a) Average firing rate for 34 LIP neurons plotted against time for all trials in which the subject chose to view the image (T2) in the neuron's response field, separated by image class. (b) Values determined for different image classes for two male monkey subjects (open and closed bars), in ms of fluid delivery time. Positive deflections indicate the subject was willing to forgo fluid to view that image class. Negative deflections indicate the subject required fluid overpayment to choose that image class. The category "hindquarters" refers to the perineal sexual signals of familiar females; "dominant" and "subordinate" to the faces of familiar dominant and subordinate males; and "gray" to a plain gray square matched for size and luminance to the other image classes. (c) Average firing rate of the population for all trials in which the subject chose to view the image (T2) in the neuron's response field, separated by fluid value relative to the nonchosen target (T1).

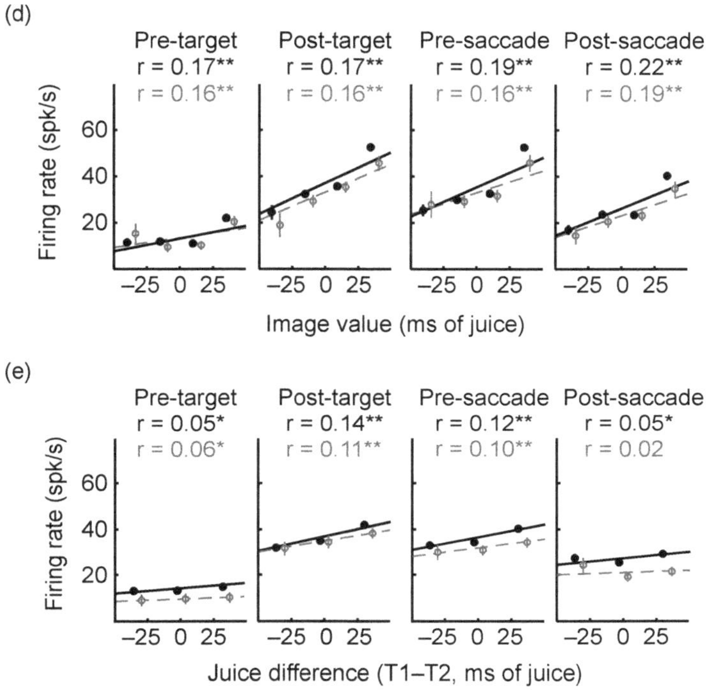

Figure 6.5 (continued) (d) Firing rates plotted as a function of image value in four 200-ms epochs. Black symbols represent regressions performed on all data in which the subject chose to view the image; gray symbols depict the same analysis restricted to trials in which the fluid payoff for choosing T1 was equal to T2. (e) Firing rates plotted as a function of the difference in fluid payoff between T2 and T1. Black symbols represent regression performed on all data in which the subject chose to view the image; gray symbols depict the same analysis restricted to trials in which the image value calculated for that block was greater than –5 and less than 5 ms. Error bars represent standard error of the mean. The data in (c) and (d) were binned for display, but all regressions were performed on raw data. $^*p < 0.05$, $^{**}p < 0.001$.

selective for faces (Kanwisher et al. 1997; Tsao et al. 2003). Neurons in other temporal lobe areas encode head direction (Perrett et al. 1985), face identity (Perrett et al. 1984), or biological motion (Oram and Perrett 1994). These populations of neurons are interconnected and feed forward from the temporal into the frontal lobe, where various modalities and features of external stimuli, including their motivational value, are integrated (Ku et al. 2011). The architecture of this network suggests that frontal decision-making mechanisms may have privileged access to social information in human and nonhuman primates alike.

Despite the presence of dedicated social perception networks in the visual systems of primates, there is little evidence to suggest that the brain regions involved in decision making, including the ventral tegmentum, basal ganglia, and frontal cortex, contain specialized subnetworks devoted to social decisions. This observation invites the hypothesis that the rise-to-threshold neural mechanism is applicable to social information gathering, as well as to perceptual and foraging decisions. Given the example in which a subordinate monkey places himself in a vulnerable position consuming a food resource, one could imagine that the presence of a potential social threat would induce a rise in firing rate in the ACC, causing the monkey to shift away from his current activity in favor of vigilance or avoidance behavior. More generally, we conjecture that a common suite of algorithms subserves both "social" and "nonsocial" decision processes, with the former differentiated from the latter by the large number of specialized structures used for detecting and processing information related to conspecifics.

Discussion

The study of ecological decision making suggests the possibility that natural selection has favored a set of simple, repeated design patterns: basic circuit elements capable of being implemented by many biological configurations. Instead of forming a single unified system for decisions, these local circuits might be capable of functioning independently for specialized subclasses of action planning and selection, as well as being recruited across regions for more complicated behaviors. Just as Gigerenzer has suggested that human decision processes draw on a well-stocked "adaptive toolbox" filled with inexpensive, approximate heuristics, so evolution appears to favor repeating algorithms, despite disparate implementations (Gigerenzer and Selten 2001).

In this view, the study of a decision-making problem like patch leaving is valuable not only because it is ubiquitous, but also because the algorithm used to solve it—comparison of local returns to a fixed threshold—represents one of the simplest forms of a single-input/single-output control system, in which the controller implements a binary threshold operation (Brogan 1985). Any neural system capable of implementing such a circuit is likewise applicable to an extremely wide class of problems, one that extends far beyond foraging. In fact, a more general version of such a system, the Kalman filter, is known to be an optimal solution to the problem of predicting returns under fairly general assumptions, and thus for fine-tuning behavior in response to changing environmental conditions (Anderson and Moore 1979; Brogan 1985). That such a system is linear, that it requires only a simple architecture, and that it is robust against noise all lead us to expect that it will not only appear in a wide diversity of species, but that it may be repeated and repurposed within a single brain to solve seemingly unrelated problems. In fact, such observations motivate a neural engineering viewpoint in which the unique classes of problems faced by an organism become primary, followed by the algorithms used to solve them, and only last their specific neural implementations (Marr 1982).

This does not mean, however, that comparative biology or neurophysiology become irrelevant. On the contrary, this evolutionary viewpoint suggests that algorithms implemented in more conservative nervous systems are more applicable than we might have thought. Just as in vision or olfaction, the insights gained from studying flies or worms suggest possibilities at the algorithmic level in birds or primates. Though the details may differ as organisms become capable of more generalized and flexible behaviors, the same simple biological components, coupled-like circuit elements, may likewise be expected to give rise to startlingly sophisticated generalizations (Brogan 1985).

Many models of decision making, particularly those derived from economics, describe the decision process as a linear sequence of first estimating abstract utility of several possible outcomes or behavioral plans, directly comparing these utilities, and finally selecting the goal or behavior associated with the highest utility (Glimcher 2004; Glimcher et al. 2005; Sugrue et al. 2005; Lee 2006; Padoa-Schioppa 2007). In some cases, such models leave the exact nature of the abstract utility undefined. Here, however, we have presented a "bottom-up" as opposed to "top-down" perspective, in which simple, reusable decision rules substitute across taxa for what are often considered outputs of a single decision-making system. We argue that these design patterns, implemented by diverse suites of neural hardware, should nonetheless prove ubiquitous on evolutionary grounds, and that their simplicity and robustness should favor them both for convergent evolution and conservation within taxa. Such claims represent a new opportunity for both systems theory and comparative biology, since the view of decision systems as evolving primarily to solve ecological problems demands renewed interest in both engineering disciplines and animal behavior. Indeed, the search for reusable design patterns in neural systems may provide a unifying framework for biological decision making in much the same way it has for vision and motor control.

First column (top to bottom): Ed Hagen, Daniel Nettle, Graham Loomes, Alasdair Houston, Dave Stephens, Alex Kacelnik, Tobias Kalenscher

Second column: Nick Chater, Dave Stephens, Alex Kacelnik, Tobias Kalenscher, Daniel Nettle

Third column: Randy Gallistel, Graham Loomes, Nick Chater, Danny Oppenheimer, Randy Gallistel, Danny Oppenheimer, Ed Hagen

7

Decision Making

What Can Evolution Do for Us?

Edward H. Hagen, Nick Chater, Charles R. Gallistel,
Alasdair Houston, Alex Kacelnik, Tobias Kalenscher,
Daniel Nettle, Danny Oppenheimer, and David W. Stephens

Do I contradict myself?
Very well then I contradict myself,
(I am large—I contain multitudes.)
—Walt Whitman, *Song of Myself* (1855)

Abstract

This chapter examines the contributions that evolutionary theory can make to an integrated science of decision making. It begins with a discussion of classical decision theory and analyzes the conceptual and empirical failures of this approach. Mechanistic explanations, which do not explicitly invoke evolutionary arguments, are presented to account for these failures. Thereafter, evolutionary approaches to decision making are examined and the failures revisited in light of evolutionary theory. It is shown that in some cases "irrational" behavior might be adaptive. The chapter concludes by exploring the open questions, levels of analysis, and policy implications that an evolutionary approach can bring to decision making.

The Bicycle

Imagine a group of aliens were to come to Earth, commissioned by their superiors with producing an intelligible account (or explanation) of a class of objects that Earthlings classify as "bicycles." The aliens could pick a set of such items and measure their thermodynamic properties, their conductivity, and other physical features, but unless they consider the purpose for which the bicycle was designed, the analysis will almost certainly fail to capture the essential elements needed to produce a coherent explanation of why such things as foldable metal bicycles, rigid plastic bicycles, racing bicycles, and mountain bicycles are all called by the same name.

As this example illustrates, considering the process by which the object of study came about and the criteria for its design (the "purpose" of the object) greatly aids the discovery of answers to certain questions. In any science that uses biological organisms as objects of study, evolution by natural selection is that process. Decision making is a functional property of organisms. Comprehensive accounts of decision processes in organisms as varied as amoebas and humans are thus aided by consideration of evolution by natural selection.

There is an important caveat to this argument for an integration of evolutionary theorizing in the study of decision making and behavior. Returning to our alien narrative, imagine that the aliens' mission was not to understand "bicycles" but "conductivity." They may well choose to study bicycles because bicycles tend to contain metal bits, rubber bits, and plastic bits, which might make it a great system for exploring how conductivity operates across multiple materials. They might even succeed in producing a coherent theory of conductivity across materials, and for them the fact that bicycles have a purpose is now irrelevant to the success of their research program. They are not interested in the *bicycleness* of their objects of study.

Within the study of decision making there are topics that are analogous to each of these different kinds of alien missions. A researcher who designs financial products may notice that private investors systematically prefer products that have lesser variance in outcome, and this suggests to her that she can make money by selling products that take advantage of this tendency. For her, the goal is to understand behavior with respect to money, and there are many ways to go about this, not necessarily involving asking why, in an evolutionary sense, investors behave as they do. Forays into *prospect theory* (Kahneman and Tversky 1979) have made advances in identifying regularities by which humans make decisions (e.g., reference points and diminishing returns for both gains and losses) without appealing to evolutionary principles.

Progress in decision science is possible without reference to evolution, but given that many of the questions that decision scientists study involve biological agents (including, of course, human beings), it is greatly aided by it. There are two major advantages of taking evolution into account. First, evolved functionality is a suitable source of candidate hypotheses for decision mechanisms. Evolved functions, which biologists term *adaptations*, are a tiny and special subset of all conceivable functions. Adaptations evolved by natural selection, which means that they must have increased the reproduction of the organism (i.e., increased biological *fitness*). Bicycles, although a highly functional form of human-powered transportation, are *not* adaptations.

For a more pertinent example, it is clear that evolution will not produce decision mechanisms that lead to maximization of lifetime accumulation of resources per se. Natural selection favors the accumulation of resources only insofar as resources contribute to fitness. If a given resource (say money) has a nonlinear relation to fitness, then there is no reason to expect that decision processes will lead to maximization of lifetime wealth accumulation (McNamara and Houston 1992). This may contradict intuitive expectations of people raised in capitalist societies, but is immediately apparent when we take evolution seriously.

A second important advantage is that since evolution acts on the norm of reaction of organisms as a whole and within specific social and environmental scenarios, the integration of decision studies across sciences such as anthropology, economics, psychology, ecology, and neurophysiology is a natural contribution of evolutionary theorizing.

In this chapter, we examine the contributions that evolutionary theory can make to an integrated science of decision making. We begin with a brief introduction to classical decision theory, followed by a discussion of its conceptual and empirical failures. We then examine some mechanistic explanations for these failures that do not explicitly invoke evolutionary arguments. Thereafter, evolutionary approaches to decision making are introduced. We revisit the failures of axiomatic decision theory in light of evolutionary theory, revealing that in some cases "irrational" behavior might be adaptive. In conclusion, we explore open questions, levels of analysis, and policy implications of an evolutionary approach to decision making.

Axiomatic Decision Theory

Decision theory aims to understand how agents—usually humans and nonhuman animals, but also microorganisms, plants, and artificial life—pursue goals in the face of options. Examples of goals include maximizing happiness, wealth, or calorie intake, and examples of corresponding options include choosing among different careers, investment opportunities, or berry bushes.

The foundations of decision theory were laid in the 17th century in a series of letters between Blaise Pascal and Pierre Fermat, who discussed the problem of dividing stakes between two gamblers whose game is interrupted before its close. To illustrate the problem and its solution, imagine a game with two players, Peter and Paul, who have staked equal money on being the first to win 3 points by tossing a fair coin. Peter wins a point if the coin lands heads, and Paul a point if it lands tails. How should the stakes be divided if the game is stopped when Peter has two points and Paul one?

Because the coin is fair, the players have an equal chance to win the next point. If Peter won the next point, he would win the entire stakes, so he is entitled to at least half the stakes. If Paul won, the players are tied, so each would have an equal chance to win the entire stakes on the next toss; in that case, each is entitled to half the stakes, meaning Peter is entitled to 3/4 of the stakes overall. In other words, the value of the unfinished game to each player is the sum over the values of each possible outcome (i.e., the stakes), each multiplied by the chances that it will occur. A key generalization that emerged from this discussion is that agents should maximize *expected value* (EV):

$$EV = \sum p_i x_i, \tag{7.1}$$

where p_i and x_i are the probability and amount of money, respectively, associated with each possible outcome ($i = 1,...,n$) of that option.

A Concave Utility Function

However elegant, the prescription to maximize expected value raises the so-called St. Petersburg paradox, famously posed by the mathematician Nicolas Bernoulli. Consider the following game of chance: to play, the player pays a fixed fee up front, x, and then repeatedly tosses a fair coin until the first head appears, receiving $2^n x$ if a head comes up on the nth toss. The expected value of the game is therefore:

$$\begin{aligned} EV &= \frac{1}{2}\cdot 2x + \frac{1}{4}\cdot 4x + \frac{1}{8}\cdot 8x + \frac{1}{16}\cdot 16x + \ldots \\ &= x + x + x + x + \ldots \\ &= \infty. \end{aligned} \tag{7.2}$$

Hence, players should be willing to pay any amount to play, yet common sense suggests that most people would only pay a few dollars to play. The most influential solution to this paradox was offered by Nicolas' cousin, Daniel Bernoulli, who argued that the value of an outcome should not be judged on monetary value, but instead on a concave function of this value, termed *utility*, which reflects the diminishing marginal utility of money (see Figure 7.1). In Bernoulli's (1738/1954) words:

> [T]he determination of the value of an item must not be based on its price, but rather on the utility it yields. The price of the item is dependent only on the thing itself and is equal for everyone; the utility, however, is dependent on the particular circumstances of the person making the estimate. Thus there is no doubt that a gain of one thousand ducats is more significant to a pauper than to a rich man though both gain the same amount.

The diminishing marginal utility of money reduces the values of the later terms of Equation 7.2 so that expected utility converges to a finite number, resolving the St. Petersburg paradox. It is also important to point out that individuals with concave utility functions are risk averse. (There have been many attempts to resolve the St. Petersburg paradox; for a review, see Martin 2008.)

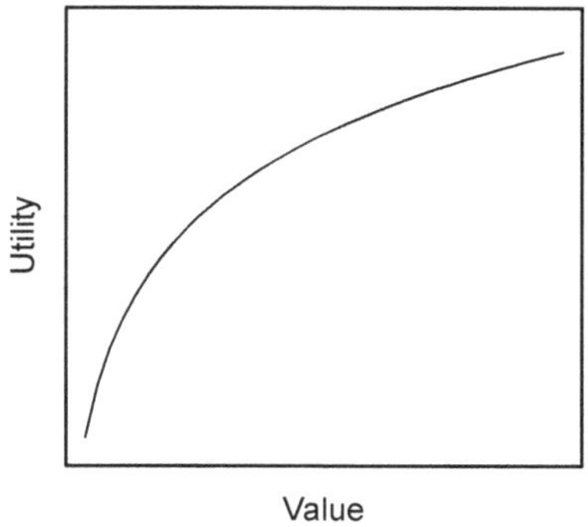

Figure 7.1 A typical utility function.

The von Neumann and Morgenstern Axioms

The ideas of Pascal, Fermat, and Bernoulli were axiomatized by von Neumann and Morgenstern (1947) as follows. Agents face mutually exclusive outcomes A_i. A lottery L is:

$$L = \sum p_i A_i, \tag{7.3}$$

where p_i represents probabilities that sum to 1. Agents choose lotteries (not outcomes!). A preference for lottery L over lottery M is denoted $L \succ M$. There are four axioms:

1. Completeness: for any two lotteries (L, M) exactly one of the following holds: $L \succ M$, $L \prec M$, or $L = M$. In other words, when comparing two lotteries, agents are never undecided about their preference (although they might be indifferent).
2. Transitivity: If $L \preceq M$ and $M \preceq N$ then $L \preceq N$.
3. Continuity: If $L \preceq M \preceq N$ then there exists a probability $p \in [0,1]$ such that $pL + (1 - p)N = M$.
4. Independence: If $L \preceq M$, then for any N and $p \in (0,1]$, $pL + (1 - p)N \preceq pM + (1 - p)N$. In other words, the preference for M over L is unaffected by the inclusion of N.

In the classical view, widely adopted in economics, agents whose decisions conform to these axioms are, by definition, *rational*.

States and Strategies

Agents' decisions will usually depend on the state of the agent (Mangel and Clark 1986; McNamara and Houston 1986). State, in this context, might include psychological variables representing information that the agent has about the world (including information about past outcomes); physiological variables such as energy reserves; and morphological variables such as body size.

A strategy is a rule that specifies the action that should be performed in each state. It is also possible that, for a particular state, one of several behavioral options is chosen randomly. The sequence of behavior generated by a strategy might also be stochastic because the organism's state might not be exactly determined by behavior. For example, the strategy in a particular state of energy might be to forage, but the consequences of foraging, and hence the next energetic state, are stochastic.

This definition of strategy does not specify the process that generates strategies, nor does it even contain the idea of maximizing utility or fitness, or otherwise "doing well." What a strategy is good for depends on the process that generated it. In axiomatic decision theory, agent strategies are generated by "rational thought" with the aim of maximizing subjective utility. The mathematical analysis of such strategies is termed *game theory* (von Neumann and Morgenstern 1947).

Mapping Needs to Utility

The Need for a Common Currency

Agents typically have numerous, constantly recurring needs, such as food, water, and safety. Under the axioms of decision theory, there exists a utility function mapping preferences for lotteries to the real numbers. The need for such a common currency arises because decisions must be made whose consequences cannot compensate for the lack of other consequences. For example, no amount of nutrient ingestion will remedy a water shortage nor pass on genes to the next generation; no amount of water ingestion will compensate for a nutrient shortage (starvation) nor pass on genes to the next generation; and no amount of copulation will restore either a water deficit or an energy deficit.

Most models of decision making assume that the decision is based on whether a decision variable of some kind exceeds one threshold or another. The decision variable represents the strength of the evidence or the strength of the need. When evidence is commensurable (i.e., when evidence of one kind can be weighed against evidence of another kind), then combining the evidence into a single decision variable is not problematic. However, evidence of a need for nutrients cannot be weighed against the evidence of a need for water. In animals, including humans, there are many neurons and neural systems involved in decision making, and these neural systems engage in different tasks and may come to different conclusions (Kalenscher et al. 2010). Animals nonetheless must, and do, decide between going to the river and going to the orchard.

One way in which such agents could do so is by mapping all needs into a common currency, usually referred to as *subjective value* or *utility*. Transforming the different options into utility allows them to be represented on a common scale and rank-ordered according to value/utility. The greater the difference between the values of the available options, the higher the propensity of an agent to choose the option with the highest value. If the analysis of the options each reproducibly (i.e., without noise) generated an output independent of the other options in the common currency (± some amount of subjective utility), then simply summing the outputs will yield a real-valued decision variable that is guaranteed to satisfy basic considerations like transitivity because the ordering of the real numbers is transitive.

Despite the theoretical advantages of a common currency, it is far from clear whether this is physically implemented in most animals (cf. Shizgal 1997) or other decision-making organisms.[1]

Old and New Approaches to Mapping Needs to Utility in Economics

In economics, there have been several approaches to mapping agents' needs to utility, which we broadly characterize as "old" and "new." "Old" involved the Jeremy Bentham-ite idea that each possible outcome has a hedonic (pain/pleasure) value which has a magnitude and can be weighted by its probability, and with the decision rule being to choose whichever is most positive (pleasure) or least negative (pain) (Bentham 1789). The Benthamite individual thus has some kind of "hedonometer" to convert outcomes into this common currency for comparison.

Because of tremendous difficulties in fitting value functions to the great variety of choices expressed by different actors, Paul Samuelson (1938) proposed to refrain from making any assumptions about the content of utility functions (i.e., what people actually value and to what degree) but to infer individual preferences, whatever they are, from observed choices. Within this "revealed-preferences" approach (which we term "new"), people behave as if they maximized a utility function, whatever this is, as long as they meet certain consistency requirements, including transitivity, independence, and completeness. In short, the idea was to forget about how the individual did the processing and weighing up of options but focus instead on how a rational individual's actual behavior could be represented by substituting "utility" indices for outcomes and weighting these by (subjective) probabilities such that if *A* is observed to be chosen over *B* the indices and weights would give a higher expected utility for *A* than for *B*.

---

1 It is even conceivable that "decision making" by plants employs a common currency. There has been a recent explosion of research on "computation" and "intelligence" in plants. Like animals, plants have sophisticated hormonal signaling networks, such as the ethylene, jasmonic acid, and salicylic acid pathways, which regulate growth and development and mediate responses to environmental stressors. These and many other signals must be integrated by the plant to reach "decisions" (Gilroy and Trewavas 2001; Trewavas 2005). It is not out of the question that, e.g., some signal molecule acts as a common currency in plants.

It is impossible, at least within the "new" framework of revealed preferences, to make any statements about utility in someone who does not adhere to these consistency requirements. Hence, in economics, consistency of choice, as indicated by transitivity, completeness, etc., is the gold standard against which the quality of a decision can be evaluated; it is the hallmark of economic rationality.

Irrationality and Exploitation

The worry about decision-making mechanisms that violate basic principles of rationality, such as transitivity, is that they expose the decision maker to potentially catastrophic exploitation. To illustrate, suppose agent X had intransitive preferences $A > B > C > A$, and agent Y possessed A. Agent Y could then sell A to agent X for $B + \varepsilon$, then sell B to agent X for $C + \varepsilon$ and then sell C to agent X for $A + \varepsilon$ (where ε is some small amount). Agent X has now given 3ε to agent Y and received nothing in return, a form of exploitation known as a "Dutch book."

Failures of the Axiomatic Approach

Despite the mathematical elegance of axiomatic decision theory, and despite the vulnerability to exploitation of agents that violate these axioms, the axiomatic approach fails on both conceptual and empirical grounds. These failures, described next, threaten those behavioral sciences that have grounded their disciplines in axiomatic decision theory, with economics being the prime example. Our principal goal in this chapter is to assess the extent to which evolutionary theory can provide an accurate, formal theory of human (and nonhuman) behavior to replace axiomatic decision theory, a task taken up below (see section on The Evolutionary Theory of Decision Making).

Conceptual Failures

Axiomatic decision theory makes several problematic assumptions about decision-making agents (at least if those agents are taken to be humans or nonhuman organisms) and has important gaps. First, although agents are assumed to have complete, transitive preferences, the theory does not explain why agents have the preferences they do, or even why they have any preferences at all. Second, agents are assumed to be able to maximize their utility functions rapidly under all conditions. The theory thus ignores practical limits on time, information, and computational power.

Axiomatic decision theory also ignores the fact that decisions are typically made in a highly structured environment, and this structure can be exploited by agents to simplify the problems they face, thus allowing them to make "good" decisions with minimal computation. Numerous experiments reveal decision-making biases that are often interpreted as errors. People are more likely to judge a statement as true if they have heard the statement before, for instance, regardless of the actual truth of the statement, a phenomenon referred to as the *truth effect* or *reiteration effect* (for a review, see Dechêne et al. 2010). Yet it is possible that such biases instead improve decisions by exploiting environmental structure.

Schooler and Hertwig (2005) and Reber and Unkelbach (2010) argue that the truth effect is mediated by cognitive fluency, such as retrieval fluency (the ease with which an object is remembered): repeated exposure to a statement increases the ease with which that statement is processed and this, in turn, increases the perception that the statement is true. The latter increase occurs, according to these authors, because statements which are true are more likely to be encountered than those that are false. Hence, cognitive fluency as a cue of truth is epistemically justified (for a review of the effects of fluency on decision making, see Oppenheimer 2008; for an argument that, for gossip statements, reiteration reduces error, thus increasing believability, see Hess and Hagen 2006).

Simon (1956, 1990) dubbed this alternative view of decision making *bounded rationality* and made an analogy with a pair of scissors: one blade represents the cognitive limitations of the decision maker and the other the structure of the environment. Understanding decision making requires understanding both blades (for further examples and a discussion of the relationship between decision making and environmental structure, see Gigerenzer et al. 1999).

Empirical Failures

Perhaps not surprisingly, there is abundant evidence that humans (Tversky 1969; Loomes et al. 1991; Grace 1993; Tversky and Simonson 1993; Kalenscher and Pennartz 2010; Kalenscher et al. 2010) and nonhuman animals (Navarick and Fantino 1972, 1974, 1975; Shafir 1994; Waite 2001; Shafir et al. 2002; Bateson et al. 2003) systematically and predictably violate transitivity and other assumptions of axiomatic decision theory. (Rieskamp et al. 2006 reviews much of this evidence and the extent to which it might be consistent with relaxed consistency assumptions.)

Moreover, the same neural systems that have been implicated in representing economic utility, identified by assuming consistency of choice (a manifestation of the utility function: "as-if becomes as-is"), are also implicated in representing local intransitive value in people making intransitive choices (Kalenscher et al. 2010). This suggests that the neural systems containing value signals do not necessitate transitivity to represent the attractiveness of one commodity over another; they thus do not work according to the requirements of a utility function.

To provide a detailed example of the empirical failure of axiomatic decision theory, we discuss intertemporal choice. We then describe mechanistic and state-dependent explanations of apparent rationality violations, followed by an introduction to evolutionary decision making theory and a reanalysis of these violations in light of it.

Intertemporal Choice

Starting with Paul Samuelson, economists in the 20th century came up with a prescriptive theory on how choices should be made between future outcomes (Samuelson 1937; Koopmans 1960; Lancaster 1963; Fishburn and Rubinstein 1982). This framework, *discounted utility theory*, posited that a decision maker behaves as if she maximized discounted utility (DU), with DU being the sum of the discount-factor-weighted utilities of all possible final states. DU assumed that the discount function, by which the (undiscounted) utilities of the outcomes are multiplied, decreases exponentially with time (Samuelson 1937):

$$f(t) = e^{-rt}, \tag{7.4}$$

where r is the discount rate. One of the important implications of exponential discounting is that the rate by which future rewards are devalued will be constant over time:

$$\frac{f(t+\Delta t)}{f(t)} = \frac{e^{-r(t+\Delta t)}}{e^{-rt}} = e^{-r\Delta t}. \tag{7.5}$$

In other words, if a glass of beer is valued half as much tomorrow as it is today, then, under exponential discounting, it is valued half as much a year and a day from now as it is a year from now.

Constant discounting has important implications for economic rationality and time-consistency of preference. According to DU, it is not irrational or nonoptimal per se to prefer small, short-term over large, long-term rewards, even if the preference for immediacy results in an overall reduced net gain over time. However, DU requires consistency over time. That is, if an individual prefers a small, short-term reward over a large, long-term reward and both rewards are shifted in time by an identical time interval, then the preference for the small, short-term reward should be preserved because both rewards should be discounted by the same rate. For example, it could be perfectly rational for a rock star to live fast and die young if he really accepts this consequence of the early deathbed. In contrast, DU would label behavior as time-inconsistent if a decision maker fails to act in accordance with his long-term interest. For instance, the failure to make appropriate retirement provisions would be irrational if the agent actually wishes to have a good and healthy lifestyle at old age.

Time-constant discounting, or more precisely, the stationarity axiom in DU (Koopmans 1960), predicts that the ranking of preferences between several future outcomes should be preserved when the choice outcomes are deferred into the future by a fixed interval because the two outcomes should be discounted by the same fraction. A wealth of empirical studies suggests that all species tested, including humans, monkeys, pigeons, rats, mice, leeches and dragonflies, violate the principle of constant discounting and other implications of DU.

In a paradigmatic test of this prediction, human or nonhuman subjects first choose between a smaller, sooner reward (SS) or larger, later reward (LL), such as receiving $1 today or $2 tomorrow. A common delay is then added to both options, such as receiving $1 in seven days or $2 in eight days. An exceptionally reliable finding across humans and nonhuman animals (Chung and Herrnstein 1967; Rachlin and Green 1972; Ainslie 1974, 1975; Green et al. 1981; Thaler and Shefrin 1981; Logue 1988; Benzion et al. 1989; Loewenstein 1992; Green et al. 1994; Kirby and Herrnstein 1995; Green et al. 1997; Bennett 2002; Frederick et al. 2002; McClure et al. 2004; Rohde 2005; McClure et al. 2007) is that the tendency to choose SS declines dramatically with the increasing length of the added delay: although many people choose $1 today over $2 tomorrow, these same individuals pick $2 in eight days over $1 in seven days. This suggests that the prolongation of the delays resulted in a preference reversal even though the difference in delays remained identical (immediacy effect; Thaler and Shefrin 1981; Benzion et al. 1989).

If these experimental results are taken at face value, human and nonhuman agents appear to add extra value to immediate outcomes, a behavior best approximated by nonconstant discount functions, such as hyperbolic (Mazur 1984, 1988) or quasi-hyperbolic (Laibson 1997) functions. A hyperbolic discount function is of the form:

$$f(t) = \frac{1}{1+rt}, \tag{7.6}$$

where r is the discount rate. In contrast to exponential discounting, under hyperbolic discounting the rate by which future rewards are devalued is *not* constant over time. To illustrate this, consider that when $t = 0$ (i.e., the present):

$$\frac{f(t+\Delta t)}{f(t)} = \frac{1/(1+r(t+\Delta t))}{1/(1+rt)} = \frac{1}{1+r\Delta t}. \tag{7.7}$$

On the other hand, when t >>Δt (e.g., the distant future):

$$\frac{f(t+\Delta t)}{f(t)} \sim 1. \tag{7.8}$$

Under hyperbolic discounting, despite the fact that a glass of beer is valued half as much tomorrow as it is today, its value a year and day from now is (almost) equal to its value a year from now.

As noted earlier, because agents who use nonconstant discount functions and exhibit other violations of rationality assumptions are vulnerable to exploitation, and because such violations challenge axiomatic decision theory, considerable effort has been made to explain these violations.

One possibility, of course, is that some such violations are artifacts of the experimental procedures. For instance, experimental procedures might inadvertently induce changes in state, at least in nonhuman animals, producing apparent violations of transitivity. This could happen if training procedures meant to teach the animal about choices also altered feeding rates, and hence the animal's energetic state. Choices, which seem to reveal intransitive preferences, result instead from choices made in different states, which do not violate transitivity, and has been demonstrated in a study of starlings (Schuck-Paim et al. 2004).

Despite the possibility that some violations are actually experimental artifacts, the empirical evidence against axiomatic decision theory is strong enough to compel us to seek an alternative. We turn now to explanations of irrational decision making that do not explicitly invoke evolutionary arguments, such as those involving the mechanisms of decision making, followed by an explicitly evolutionary approach to decision making which, in some instances, identifies an adaptive logic underlying seemingly irrational decisions.

Mechanisms of Decision Making

Some of the observed failures of axiomatic decision theory could be the result of the cognitive architecture of decision making.

Voting Can Produce Intransitivity

One suggestion is that intransitivities are the consequence of a system where multiple independent decision mechanisms in the brain "vote" for a choice, and the winning choice becomes the agent's decision. Condorcet's voting paradox tells us that when decisions are made by aggregating over votes (rather than, e.g., summing real-valued outputs), and where there are three or more voters (in this case, decision-making mechanisms), the revealed preferences of the system as a whole may very well be intransitive, even though every voter has a transitive ordering of the options. When choosing among two of the three decision options by voting, the preferences (see Table 7.1) will yield $A > B > C > A$.

Kenneth Arrow (1950) generalized Condorcet's paradox in his *impossibility theorem*, which proved that no voting system can be designed that satisfies three fairness criteria:

Table 7.1 Preferences illustrating Condorcet's paradox, where *A*, *B*, and *C* are decision options; I, II, and III are decision-making agents; and 1st, 2nd, and 3rd represent the ordinal preferences of each agent.

	I	II	III
A	1st	3rd	2nd
B	2nd	1st	3rd
C	3rd	2nd	1st

1. If every voter prefers alternative *X* over alternative *Y*, then the group prefers *X* over *Y*.
2. If every voter's preference between *X* and *Y* remains unchanged, then the group's preference between *X* and *Y* will also remain unchanged (even if voters' preferences between other pairs like *X* and *Z*, *Y* and *Z*, or *Z* and *W* change).
3. No single voter possesses the power to always determine the group's preference.

Hence, some behavioral tendencies might simply be by-products of the neural architecture (for a discussion, see Schneider et al. 2007).

Representation

Constraints on representation may be another central way in which cognitive architecture shapes decision making, including exhibiting intransitive preferences and other irrational behaviors. The concept of representation encompasses different phenomena according to different authors, so it is worth prefacing any discussion or example of its role in decision making with a minimum attempt at semantic precision.

By representation, we mean a system of symbols that is homomorphic[2] to another system. A representation has three essential properties (Gallistel 2001):

1. Reference: the symbols in the representing system must be causally connected to that to which they refer.
2. Homomorphism: operations on the symbols in the representing system must be homomorphic to processes and relations that obtain between their referents.
3. Functionality: the operations on the symbols in the representing system must at least occasionally direct behavior in ways consistent with the state of that aspect of the represented system to which the symbols refer.

Neurons involved in yaw correction in the housefly provide an example of all three properties (yaw is rotation around an axis perpendicular to the fly's wings). First, large-field image motion (which under natural circumstances is a

[2] A homomorphism is a "structure-preserving" transformation.

measure of yaw) alters spike trains in two axons in the visual system of the fly, establishing causality. Second, a linear operation on the spike train data recovers an accurate representation of the yaw waveform, establishing homomorphism (Rieke et al. 1997). Third, there is good evidence that the spike trains in these neurons drive the yaw correcting responses of the flying fly, establishing behavioral function.

Whether "thoughts" (higher-level cognitive processes), in general, are best understood as computations over representations, or as emergent phenomena of the simultaneous operation of large numbers of simple, interconnected processing units (neurons, presumably, or groups of neurons) is, of course, a classic debate in the fields of cognitive science and the philosophy of mind. If the symbolic view of cognition is correct, then it has important implications for decision making in humans. Insofar as information-processing systems in nonhuman organisms, including microorganisms and plants, are also symbolic, the implications we describe next apply to them as well; in fact, some of the most compelling examples come from insects.

A decision is made between alternatives: between alternative possible states of the world or between alternative actions. The representation then specifies the set of alternatives. Thus, it constrains the decision. No decision can be taken on an alternative not specified in the representation. Consider, first, decisions as to which, among a range of possible states of the world, is in fact the currently prevailing state, a problem often solved with Bayesian analysis. A Bayesian decision maker has a representation of the possible states of some aspect of the world, which precedes and constrains the outcome of the decision process. Consider, second, decisions as to which action to take. Again, the animal cannot choose an action that it cannot represent. In reinforcement learning, the values that are computed attach to the symbols that represent the possible actions. No decision can be taken on an action that has not been symbolically represented.

This is a very general definition of the concept of representation as it applies to choices, one that encompasses a broad range of views regarding the veridicality of representations. At the "minimalist" extreme, a representation is simply a member of a set over which a nonzero probability distribution has been defined. The set and its members need not have much, if any, structure. The primary requirement is that, in the environments in which an agent typically finds itself, the set of representations helps generate functional behavior.

At the same time, this definition of representation can also accommodate highly structured representations of the world. If an agent solves a problem like finding its way in space, then it must somehow represent space, in the sense that it must compute solutions using "quantities" that represent attributes of the way-finding problem. For an agent that navigates along a chemical gradient, the representation of space could be as simple as "higher concentration" and "lower concentration." At the other extreme, an agent that must achieve pinpoint accuracy in order to navigate long distances, such as a migrating bird, might have a representation of space whose structure closely approximates the structure of Euclidean space (i.e., $\mathbb{R}^3$ with the standard metric). Much ink has been spilled on the question of whether nonhuman animals possess such a "cognitive map," and although there are many threads to this story—from "short-cut style experiments" to the analysis of so-called place cells in the mammalian hippocampus—there does not seem to be have been a satisfying resolution of this question (see, e.g., Collett and Graham 2004; Cruse and Wehner 2011; Dyer 1991; Gallistel and Cramer 1996; Gould 1986; McNaughton et al. 2006; Menzel et al. 2005; Tolman 1948).

State-Dependent Valuation Learning

The foregoing theoretical arguments make it clear that agents' representations could shape their decision making, but they do not address the issue of which aspects of experience are first encoded and then recalled to drive choices. Given a certain experience (say an action followed some time later by access to a food item), the subject can encode perceptual information about the food item or the time that has lapsed between action and outcome, as well as encoding internally generated information such as changes in energetic state consequent on ingestion. It is possible to explore these issues experimentally.

Consider a subject that experiences two stimuli: red and blue. Responding to red leads to outcome A whereas responding to blue leads to outcome B. In one experiment, these stimuli are randomly assigned to two internal states, so that, for example, red is experienced when the individual is hungry (H) and blue when it is not hungry (N). We start with a case where the sizes of A and B are equal. Once the subject is acquainted with each stimulus and its consequences, choices between them are introduced in both states (H and N) and preference is measured. The questions involved are: Which is preferred (if any), and what kind of representation may explain such preference? The problem is described in Figure 7.2 (Aw et al. 2011).

Focusing on Figure 7.2b, let us consider several putative criteria for choice. The subject may:

1. Use representations of outcome sizes, or equivalently of the change they cause in energetic state, shown as ΔR_A and ΔR_B. Since these are equal, the subject would be indifferent.
2. Form distorted representations so that A is encoded as being bigger than B. This is shown in the cartoon representing the bird's memory, where ΔR_A appears larger than ΔR_B. This may happen because the significance of the change in state is greater when the animal is hungrier. If this happens, the bird will prefer A (or the red stimulus that causes it in our example).
3. Use representations of the significance, or value, consequent of the changes, shown as ΔH_A and ΔH_B, leading again to preference for the

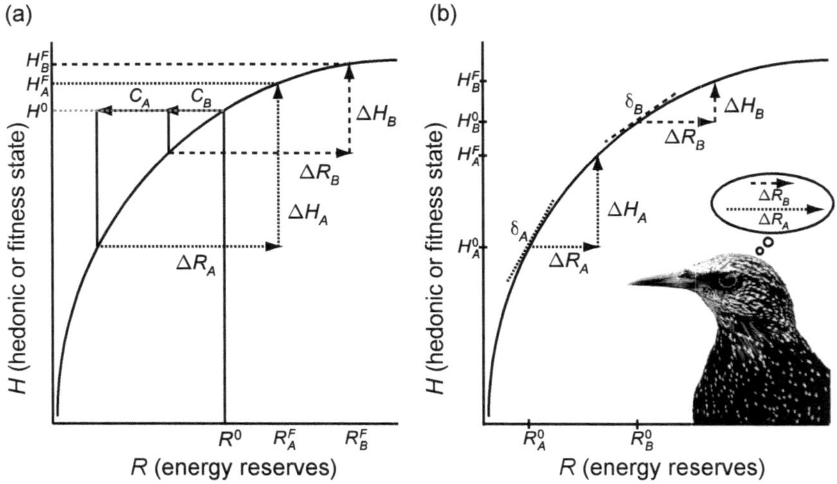

Figure 7.2 A nonlinear relation between state and a measure of subjective value. (a) Depiction of a situation where an animal in a given state, R^0, pays different costs C_A and C_B for equally sized rewards, *A* and *B*, which then cause increases in reserves, ΔR_A and ΔR_B of equal magnitude, but with different consequences, ΔH_A and ΔH_B. (b) Depiction of a situation where the encounters with *A* and *B* occur when the subject's state is manipulated so that *A* is encountered in R^0_A and *B* in R^0_B, where the former refers to stronger hunger. In both cases, because of the concave shape of the relation between state and consequences, *A* and *B* differ in value gain despite being equal in size. After Aw et al. (2011).

stimulus causing *A*. This differs from second hypothesis because percepts are encoded here veridically; their different affective consequences are encoded and then used. Hypotheses 2 and 3 can be differentiated by designing experiments so that the subject is induced to reproduce its memory for each outcome rather than choosing between them.

4. Use representations of the slope of the value versus state function at the time of encounter with each stimulus ("red gives more value per unit of reward"). This appears as δ*A* and δ*B*. In this case, it would also lead to preference for *A*, but could be differentiated from the other hypotheses by using protocols where the magnitudes of *A* and *B* are not equal. Under hypothesis 4 the magnitude of the outcomes exerts no influence.

As we see, one hypothesis leads to indifference and three to preference for red; namely, the stimulus associated to greater hunger.

Such experiments were conducted in starlings, fish, and locusts (Aw et al. 2009, 2011; Marsh et al. 2004; Pompilio et al. 2006; Pompilio and Kacelnik 2005). In all three species, the subjects preferred the stimuli associated with a state of greater need, refuting the possibility that direct representations of the metrics of the outcomes, without any additional encoding of value, may be sufficient to explain choice.

Further experiments permit some level of differentiation between the surviving three hypotheses. If the fourth hypothesis were true, the subjects would be indifferent to the size of the outcome. We can discard this in the starlings because when the outcome metrics are manipulated, so that the stimulus associated with hunger takes values below that of the alternative, preferences for it decline and eventually reverse. This sensitivity to outcome magnitude eliminates (at least for the only species where it was tested) hypothesis 4.

Hypotheses 2 and 3 can also be differentiated for starlings. In experiments where outcomes *A* and *B* differed in the delay in delivery of food, it was shown that the birds' preferences were independent of having accurate representations for the delays in both options (Pompilio and Kacelnik 2005). They sometimes preferred responding to receive outcome *A* in spite of knowing accurately that its metrics were worse (longer delay) than in outcome *B*. For locusts, however, preexistent neurobiological evidence indicates that it is likely that the gain of receptors is adjusted, so that they would perceive the stimulus associated with hunger as being more salient than the alternative. Admittedly this last matter is not as categorically sorted as the former; however the point here is to illustrate how the nature of the representation and its effect on choice can be unconfounded with suitable experimentation.

From an evolutionary standpoint, these studies provide an example of convergence to the same choice mechanism (i.e., state-dependent valuation learning) in a set of very distant organisms, raising the question of its adaptive significance (for an evolutionary model, see McNamara et al. 2012).

Multiple Attribute Problem

Choosing between options with many attributes is a fundamental problem of decision making and provides another illustration of the impact of representation. Broadly speaking, there are two approaches. One approach attempts to measure each option directly through a common currency. Suppose that we are attempting to decide which of two apples to eat, and each varies in both size and taste. Very crudely, according to this approach, we attempt to value the apple based on, say, size and taste, as having a certain utility (say 23 vs. 25 "utils"). The apple that is assigned the highest value is preferred. If people could stably associate such complex objects with values using such an internal currency, then choices would be highly stable and transitive. However, the problem with such "object-based" comparisons is that it is extremely difficult to know how to map complex objects onto utilities; indeed, when people are asked to do this explicitly using evaluations, the results are notoriously unstable (e.g., Hsee and Rottenstreich 2004).

Attribute-based comparison is an alternative approach. Here, the agent does not evaluate each apple separately. Instead, the apples are compared according to size (e.g., apple *A* is slightly bigger than apple *B*) as well as taste (apple *A* is a very much less tasty variety than apple *B*). Each attribute thus provides a potential argument for or against the choice of apple A or B. The agent then attempts to weigh these "arguments": the big difference in taste may overwhelm the small difference in size, and result in apple being chosen.

Almost all versions of attribute-based decision making will not be equivalent to any object-based approach. For example, if one object is valued slightly better on one dimension, and yet identical on all others, it will be chosen nearly always. From the attribute-based view, this is easily exploitable, because there is just one "argument" for one object, and none against it. By contrast, the choice between two objects, such that one is better on attribute f while the other is better on attribute g, will be highly unstable, because the agent has two almost equally good arguments to trade off (Loomes et al. 2012). Moreover, if the object has three attributes, which are appropriately arranged, then Condorcet's paradox may arise (see earlier discussion).

Internal Conflict

Some decidedly irrational behaviors in humans and other animals appear to reflect "conflict" in the decision-making machinery. Rats offered both food and shock at the end of an alley, for instance, oscillate at a certain distance from them; this behavior could indicate conflict between approach and avoidance mechanisms. Human alcoholics willingly take drugs that will make them sick when they next consume alcohol. Livnat and Pippenger (2006) argue that if behavior is generated by a decision-making system that is subject to computational constraints, if conflict is defined in a particular way in terms of utility functions, and if parts of the decision-making machinery can be assigned utility functions based on information theoretic considerations, then even an optimal system designed for a single purpose can comprise agents that are in conflict with one another and may occasionally produce maladaptive behavior. In the rat case, this would involve approach and avoidance mechanisms that "selfishly" optimize their respective goals, with computational constraints preventing an optimal trade-off between the two goals; this results in maladaptive oscillations under certain parameters of food and shock.

The Evolutionary Theory of Decision Making

The synthesis of evolutionary theory and decision theory is achieved primarily by using *fitness*,[3] rather than subjective utility, as the common currency to compare options (e.g., McNamara and Houston 1986), with fitness optimized by natural selection rather than rational thought. What evolves are strategies (not preferences!) that maximize fitness across the statistically variable sets of environments encountered by members of a population over evolutionary time. The

[3] The number of descendants left far into the future.

mathematical analysis of such strategies is termed *evolutionary game theory* (Maynard Smith and Price 1973; Maynard Smith 1982; for reviews of the strategy concept in biology, see Hammerstein et al. 2006; Laubichler et al. 2005).

The substitution of fitness and natural selection for subjective utility and rational thought, respectively, overcomes many of the conceptual limitations of axiomatic decision theory. First, fitness is a core concept in the theory of natural selection. Unlike subjective utility, it has an exceptionally strong theoretical justification and applies to all forms of life, including viruses, bacteria, plants, and animals. Second, because fitness is optimized by natural selection in a population over many generations, and not by cognitive processes within the agent, agents can behave optimally even with extremely limited computational abilities. In fact, the strategy concept can be applied not only to behavior but also to ontogeny and morphology (e.g., Hagen and Hammerstein 2005). Third, because natural selection acts to adapt agents in a population to their environment, there is a fundamental relationship between decision making and environmental structure (for a review of the increasingly productive interaction between evolutionary biology, economics, and other social sciences, see Hammerstein and Hagen 2005).

This synthesis has implications that are not as widely recognized as they should be. In axiomatic decision theory, rational agents are characterized by their complete transitive preferences. The strategic behavior of agents in a particular context is the result of optimizing computations in the agents' brains, which takes place more-or-less in real time. In evolutionary game theory, in contrast, the agent is a strategy executor rather than a utility optimizer. In particular, evolved agents do not have a fitness utility function coded in their brains and do not compute how to optimize their fitness (the distinction between agents that optimize over preferences vs. those that simply execute strategies might be critical for interpreting the seemingly anomalous results of some behavioral economics experiments; Hagen and Hammerstein 2006).

Ecological, or Substantive, Rationality

We do not wish to infer that, in evolutionary models, agents never optimize in real or ontogenetic time. When they do, however, it is in a circumscribed socio-ecological domain (e.g., foraging or mate choice) that was important to fitness over many generations. This requires a mechanism that makes some (possibly erroneous) commitments to the nature of the reality about which decisions and choices are being made, an approach to decision making dubbed *ecological rationality* (Simon 1956; Gigerenzer et al. 1999) or *substantive rationality* (Chater, this volume).

The problem of characterizing the environments in which decision mechanisms evolved, and the environments in which they operate (and the former could well differ from the latter), comes up in many different guises. For example, in understanding the performance of a heuristic, one clearly needs to know something about the range of situations in which the heuristic is applied. Moreover, a common and reasonable explanation of why we fail to observe globally optimal decisions is that the decision mechanism in question evolved to make decisions in a specific environment, and we would therefore hypothesize that this mechanism should perform at a high level in that environment even though it performs poorly in other environments—a problem often referred to as an "evolutionary mismatch." For instance, a popular hypothesis holds that human obesity occurs because humans evolved in conditions of relative scarcity, which, the argument goes, selected for a tendency to favor foods high in fat and carbohydrate, leading to overeating in the modern environment where these foods are plentiful (e.g., Nesse and Williams 1996). In such cases, the behavioral anomalies studied provide insight into the actual evolutionary pressures to which the decision mechanism was adapted.

Despite its importance, characterizing the environments in which a mechanism evolved versus those in which it operates is often challenging and very seldom even attempted. Indeed, in most decision-making studies, consideration of "the environment" never goes beyond the specific situation presented to subjects.

The research program suggested here has three components: (a) selecting the environment to characterize, (b) choosing the relevant dimensions of the environment to measure, and (c) evaluating and interpreting the performance of a hypothesized mechanism in the now characterized environment. Each presents a different set of challenges.

First, there is the problem of selecting the appropriate environment to characterize. Should we study the current environment in which the animal lives or the environment it experienced in its selective past? From an evolutionary perspective, one is obviously most interested in the environment in which selection has acted, but identifying this could be quite difficult. In most cases, one would have to make some type of stationarity assumption (e.g., that the current environment fairly represents the "adaptation-relevant environment"). Stationarity could, of course, fail in several ways. A given lineage may have experienced a cyclic environment, for example, or its distribution might have covered a large geographic distribution such that the lineage has sampled a wide range of conditions.

Second, it is difficult to choose which attributes of the environment one needs to characterize. To make headway, we probably have to focus on a particular class of decisions. The best example stems from the study of the visual environment. In these studies, investigators have collected thousands of images of the human visual environment and characterized the properties, such as brightness and color distributions (Olshausen and Field 2000), allowing vision scientists to interpret the neural mechanisms of visual process in terms of the statistical regularities of the visual environment. This is the best developed example that we know, and we suggest that it should be a model for future studies of the decision environment. We should, perhaps, point out that this "best example" was developed to study the mechanisms of visual perception rather than decision making.

A hypothetical example that is directly applicable to decision making is the problem of the foraging environment. Suppose we consider a classic foraging problem like patch exploitation (Charnov 1976; Stephens and Krebs 1986). We would want to know the probability distribution of patch types, the gain functions associated with these types, and distributions of travel times associated with these types. In general, we would not expect these properties to be independent of each other, so we would actually want to know, for example, the joint distribution of gain function properties and travel times. Moreover, it is clear that regularities, in the form of correlations between economic properties, are exactly the type of thing that might influence the evolved form of decision rules.

Assuming that one can solve the problems described above (and these problems should not be minimized), the problem of evaluating a decision mechanism, given a well-characterized selective environment, seems relatively straightforward. Yet, even here one must make decisions about the appropriate level of analysis. Crudely using this information to understand a hypothesized neural mechanism requires a different tool box than evaluating the performance of a simple patch-leaving rule.

Even when there are mechanisms, such as reinforcement learning, that solve many different classes of problems, there is a need for domain-specific adaptations. Reinforcement learning can only be deployed where selection has made relevant end-states rewarding, and prepared constraints on the types of actions or stimuli which can become associated with those rewards.

Clear examples of the role of preparedness come from conditioned taste aversions in omnivores (Garcia and Koelling 1966; Rozin and Kalat 1971; Holder et al. 1988; see also Hammerstein and Stevens, this volume). The relevant learning mechanisms make strong species-specific commitments to the nature of the cues that are likely to prove reliable in predicting the nutritive and pathogenic properties of the different foods that they must decide to ingest or not to ingest. Specifically, animals readily form associations between novel flavors and gastrointestinal malaise but fail to do so when it the color of the food predicts ingestibility.

Interestingly, vampire bats do not readily form conditioned taste aversions (Ratcliffe et al. 2003), although they are presumably able to learn associatively in other domains. Given that they do not encounter variation in the level of putrefaction of possible foods (since they ingest blood directly from healthy animals), this illustrates the role of natural selection in maintaining particular associative learning pathways for specific domains in particular lineages.

In another well-studied example, many animals, including insects, use the Sun as a directional referent. To do this, they must learn the solar ephemeris, the compass direction of the Sun as a function of the time of day. The mechanism that learns this makes strong a priori commitments to the form of this function (Dyer and Dickinson 1994).

In studies of humans, economists routinely take for granted that people making economic decisions should and do assume that their environment contains other decision makers that have accurate knowledge of the relevant aspects of the world; decision makers who reason and make decisions in accord with principles of rationality identical to the principles that the decision maker herself employs.

Irrationality Revisited

Evolutionary decision theory provides powerful tools to help resolve the many apparent violations of rationality assumptions in humans and nonhuman animals. Perhaps foremost among these tools is the comparative approach: decision making can be studied in numerous, often distant, branches of the tree of life, including, in principle, organisms without nervous systems, such as plants and bacteria. If a violation of rationality assumptions, such as hyperbolic discounting, occurs systematically across taxa as well as across ecological contexts within taxa, this provides a vital clue to the nature of evolved decision mechanisms.

Below we revisit some apparent violations of rationality assumptions discussed earlier in light of the evolutionary theory of decision making. In some cases, we find that behavior which violates assumptions of axiomatic decision theory is, from an evolutionary perspective, adaptive.

Intertemporal Choice

How is it possible that evolution has shaped a behavior—hyperbolic discounting—that is so ubiquitous across domains and species, yet so apparently and dramatically maladaptive? Optimal foraging theory (Stephens and Krebs 1986) might provide some insight.

Optimal foraging theory makes predictions about the behavior of an animal foraging in an environment with patches of food that vary in location, density, quality, and other variables. The animal needs to decide whether to put its background activity—foraging—on hold to enter a food patch when encountered. Once the animal enters a patch, the marginal rate of energy return decreases with time because the patch is progressively depleted. The animal, therefore, needs to decide when to leave to seek the next patch.

When making such a decision, the animal has to take into account the current (decelerating) energy gain rate, and thus the time between energy unit consumptions in the current patch and the time (and effort and risk) needed to reach the next patch. Within the classical optimal foraging framework, animals have to trade off, among other things, delays and reward magnitudes. Hence, they face an intertemporal choice.

Optimal foraging theory proposes that animals maximizing Darwinian fitness should use foraging strategies that maximize the net energy gain per time unit over the long term. In formal terms, it is assumed that organisms maximize, at least in the long run, the ratio of food intake and the time needed to obtain and consume the food (Stephens and Krebs 1986). In a choice between large, delayed rewards and small, short-term rewards, rate maximization predicts that animals prefer large rewards when the ratio of reward amount per time unit is higher for the large than for the small reward.

To illustrate the implications for putative consistency violations, assume an animal chooses between (a) two food items delivered after 2 seconds (rate: one item per second) and (b) four food items delivered after 8 seconds (rate: 1/2 item per second); hence (a) > (b). Let us now play the economic game presented above and add a common time interval to both options: if both rewards were delayed by 10 seconds, the energy rate for option (a) would change to 0.17 items per second (two food items after 12 seconds) and for option (b) to 0.22 items per second (four food items after 18 seconds). Thus, in contrast to DU, optimal foraging theory predicts a preference reversal: the animal should prefer (a) over (b), but (b′) over (a′). The same behavior that is labeled irrational in economics may be considered well-adapted in biology (e.g., Fawcett et al. 2011).

Experimental work by Stephens corroborates this notion. He has shown that the same animals that perform poorly in traditional intertemporal choice tasks, similar to the ones described above, perform remarkably well in economically equivalent tasks whose structure, however, resembles more the patch-like environment under which the intertemporal choice policies presumably evolved (Stephens and Anderson 2001; Stephens 2002, 2008; Stephens et al. 2004).

Apparent Violations of Transitivity

Houston et al. (2007b) use the foregoing evolutionary strategic framework to obtain *apparent violations of transitivity*. The term "apparent" is used because the violation appears if an observer takes the organism to be choosing between two options, ignoring the possibility that the organism is able to make *repeated* choices between options. The model involves a single state-dependent rule for choosing between two options, given that these options will always be available to the organism in the future.

Specifically, it is adaptive for a foraging animal to consider the alternative food sources when determining the value of a given source. If the value of one food source depends on the nature of the alternative food sources available, intransitive choices can occur. Assume an animal is foraging in an environment containing food sources that can be characterized, among others things, by reward magnitude, reward probability, and predation risk. For example, a starving animal may prefer a rich, but risky (high predation risk) food source (source *A*) over a poorer, but safer food source (source *B*), and may prefer *B* over a rich, safer source that has, however, a low reward probability (source *C*; thus preferring $A > B$ and $B > C$). Under certain circumstances, this animal may prefer the rich, safer source *C* with the low reward probability over the rich, but risky source *A*, thus preferring $C > A$, because the availability of the high-risk source *A* acts as an insurance against the outcome variance of option *C*. The animal could first try out the low-risk patch *C*, and, if no reward is found, it could still opt for source *A*, which yields a large reward with a high probability, albeit at a high predation risk.

At the level of these strategies there is no intransitivity; for any pair of options, there is a single best state-dependent strategy. Even if experiments do not involve repeated choices (as in this model), animals might use rules that evolved to cope with environments in which the options that are available to a decision maker persist into the future. Migrating birds are seen to use current food availability to predict future food availability (Houston 1997), for example, and humans might also behave as if interactions will be repeated (e.g., Hagen and Hammerstein 2006).

Multiple Attribute Problem

Russo and Dosher (1983) maintain that, in multiattribute choice (e.g., a decision between gambles varying in reward probability and magnitude), an agent comparing the levels of the attributes separately will fail to treat each option as an integrated whole, but will evaluate the available options much faster and with fewer errors than an agent integrating probability- and reward-representations into a single utility representation for each option. Assuming that the world is structured in such a way that attribute-based comparisons generally result in recommendations similar to utility-based comparisons, then attribute-based policies will be favored by evolution because of their supremacy over utility-based policies in terms of processing speed and accuracy, which, however, comes at the (possibly small) cost of occasional intransitivities.

These are not the only optimality accounts of intransitive choice (Houston 1991, 1997). They do, however, illustrate the gist of the idea that evolution may not have favored choice consistency, but the development of mental policies that were adapted to the environment in which they evolved.

Two Open Questions

Specificity versus Generality of Decision Making

The *heuristics-based approach* (known in behavioral ecology as the rules-of-thumb approach) correctly and usefully separates the functional from mechanistic aspects of decision processes, yet it tempts theorists to propose overly specific heuristics. For instance, there have been a number of heuristics proposed for whether people find a message persuasive. These include the "audience response heuristic" (Axsom et al. 1987), the "consensus heuristic"(Giner-Sorella and Chaiken 1997), and the "endorsement heuristic" (Forehand et al. 2004), all of which suggest that the more positively the message is received by others, the more positively it will be received by the target (but which vary based on the specific nature of who the "others" are). There is also the "likability heuristic" (Chaiken 1980)—people are more persuaded by likable speakers—and the "expertise heuristic" (Ratneshwar et al. 1987) which suggests that people are persuaded more by experts.

Instead of having a repertoire of extremely specific heuristics, it might be that there is a more general strategy that the highest validity, positively valanced information available in a given context is what determines persuasion. More generally, the most valid information to a given task is used to solve that task. Indeed, Shah and Oppenheimer (2008) have argued that there may be only five general forms of heuristics, and that other heuristics can be created as specific instantiations of a combination of these elemental heuristics.

Kacelnik (this volume) suggests that one robust, heritable, broad-domain algorithm might be when facing a novel problem, display an inherited broad-domain behavior, and then use outcomes to modify the response in the appropriate direction (basically, a loose version of reinforcement learning). As an example, he considers a puppy that is trying to reach a moving ball; initially it runs directly to the ball, thus having to correct its direction constantly and being highly inefficient. With experience, the puppy starts to anticipate these corrections and may converge to a constant angle of gaze control in its chases. The point here is that there was no need to postulate that a heuristic for constant gaze was selected through evolution. What was selected in this case was the very broad "learning through consequences" mechanism, which the puppy uses for such different situations as identifying which human is more likely to deliver a treat versus a kick, which plants sting, what size of puddle is jumpable, or how to catch a moving object.

By developing more general principles rather than a laundry list of heuristics, we have more broadly predictive theories as well as more plausible targets for natural selection.

Substantive versus Formal Rationality

Two types of mechanisms that might be present in the cognitive systems of humans and nonhuman animals can be distinguished: those which embody *substantive* and *formal* rationality. Mechanisms involving substantive rationality assist the organism in reasoning or decision making, by embodying contingent aspects of the world or fitness-relevant goals. For example, as we mentioned earlier, Olshausen and Field (2000) measured spatial statistics of natural images, and argued that the receptive fields in the early cortical visual processing area V1 may arise because they are optimized to encode the image using the minimum number of active neurons. (Substantive rationality is similar to ecological rationality; we use the former term because we want to remain agnostic about some of the theoretical commitments of Gigerenzer and colleagues, with whom the latter term is closely associated.)

Substantive information may be more abstract, such as the apparently strong constraints on how bees learn to use the movement or the Sun in navigation (Wehner and Rossel 1985). It also may include procedures, such as nest-building or information concerning fitness-relevant goals of the organism, such as food and sex. This substantive information contributes to the fitness of the organism, in the context of its actual environment and ecological niche.

In contrast, there might be cognitive mechanisms that embody formal rationality; that is, mechanisms that do not themselves make commitments to the nature of reality but instead impose a priori consistency constraints that are critical to the adaptiveness of decisions. For example, virtually all decision making must take account of the uncertain (probabilistic) nature of the input on which the decision is based. Bayesian inference, the normative form of probabilistic inference, provides the formal rationality constraints for this problem and therefore yields optimal results. Insofar as one believes that evolution optimizes the properties of critical mechanisms, a biological decision-making machine should have mechanisms that implement Bayesian inference.

For Bayesian reasoning, the problem specificity, or substantive content, is found in the "support" (in the statistical sense of the word); that is, in the representation of the possibilities. Prior probabilities (or probability densities) and likelihoods are defined over this support. For example, in deciding where one is, the support is the vector space of one's possible locations. In deciding on what course to follow, the support is the points on the unit circle (the possible compass directions in which one could head). In deciding on whether one is within the radius around one's goal within which one should stop the straight run toward the goal and initiate the search pattern (Wehner and Srinivasan 1981), the support is the possible distances from the goal. In deciding whether to respond to a cue that may or may not predict the time of occurrence of an event of interest, such as the onset of food availability, the support is the measure of how well the onset of the predictor predicts the onset of the predictee. This support is the interval on the entropy dimension between 0 and the source entropy (because the mutual information between the predictor and the predictee cannot be negative and cannot be greater than the source entropy, the amount of uncertainty about when the next predictee might occur; Balsam and Gallistel 2009).

The prior probability distribution specifies the probabilities (or probability densities) of the different possibilities in the light of both analytic considerations and evidence that has already been processed. An appeal of the Bayesian formulation is that it naturally melds information instilled in the genes through evolutionary time (McNamara and Houston 1980) with information acquired by the animal. The likelihood function specifies the likelihood of the possibilities in the light of the latest input to the processor (latest signal, newest data). The support for the likelihood function is the same as the support for the prior distribution; namely, the representation of the possibilities. The posterior likelihood function is the point-by-point product of the prior distribution and the likelihood function. It specifies the relative likelihood of different possible states of the world "all considered." When normalized, it is the posterior probability distribution.

Some cognitive scientists and neuroscientists have suggested that computational mechanisms which carry out (some approximation to) Bayesian inference may be implicated across perception, language processing, and inference (e.g., Chater et al. 2006; Knill and Pouget 2004). Here we illustrate how formal Bayesian inference procedures could decide on behavioral outcomes. A given behavioral option will have various outcomes associated with it. Each outcome can be associated with a value which may be subjective or, in the case of evolutionary explanations of behavior, is its reproductive value. Once the prior information is combined with available information to form the posterior distribution, the posterior probability of each outcome can be computed. The expected (mean) value of the behavioral option can then be computed. This is the average value of the outcome, where the average is formed by weighting each value by its corresponding (posterior) probability of occurrence. When there are a range of possible behavioral options, the decision process will select the option with the greatest expected value.

Another foundational mechanism that operates across all domains is a system of the arithmetic processing of quantities; it enables the elaboration of the representations upon which decision making is based (Gallistel 2011). That is, arithmetic manipulation is necessary for constructing the model that mediates the decision, no matter what the substance of the model (Gallistel 2011). Navigational computations use arithmetic to establish vector space representations from which ranges (distances) and bearings (directions) may be computed. Computations of the mutual information between events distributed in time use arithmetic as well, but they do not use it to create vector spaces. Instead, they use it to compute entropies (amounts of information). In both domains, arithmetic is used to carry out the Bayesian inference (i.e., to take the product of the prior distribution and the likelihood function to marginalize or integrate the result) to take the ratio of different likelihoods (i.e., to form Bayes factors) and to multiply Bayes factors by prior odds.

Similarly, consistency conditions for linking beliefs, values, and actions are provided by various types of decision theory. Indeed, statistical decision theory has proved to be a powerful framework for understanding perceptual motor control and simple perceptual choice tasks (e.g., Bogacz et al. 2006).

The distinction between substantive and formal rationality raises a number of questions: Which organisms have mechanisms that capture different aspects of substantive and formal rationality? Which, if any, organisms embody mechanisms for formal rationality at all? How might the "tinkering" of evolution

(Jacob 1977), and perhaps also learning mechanisms, apply substantive or formal mechanisms outside their original domain. Could there be general principles underlying the inferential "machinery," and their implementation in neural circuitry, which might be carried across many aspects of brain function? When should we attribute an organism with specific substantive rationality (e.g., concerning particular algorithms for navigation), and when is it appropriate to postulate, in addition, principles of formal rationality concerning, for example, the principles of Euclidean geometry, which might underpin such navigational strategies? Is it necessary that we postulate that substantive or formal information is represented? What neural or behavioral evidence might help in answering such questions?

Policy Implications

We have argued that consistency is not the only standard against which to evaluate the quality of a decision since what is irrational, from the classical point of view, could, from a biological perspective, be adaptive. However, we cannot escape the conclusion that time-inconsistent preferences are a problem for the individual and society, even if they make sense from an evolutionary perspective: if everyone were slave to their present bias, and there is evidence that a significant fraction of the population is, we would live in a society facing severe problems such as old-age poverty (as a result of the failure to take appropriate retirement provisions), severe health issues (as a consequence of the failure to pay regular health insurance premiums when not being ill), eating disorders (as a consequence of succumbing to the lure of tempting, yet unhealthy food), and financial illiteracy (as a consequence of the inability to deal with loans and credit cards).

Hence, to generate policy recommendations, society needs a normative standard against which the quality of a decision can be evaluated. The problem is that a theory which performs well in explaining, describing, and predicting behavior cannot tell us what people ought to do; evolutionary theory does not make policy suggestions. Thus, although we propose that all disciplines concerned with the explanation, description, and prediction of decision making (e.g., economics, psychology, behavioral ecology, and ethology) should eventually converge to a unifying framework on how decisions are made, this would not displace the need for a modern normative standard. A bold and certainly controversial proposition would be that the normatively flavored approaches in neoclassical economics, such as revealed preferences, are nothing but normative (i.e., they have no descriptive, explanatory or predictive value), and that other, more evolutionary flavored approaches replace them. If we adhered to this proposition, we would have two sets of theories: one that does well in describing, explaining, and predicting human (and animal) behavior, and one that prescribes choice by telling us what we ought to do. Any discrepancy could then be used for policy intervention.

Caveats

To truly advance the science of decision making, an evolutionary approach will need to do a number of things. First, it must offer more than the concept of optimization, which is already broadly used in psychology, but without the evolutionary constraints (e.g., Anderson 1990). Second it must offer more than comparison with other species, which again, already influences decision theory. Third, it must go beyond simply documenting between-species differences in decision making by making testable predictions about both cognitive mechanisms and behavior that apply to a particular species, especially humans. Fourth, it needs to provide constraints that are generally instructive. An important theme of this chapter is that decision-making mechanisms were shaped by, and might take advantage of, the structure of the environments in which they evolved. When it comes to humans, some aspects of the ancestral environment are certain. For instance, women got pregnant and men did not—a fact that undoubtedly influenced the evolution of mating and parenting psychology. Other aspects of the ancestral human environment, such as sex differences in social status, are unknown and might never be known. A related challenge is that it is often difficult to distinguish between behaviors or cognitive patterns and strategies that have been learned versus those that evolved by natural selection.

Finally, even if evolutionary perspectives advance psychological theorizing, it is important to note that there are many other approaches that can also be generative. As such, we would not want evolutionary theorizing to replace or subsume other, useful approaches, but rather it should add to the theorist's toolbox.

Concluding Remarks: Levels of Analysis

As in any science, we seek a compact account of decision-making processes in evolved systems that is as simple as possible, but no simpler—one that is parsimonious, yet able to capture everything that is systematic in biological decision making. Such a theory would enable us to understand and anticipate the properties of decision-making mechanisms not yet investigated. It would enable us to understand the many seemingly paradoxical aspects of human decision making in personal, interpersonal, societal, and economic contexts.

The account we seek should extend across the levels of analyses of cognitive processes delineated by Marr (1982). It should describe the representations on which decisions are based, because, by specifying the alternatives or options between which the decision is to be made, they strongly constrain the results of the decision. It should offer guidance on what constitutes an appropriate and functionally adequate representation of the alternatives. It should describe how the representations are computed and the computations that mediate the decision itself. This is Marr's algorithmic level of analysis. Finally, it should include a specification of the (mostly neural) mechanisms and processes that implement these algorithms.

We advocate an evolutionary account of decision making. Evolutionary thinking emphasizes function: many organismal structures are best explained as effective means to some end. Evolved decision-making mechanisms would therefore be designed to optimize some resource or outcome that was important to fitness (e.g., energy, mates, and offspring number and quality). Functional thinking, in turn, requires a focus on the structure of the organism's environment: the mechanisms which have been optimized by natural selection to function in particular environments. For decision-making mechanisms, this environmental structure would include the risks, time delays, and threats typical of the environment in which the organism evolved. Evolutionary explanation also involves the gradual modification of a biological lineage, and conserved structures from this lineage, from reflexes to computational patterns in neural circuits, might potentially have profound effects for modern human decision making.

Such a theory will provide a conceptual unification, bringing together the currently very disparate bodies of work in the many disciplines that study decision-making processes and mechanisms.

Acknowledgment

This chapter benefited greatly from discussions with Graham Loomes.

Robustness in a Variable Environment

8

Robustness in Biological and Social Systems

Jessica C. Flack, Peter Hammerstein, and David C. Krakauer

Abstract

Defined as the invariance of system structure or function following a nontrivial perturbation to one or more important system components, robustness is a characteristic property of all adaptive systems. This chapter reviews the theory of robustness in biology, the design of experiments used to assay robustness (including the functional behavior or outputs of a system), and the adaptive response of those parts or components which are compromised by a perturbation. Emphasis is given to a rigorous logic of measurement that carefully factors apart the many causal contributions to robust function. Insights from the study of robustness in biology are applied to the social and decision-making domains, and modifications of experimental design and theory are proposed to account for challenges unique to human agents.

Introduction

Due to the large number of parts and interactions that are characteristic of biological and adaptive systems, one could imagine that a system would be vulnerable to failure in many ways. The number of defective individuals or mutants observed in nature, however, remains relatively low. One trivial reason for this is that mutants are rarely viable, and thus are eliminated before they are observed. A second reason is that biological systems have evolved the means to correct malfunctions and errors, and are consequently said to be robust to mutation and other kinds of perturbations (de Visser et al. 2003; Krakauer and Plotkin 2004; Lesne 2008; Wagner 2007). A system can be said to be robust if one or more of its structural or functional properties remains invariant despite exposure to nontrivial perturbations.

Robustness has been well studied in biology, ecology, engineering, and computing. In each of these disciplines, the importance of maintaining performance despite a changing or unpredictable environment, component error and failure, and component conflict is well recognized. As early as 1956, for example, von Neumann suggested that components of computer systems which are individually inaccurate could in aggregate perform reliably (von Neumann 1956). In addition, a fundamental goal in biology and ecology has been to account for the origins of complex structures. The layering of robustness mechanisms in "anticipation" of fault and failure has been proposed as an explanation for hierarchical complexity (sometimes called robust-over-design; Krakauer and Plotkin 2004; Hammerstein et al. 2006) and has generated many studies on the relation between robustness, complexity, and evolvability (Gould 2002).

More generally, studies on social evolution and social systems have lagged behind biology in addressing robustness issues. In social evolution, the focus has been on trait economics; that is, the cost and benefit scenarios which support the evolution of alternative behavioral strategies (e.g., Bourke 2011; Frank 1998). This has often happened at the expense of studies that address how social systems and their constituent strategies persist in time and related mechanistic questions. In addition, there is the recognition that the rigorous study of robustness requires carefully controlled systematic experiments, which are difficult in social systems.

In this chapter we discuss how ideas of robustness can inform our understanding of social evolution and social decision making, and how robustness can be investigated experimentally and theoretically. We provide a basic introduction to robustness concepts and introduce several generic robustness principles which come primarily from the biological and engineering perspective. We conclude with a discussion on how these principles might shed light on trait economics and the evolution of individual strategies and decision-making rules. To ground these ideas, we begin with a brief history of the robustness concept as developed in the study of living systems.

Brief History of Robustness

The study of robustness in biology finds it origins in the study of the phenotype. There have been two major trajectories in evolutionary theory with regard to the evolution of phenotypes. Although an oversimplification, we might term one the functional perspective and the other a developmental or constraints perspective. The functional perspective, consolidated during the Modern Synthesis, has focused on selective or mutational processes which ignore many details of the organism. Here the emphasis is on the origins and maintenance of genetic and phenotypic diversity, including the key concepts of dominance and epistasis, as well as on adaptation, either to an environment that changes slowly or to rival strategies in a population that can change quickly (frequency dependence: Mayr and Provine 1998). It is from this perspective that sociobiology, evolutionary game theory, behavioral ecology, and eventually evolutionary psychology have grown.

The developmental or constraints perspective focuses on *self-organization*; that is, the interaction between the dynamics of metabolic and physiological processes with the developmental process (Laubichler and Maienschein 2009). Within this perspective there are three broad traditions: one is associated with D'Arcy Thompson, another with C. H. Waddington, and a third with Theodore Boveri and Richard Goldschmidt. Intellectual descendants of D'Arcy Thompson focus on the constraints of phenotypic space imposed by physical laws; the scaling work of West, Brown, and Enquist (1997) provides a modern example. The "Waddington camp" is interested in what might be called evolved constraints. This includes the study of canalization—a developmental strategy for addressing uncertainty in which an organism essentially renders itself impervious to many fluctuations in the environment (Waddington 1942)—and the study of frozen accidents or contingencies which arise because of genetic entrenchment and pleiotropy—processes that lead to evolutionary conservatism (Wimsatt 2001). Goldschmidt-Boveri descendants pursue the idea that organismal transformations, and hence changes to phenotypes, result from the reorganization of genes and present as combinatory and regulatory mechanisms (Laubichler and Maienschein 2009). The questions of evolved constraints, canalization, and regulation all influence issues of robustness.

In general, the constraints perspective has been in the domain of the mechanistic and experimental biologist, whereas the functional perspective rests in the domain of the evolutionary biologist. Two important exceptions to this norm are the macro-evolutionary biologists and the evolutionary developmental biologists, who have been deeply concerned with the role of constraint in evolutionary change (e.g., Gould 2002; Laubichler and Maienschein 2009). With respect to robustness studies, these two groups began to converge in the 1990s and early 2000s, as students of evolutionary theory recognized that robustness mechanisms were one of the bridges connecting the dynamics of ontogeny with the dynamics of phylogeny. This is because robustness mechanisms limit phenotypic variation and thereby the strength of selection, and simultaneously provide a means of exploring alternative genotypes without compromising the phenotype by promoting neutrality (Krakauer 2006; Wagner 2007).

Robustness: Basic Issues

The term *robustness* has many definitions. Most include some mention of (a) mitigation of modification to a "phenotype" following a large perturbation to the genotype, (b) undetectable or minor modification to a phenotype following a large perturbation to the phenotype from the environment, and (c) undetectable or minor modification in function following a large perturbation to the genotype or phenotype with or without a correlated change in the phenotype (Krakauer 2006). The important distinction between genotypic and phenotypic or environmental robustness is that in the former, perturbations are inherited whereas in the latter they are not. An additional complexity is that functional robustness can be achieved either through phenotypic invariance (where the phenotype resists change) or phenotypic plasticity (where the phenotype tracks changes to the environment). Genotypic and environmental robustness are often measured through the environmental (V_e) or mutational variance (V_m) of a trait, whereas functional robustness can only be measured as the variance in geometric mean fitness. Often a single mechanism leads to all three forms of robustness, in which case we observe what has been called *plasto-genetic congruence* (Ancel and Fontana 2000).

We began by stating that a system can be said to be robust if one or more of its structural and functional properties remain invariant despite exposure to nontrivial perturbations. This can be stated more formally: system property *S* is robust if it remains invariant under a perturbation *P* that removes, disables, or otherwise impairs the performance of adaptive component *C* (Ay et al. 2007; Ay and Krakauer 2007). This statement, which captures the essence of what most researchers mean when they employ the term "robustness," is applicable to most of the examples given in the first paragraph of this section. Below we scrutinize each aspect of this statement, giving examples to make the definition more concrete. In doing so, we provide a basic introduction to a few of the issues that arise in the systematic, quantitative study of robustness.

Perturbation

Perturbations are exogenous or endogenous factors that increase disorder (entropy) within a specified system. The range of perturbations can include environmental factors, like UV radiation, and internal factors, like the propensity of a cell to mutate, generate noisy outputs, or fail. Chronic conflict between components, which arises from imperfectly aligned interests and competing objectives, is an important source of functional errors. To be considered nontrivial, a perturbation must be capable, given its frequency and magnitude, of having an effect on that system. For example, if one person throws a paper airplane from a distance of four feet at another person, the perturbation is unlikely to cause injury or even be perceived. This might be considered a trivial perturbation. On the other hand, a large rock thrown from a comparable distance is likely to be noticed. Ideally, the relative magnitude and frequency of the perturbation should be quantified and defined in relation to the system component it is perturbing.

When a perturbation has no effect on a system despite being of a causally significant scale, there are two possible explanations. First, there may be a robustness mechanism protecting the target or the function. Second, the target of the perturbation may not be causally important, as discussed below.

Knockout and knockdown are techniques used primarily in genetics and cell biology to disable parts of systems; that is, they are perturbations administered by the experimentalist (see Kühn and Wurst 2009). Often, these are

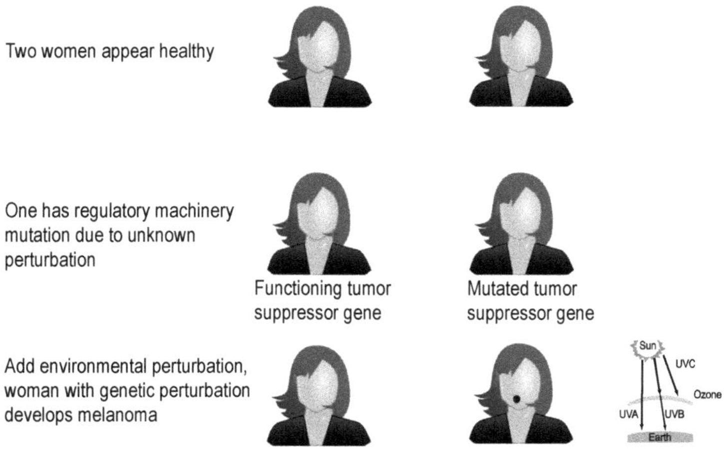

Figure 8.1 Robustness is only apparent under perturbation. If the system has evolved robustness mechanisms, minimally coincident dual perturbations will be required to reveal a lack of robustness to some specified perturbation. Here, one women possesses a robustness mechanism: a functioning tumor suppressor gene (TSG) that prevents rogue cells from proliferating and causing cancer. The second women's TSG is damaged. When exposed to UV radiation, the second woman develops a melanoma, indicating a lack of robustness against this perturbation because the robustness mechanism has failed and the UV radiation has caused a cell to begin proliferating, and this proliferation detrimentally affected some critical physiological function. An experiment with this two perturbation design can also be used to show that a posited TSG is in fact a robustness mechanism as long as it has been demonstrated independently that the second perturbation—in this case, the UV radiation—can cause cells to proliferate uncontrollably.

second-order perturbations of putative robustness mechanisms and need to be administered conjointly with a perturbation that the mechanism is thought to buffer against in "the wild." Hence most robustness experiments involve administering a minimum of two perturbations (Figure 8.1). The number of perturbations that need to be administered experimentally varies depending on whether the goal is to determine whether the system is robust to a specified perturbation or whether a posited robustness mechanism is operational.

In the context of system vulnerabilities, natural perturbations (i.e., perturbations that occur in the "wild") can be interpreted in terms of the selection pressure driving the evolution of robustness mechanisms. In other words, perturbations that have occurred multiple times over many generations are expected to leave an imprint in the genome in the form of encoded robustness mechanisms.

Functional Contribution and Exclusion Dependence

An important consideration in the study of robustness is the magnitude of the functional contribution made by a component that is the target of a perturbation.

Colloquially, the functional contribution of a component is its relevance or adaptive value to a system property when it is unperturbed and performing "normally" (Ay et al. 2007; Ay and Krakauer 2007; Ay and Polani 2008). To illustrate why this is an important concept, consider a social network in which some nodes have a degree of 0 (meaning they are disconnected from the remaining nodes), and assume that the only way a node can influence the behavior of other nodes or global network properties is through their connections. We expect a perturbation to a node with degree 0 to have no effect because it makes no functional contribution. At best the system can be said to be trivially robust to perturbations on such components.

A second critical concept is exclusion dependence; that is, how much a specified structural or functional property of the system changes in response to a perturbation that impairs functionally relevant system components (Ay et al. 2007; Ay and Krakauer 2007). For example, imagine we have shown in an independent experiment that a gene is important to DNA repair. To measure the robustness of the system to a perturbation of that gene, we might knockout the gene and ask how the DNA repair ability of the cell has changed. This change is the exclusion dependence. As shown in Figure 8.2, the extent to which a system property is robust can be measured by taking into account these two quantities: contribution and exclusion. Robustness mechanisms minimize the exclusion dependence of functionally important components.

In some cases it might seem that there is little observable difference between functional contribution and exclusion dependence because the same experimental design is used to infer both properties: experimental knockout

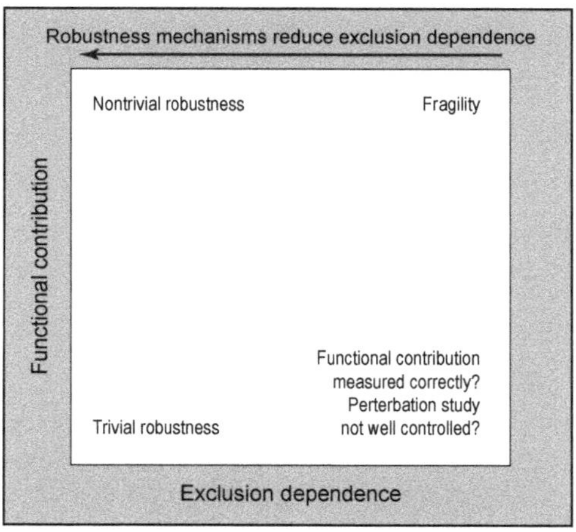

Figure 8.2 Functional contribution is the contribution of the target component to a system property or function. Ideally, this should be quantified using clamping. Exclusion dependence is how much the system function changes when the component has been perturbed. This measurement can be made using knockout and knockdown designs. Robustness mechanisms reduce exclusion dependence.

of a component and measurement of the system change. However, because a system can reconfigure upon knockout (either finding a new solution or unmasking a latent function), knockout is not a good way to measure functional contribution. Functional contribution should ideally be quantified using clamping (Ay and Polani 2008; Pearl 2010). Clamping involves holding the inputs to a target component at a constant value and evaluating system function as noise (proxy for error) is injected into the system. This allows the experimentalist to isolate the contribution of the target component for some specified function without concern that the system will reconfigure or that latent backup solutions will be unmasked.

Types of Questions

When studying robustness, biologists typically ask three questions:

1. Is X a robustness mechanism? That is, does it reduce the exclusion dependence measured for some system property S given a perturbation P that affects a functionally relevant component C?
2. Is system property S robust to a perturbation P affecting component C?
3. Is system property S robust to a perturbation P that disables a known robustness mechanism C_r?

Each of these questions requires a different experimental design, both in terms of the number of perturbations that must be experimentally administered and the timescales over which the consequences of the perturbations are assessed. For example, to determine the robustness value of a posited robustness mechanism, the mechanism must be disabled so that its robustness contribution can be inferred from the effects of its removal. The knockout must be temporary to prevent the system from adapting (i.e., from building a new robustness mechanism or activating a latent backup mechanism), which would confound the results. In contrast, to determine whether the system is robust to a specific perturbation, a "long timescale" knockout is required so that time-series data can be collected to determine whether the system destabilizes, restores function redundantly, or reconfigures to establish a new robustness mechanism.

A Word of Caution

No biological or social system is robust to all perturbations. In some cases, a perturbation that has negative consequences on one component can have neutral or even positive consequences on another. Robustness at one level of a system (the robustness of some global property) might even require sensitivity of components (a lack of robustness) at lower levels (Krakauer and Plotkin 2004). Thus in studying robustness, it is critical to be specific always about the perturbation, the component subject to the perturbation, the system property thought to be jeopardized by the perturbation, and the robustness mechanism that buffers the system against the perturbation.

Krakauer and Plotkin discuss three principles that have arisen in an effort to understand the evolutionary response to mutations: the principle of canalization, the principle of neutrality, and the principle of redundancy (Krakauer 2006; Krakauer and Plotkin 2004). They contrast these with the parameters of robustness; that is, mechanisms by which these principles are realized. The principles and parameters metaphor is derived from linguistics (Chomsky 1981), where the principles are the invariant properties of universal grammar and the parameters the local rules and practices of language. A focus on principles allows for several mechanisms to vary parametrically but employs some form of redundancy to be classed together. The same would be true for modularity and so on.

Robustness Principles

To provide a brief introduction to robustness principles that are commonly invoked in biology, we extend the Krakauer list (Krakauer 2006). Thereafter we discuss briefly the relation of robustness and canalization to niche construction and the extended phenotype, with the goal of clarifying a common misapprehension about what makes for a robustness mechanism rather than a common-garden adaptive trait.

Redundancy

A rigorous means of identifying the function of a gene is to perform a knockout experiment to remove or silence a gene early in development. By assaying the resulting phenotype, the putative function of the absent gene can be inferred. In many such experiments, there is no scoreable phenotype: the knockout leaves the phenotype in the wild-type condition. Biologists refer to a gene x on a background y as functionally redundant (Tautz 1992). This is taken to mean that the target gene is one of at least two or more genes contributing epistatically to the phenotype (Krakauer and Nowak 1999).

Removal of redundant gene x leads to compensation by other genes, the compensatory part of the genome. Let $f(y)$ be the fitness of a part y of a genome and $f(x, y)$ the fitness of this part with an additional gene x. If $f(x, y) = f(y)$, then gene x is redundant with respect to part y. In the simplest case, y consists of a single gene that is a redundant copy of x.

True redundancy might be rare, and what we observe may be "artifactual" redundancy or experimental neutrality, in which the effect of perturbation remains below an experimental detection limit (Ponte et al. 1998). Assuming that we are able to detect small changes, the degree of redundancy describes the degree of correlation among genes contributing to a single function. Models of redundancy in biology tend to focus on the evolutionary preservation of redundant components, and hence employ population genetics approaches. More recently, differential equation-based models for the dynamics of regulatory systems following structural perturbation have also been explored (Wagner 1996).

Purging and Anti-redundancy

Whereas redundancy buffers the effects of perturbation, purging acts in the opposite fashion and amplifying the effects of perturbation ensures the purity of a population by making perturbed genomes more vulnerable to selective forces. Consider a genetic background y, a wild-type gene z^+, and a deleterious mutation z^- of this gene. Let the fitness of an organism with this genetic background and the deleterious mutation z^- be denoted by $f(z^-, y)$, so that the fitness of the wild type is greater than that of the mutant: $f(z^+, y) > f(z^-, y)$. An additional gene x is called functionally anti-redundant if its presence reduces the fitness of the mutant even further: $f(z^-, y) > f(x, z^-, y)$. In the more general case, z^- can consist of more than one damaged gene.

Purging is only effective when individual replication rates are sufficiently large to tolerate the effects of removal of defective components. Thus apoptosis (programmed cell death) is a common strategy for eliminating cells upon damage to their genomes or upon infection, provided these cell types are capable of regeneration. Nerve cells and germ cells produce factors which strongly inhibit apoptosis (Matsumoto et al. 1999) and removal, in these cases, has deleterious consequences. In severe infection, it can make sense to purge nerve cells (Krakauer 2000).

Recent models that address purging-type phenomena have involved stochastic models which assume finite populations. The key insight from the study of anti-redundancy is the ability of particulate, hierarchical systems to exploit cellular turnover to eliminate and replace deleterious components from populations.

Multiple Pathways

Organisms are surprisingly robust against the loss of single genes. Somewhat surprisingly, knockout studies investigating the benefits of duplicate genes find that simple duplication accounts for substantially less phenotypic invariance than was originally assumed. These observations suggest that phenotypic invariance is not exclusively the consequence of duplication but of some other feature of the organism. One finding from the study of metabolic networks suggests an explanation: metabolic networks which evolved in fluctuating environments show high mutational robustness and are characterized by a greater number of multifunctional enzymes and independent pathways (Ihmels et al. 2007; Soyer and Pfeifer 2010). This finding suggests that robustness can be achieved as a consequence of building multiple pathways in response to selection pressures which support the ability to process different kinds of metabolites in fluctuating resource environments.

These results are consistent with theoretical studies on the evolution of canalization which suggest that networks evolving to be more highly connected show greater insensitivity to mutations to the network's functional equilibrium state (Siegal and Bergman 2002).

Mutational Robustness

There are many mechanisms associated with *mutational robustness*, defined as invariance of the phenotype in the face of mutation. A canonical example is the heat-shock protein 90 (Hsp90). In *Arabidopsis thaliana*, Hsp90 buffers phenotypic traits against genetic variation and normal development against the destabilizing effects generated by stochastic processes (Queitsch et al. 2002). Hsp90 chaperones metastabile proteins, preventing them from forming conformations prone to misfolding. Hsp90 is activated under heat and environmental stress when misfolding is more likely.

Interestingly, although mechanisms of mutational robustness are often thought to promote robustness at the expense of evolvability (i.e., genetic variation that might serve as the raw material for selection), this is not always the case. The genetic data suggest that robustness mechanisms, such as Hsp90, can decrease selection on nucleotide substitutions, thereby allowing a greater diversity of neutral or nearly neutral nucleotides to be maintained and to even accumulate (Queitsch et al. 2002). This could allow a population to move to a new local optimum without having to pass through an adaptive valley.

Neutrality and Sloppiness

It is now well known, particularly from the study of RNA secondary structure (Schuster 1999; Wagner 2008), that the relationship between sequence space and network topology, on the one hand, and phenotypic characters, on the other, is degenerate. Large regions of these spaces can be explored without any consequences for the phenotype (no variation). One advantage of neutral networks is thought to be that they provide a means of exploring alternative genotypes without compromising the phenotype and its function.

Many of the results on neutral networks are understood for discrete genotype and phenotype spaces. Recent work on biochemical networks consider robustness for continuous parameter spaces (Daniels et al. 2008). Network behavior is highly sensitive to movement in parameter space along a few directions or dimensions (where a direction is defined over multiple parameters) and is largely insensitive to movement along the remaining directions. Daniels et al. (2008) refer to this property as sloppiness and suggest that the hypotheses for the evolution of continuous neutral networks are likely to resemble those of the discrete case.

Feedback Control

Feedback control systems have three components (Emanuel 1979): a plant (the system under control), a sensor (measuring the output of the plant), and a controller (generating the plants input). A measure of performance is often the degree to which the output of a plant approximates a function of the input to the controller. In biology, a plant could be RNA or protein concentration, protein kinase activation, immune effector cell abundance, or species abundance. Inputs in each of these cases would be transcription factors, protease concentrations, chemical agonists bound to receptors, antigen concentrations, and death rates. The controllers are more often than not aggregates of several mechanisms. Feedback is a mechanism of robustness as it enables plants to operate efficiently over a range of input values. The question remains as to whether the controller is robust to variations in the plant: Does the controller provide robust stability? For example, in biology, can a single feedback controller regulate the concentrations of several different proteins?

A well-known example of feedback control in biological systems, which appears to be characterized by robust stability, is bacterial chemotaxis. *Escherichia coli* move toward chemical attractants and away from repellants using a sensory system which translates chemical concentration gradients into changes in the cell's tumbling (or movement) patterns. The cell's tumbling frequency (movement rate) can change rapidly in response to changes in the concentration of a chemical stimulant, gradually adapting back to a baseline value. This has been called *exact adaption* and has been found to be robust to variation in the cell's biochemical parameters (Alon et al. 1999).

Modularity

If a system contains networks with characteristic input-output functions that are capable of recombining or shuffling to produce new aggregate functions, this is described as *modularity*. Modularity arises in structures that balance autonomy (single and noncomposable input-output map) and integration. Within a module there is strong integration, whereas populations of modules are only weakly coupled (Slotine and Lohmiller 2001). This has also been called *near decomposability* (Simon 1962). In genetics, modularity involves a minimum of pleiotropy, in which sets of genes contributing to one module or trait (e.g., an organ system) contribute little to other complexes or traits (Goldberg 1995; Raff and Raff 2000; Raff and Sly 2000). These modular genetic systems are found in different genomic contexts performing a similar function. Of course modularity can be defined at levels of organization above that of the gene (Winther 2001)—the extent to which organs operate independently during homeostasis. The dissociability of modules provides one means of damage limitation through encapsulation and can therefore be a mechanism of robustness.

There is no consensus over the most appropriate or predictive models for analyzing modularity in biosystems. To date quantitative genetics models have been used to explore the limits to the evolution of modularity, and neural network models have been used to explore how modularity can lead to more efficient task management (Calabretta et al. 1998). For a comprehensive review of the recent literature on modularity, see Schlosser and Wagner (2004).

Spatial Compartmentalization

Compartmental systems are those made up from a finite number of macroscopic subsystems called compartments, each of which is internally well mixed. Compartments interact through the exchange of material (Jacquez 1985). The spatial compartmentalization of reactions leads to robustness by minimizing covariance (interference) among reaction components that participate in functionally unrelated processes. Thus spatial de-correlation through compartmentalization substitutes for temporal correlation in biological functions. Robustness is achieved in at least two ways: (a) interference (chemical, epistatic, or physiological) is minimized, and (b) mutual dependencies are minimized, thereby attenuating the propagation of error through a system. The study of spatial compartmentalization is particularly rich in theoretical ecology and epidemiology (Levin et al. 1997) and has been used to explore the maintenance of antigenic diversity, restrictions on pathogen virulence, and seasonal forcing. In addition, in molecular biology it has been used where proteins have been found to be compartmentalized (Prior 2001).

From a modeling perspective, compartmentalization is often approached from the perspective of metapopulation dynamics or coupled oscillators, in which space is assumed to be discrete (implicit space) and nonlocal (Hanski 2001). An alternative approach, based on continuous space (explicit space) with local interactions, employs partial differential equations to study diffusion and advection of components (Murray and Stanley 1986). A third approach assumes discrete space with local interactions and employs coupled map lattices and cellular automata. A fourth approach analyzes the statistical connectivity properties of undirected graphs and their response to node or edge elimination (Albert et al. 2000).

Distributed Processing

Distributed processing describes those cases in which an integrated set of functions are carried out by multiple, semi-autonomous units (McClelland 1990; Hertz et al. 1991). The most obvious example is that of nerve cells comprising the nervous system. Distributed processing, or connectionism, might be assumed to be a combination of modularity and spatial compartmentalization. However, distributed processing brings additional demands on structure: it requires that a single function is emergent from the collective activities of units, and that at some level integrated or correlated activity is a desired outcome.

The robustness properties of connectionist models are the ability (a) to identify incomplete patterns, (b) to generalize from a subset of learned patterns, and (c) to degrade gracefully upon remove of individual nodes. Connectionist models range from a simple application of linear algebra, to dynamical systems and Hamiltonian representations of steady states, through to the use of statistical mechanics models of frustrated systems, such as spin glasses.

Conflict Management

In most biological systems, component or individual interests are only partially aligned, and this leads to chronic conflict. Chronic conflict can jeopardize performance and the formation of stable interaction networks critical to resource extraction and production (Flack et al. 2006). Mechanisms have evolved in many biological systems to manage conflict, keeping it at levels thought to be useful for state-space exploration, which might support adaptation when the environment changes. Many conflict management mechanisms serve as robustness mechanisms that buffer resource extraction and production against the damaging effects of chronic conflict. Examples include third-party policing in animal societies, whereby individuals uninvolved in a dispute impartially intervene into the dispute, usually at low cost to themselves and low immediate cost to the conflict participants, thereby causing the conflict participants to disperse (Flack et al. 2005a, b, 2006).

Error Correction and Repair

Another strategy for buffering against perturbations is to repair the damage directly. For example, in the cells of our bodies, environmental factors and metabolic activity can directly damage DNA. This damage can compromise gene transcription or the survival of cells after cell division. Organisms have evolved means for repairing damaged DNA (Friedberg et al. 1995). These mechanisms, called DNA repair mechanisms, range from chemically reversing damage to removing damaged sections of one strand of DNA and resynthesizing the missing piece using the second undamaged strand as a template. In the case of double strand damage, this requires identifying an identical or nearly identical sequence and using this as the template.

In animal social systems, social relationships can also be damaged by conflict (de Waal 2000). A common mechanism of repair that restores the quality of relationships to baseline (preconflict) levels is reconciliation (de Waal 1993). Reconciliation typically involves fight participants showing increased social investment in each other immediately following conflict. This increased social investment presumably serves as a costly behavioral signal, the function of which is to reestablish mutual trust.

Relationship to Niche Construction, Canalization, and Other Adaptive Traits

A major challenge faced by biological and social systems is *environmental uncertainty*. Two classes of strategies exist to handle environmental uncertainty. First, when the environment is largely unpredictable (i.e., when the time series of environmental states is characterized by minimal regularity), organisms make themselves impervious to environmental fluctuations by canalizing development (Waddington 1942). In other words, a developmental program evolves along a rigid trajectory, producing an invariant phenotype, regardless of environmental fluctuations. Another way to say this is that randomness is countered with a near deterministic buffer.

Second, when an environmental process is characterized by sufficient regularity, organisms can evolve the means to intervene directly into the environmental process and modify it. This involves changing the rate or trajectory of environmental events to make them more predictable and compatible with an existing phenotype. One benefit of this strategy is that it permits an organism's fit to the environment to be more finely tuned, resulting in a higher exchange of information between the organism and environment (increased adaptability). Strategies of this type include extending the phenotype (Dawkins 1982) and niche construction (Odling-Smee et al. 2003), whereby individuals control, to some extent, the selection pressures to which they are subject by modifying the environment. Examples would include bird nests, termite mounds, and spider webs (Odling-Smee et al. 2003; Stamps 2003), or social structures like power structures (Boehm and Flack 2010; Flack et al. 2005b, 2006).

Canalization and niche construction are strategies for reducing variance, and represent two extreme measures for reducing uncertainty over a lifetime. Whereas robustness mechanisms often appear to resemble canalization, most robustness mechanisms observed in nature fall between the two extremes. Niche construction, where elements of the environment are encoded in the organism's genome, can be shown in a limited way to lead to the evolution of an extended developmental program (Krakauer et al. 2009). It is often through the requirements of robustness that mechanisms of niche construction resembling developmental canalization evolve.

If robustness mechanisms include uncertainty reduction mechanisms to bridge niche construction to canalization, what kinds of adaptive strategies are not instances of robustness? What distinguishes a robustness mechanism from adaptive mechanisms more generally? All robustness mechanisms are certainly adaptive traits, but not all adaptive traits are robustness mechanisms. The critical distinction is that robustness mechanisms are adaptive traits that promote system persistence by reducing variance in performance in the face of entropic factors. For a trait to be considered adaptive, however, it must increase, over some timescale, mean performance in some well-defined functional context.

Robustness in Social Systems: From Decision Making to Organization

The question of how and why social systems persist, despite individuals dying and possessing partially aligned interests, has been addressed qualitatively by structural anthropologists and to some extent sociologists. Emile Durkheim (1895/1964) and Lévi-Strauss (1969) pursued questions on the functional value of the division of labor and ubiquitous features of culture, such as the incest taboo; Ruth Benedict (1934) addressed the role of the individual in the production of culture and the role of culture as a constraint on individual behavior. However, as anthropology devolved into post-modernism, the issue of robustness was largely abandoned. Only a few small communities of anthropologists and archeologists interested in the rise and decline of civilization or societal robustness continued to pursue related concepts (Lansing 2006; Wright 1935).

Beginning in the late 1960s and early 1970s with the rise of sociobiology, behavioral ecology, and eventually evolutionary psychology, attention turned with renewed vigor toward social behavior; this time biologists and cognitive scientists with training in evolutionary theory took the lead. Rather than focusing on the mechanics of social organization, the focus has largely been on "economics," emphasizing the cost and benefits that support one behavioral strategy over another. The overwhelming goal in this research trajectory has been to give a compelling account for when selection favors group formation and cooperative interactions in the framework of game theory (Frank 1998; Nowak et al. 2010).

Unlike evolutionary theory, which developed in parallel with mechanistic biology (i.e., with an empirical emphasis on the genetic foundations of population structure), the theory of social evolution has suffered from a relative absence of a detailed mechanical understanding of how societies arise. Both biological and social evolution have tended to underemphasize ontogenetic or developmental regulatory dynamics, such as metabolic, physiological, and related processes.

Thus, despite a growing body of research on the cognitive and neural bases for the cooperative strategies studied in game theoretic models and experimental games, there has been relatively little quantitative work on how collective properties, social structure, and other aggregate features of groups are produced collectively through the interaction of individuals adopting different strategies and decision-making rules. There has been even less research on how aggregate properties persist in the face of high levels of uncertainty and noise. Exceptions to trend include research on the rules underpinning fish schools and bird-flocking behavior (Couzin 2009; Sumpter 2006), studies of colony formation in social insects (e.g., Hölldobler and Wilson 2011; Theraulaz et al. 2003), the emergence of social power structure (Boehm and Flack 2010; Flack and Krakauer 2006), collective conflict dynamics (Dedeo et al. 2010, 2011), and the implications of robustness mechanisms for social niche construction in primate societies (Flack et al. 2006).

Next we discuss how robustness can be studied in social systems and what can be learned about social evolution by considering how societies and the decision-making rules that underpin them persist, despite frequent and often severe perturbations.

Robustness at the Societal Scale

If we are to retain the formal definition of robustness provided earlier in this chapter—system property S is robust if it remains invariant under a perturbation P which removes, disables, or otherwise impairs the performance of adaptive component C—we have to determine what is meant by a system property S in the social case.

In the biological domain, system property S is typically some aspect of the phenotype, such as the number of spines or the number of cells in the endomesoderm of a sea urchin. It can also be a highly conserved trait across lineages, including the ability to metabolize a novel substrate. Each of these examples can be thought of as a collective feature of a system insofar as they arise from the interaction of genes and proteins through a complicated regulatory process. The dynamics or logic of regulation can then be summarized in terms of causal networks or regulatory circuits (Davidson 2010).

In the social domain, the analogs to these phenotypic properties are aggregate properties that arise out of the collective behavior of individuals in regulatory social networks. Examples range from the statistically simple to the complex, including the trajectory and/or volume of schools of fish, the distribution of fight sizes in a primate society, the degree of assortative or reciprocal behavior that characterizes social interactions, and the distribution of power in animal societies. Whereas some of these aggregate properties arise from simple rules, others require integration over complex dynamics at the microscopic level.

Perturbations relevant at the social level include chronic conflict and environmental, physiological, or cognitive deficits that result in anomalous behavior or the failure of individuals to perform critical social roles. Other common perturbations are acts of predation which remove key individuals from a society. When roles and functions are institutionalized or supported by "pool" mechanisms (e.g., Sigmund et al. 2010), perturbations can take the form of a sudden lack of funds required to support the function. An example would be a market crash that might lead a community to disband their police departments. The key is to show that these perturbations remove individuals or subgroups that have critical social roles with respect to some system function, or directly impair the functions themselves. Thus, as with the study of robustness in biological systems, we must demonstrate that the component compromised by the perturbation makes a critical contribution to some measure of system performance. We must also demonstrate that the perturbation is nontrivial and operating at a relevant scale.

To make this concrete, consider the following example: In pigtailed macaque societies, third-party policing is a primary conflict management strategy (Flack et al. 2005a). This social function is performed by a small subset of individuals at low cost (quantified in terms of aggression received from conflict participants in response to their intervention). These individuals pay low cost because they sit in the tail of a heavy-tailed power frequency distribution; that is, they are perceived by group members to be disproportionately capable of using force successfully during fights and thus are rarely challenged when they intervene. The high-variance power distribution thus supports the policing mechanism because it makes the cost of policing negligible for individuals in the tail, yielding a large net benefit (no group selection required).

In an experimental study in which the policing mechanism was temporarily disabled or "knocked out" (by temporarily preventing the policers from performing this behavior), the cost of social interaction was lower and individuals were less conservative in their social interaction patterns when policing is functional (Flack et al. 2005b, 2006). Mechanistically, policing checks the escalation of conflict, thereby allowing individuals to interact at relatively low cost and build more connected and integrated social networks. Dense networks facilitate the extraction of social resources (e.g., alliance partners, information about location of food and sleeping sites). These results suggest that policing is a robustness mechanism that changes the economics of behavior by modulating the cost of interactions and allowing individuals to invest more in beneficial, socio-positive interactions.

Is the policing mechanism, however, a robustness mechanism? The system property S is the degree of integration in social networks, and this property is quantified using a variety of biologically meaningful social network metrics. The perturbation is chronic conflict, and the component or feature of the system being disrupted by the perturbation is the behavioral strategy set associated with social niche construction, or the ability of individuals to build edges in their social networks.

Robustness at the Behavioral Scale

In the policing study, the system feature being studied—degree of network integration—is a collective feature defined over nodes in the network. It is also possible to ask about the robustness of the underlying behavioral strategies themselves. These are the decision-making rules individuals use when, for example, they decide to fight or help an alliance partner or invest in an edge in a social network.

In the case of robustness mechanisms, such as third-party policing which buffers against conflict, the "system" is the social system. The "system" in the case of decision-making rules or strategies is the individual's cognitive system, which encompasses its neurophysiological and behavioral activity. The strategy is the output of the cognitive system. The system property S is a statistical feature of the strategy. The perturbation could include environmental factors, either ecological or social, or factors internal to the organism that increase, for example, entropy in neural firing patterns. The component disrupted by the perturbation might be the firing pattern of a single neuron, the firing pattern defined over a network of neurons, the degree of blood flow to a brain region, or any other neurophysiological behavior that makes a functional contribution to the cognitive/behavioral output of interest.

Let us say, for example, that we are interested in determining if the behavioral strategies which individuals use to build edges in their social networks are robust to perturbations such as conflict. This is a very different question than the one asked in the policing study. How might we address it? We first need to define the space of strategies (e.g., maximize partner number and diversity). We would then want to determine what cognitive component of these strategies (which we might interpret as the "system function") is being targeted by the perturbation (conflict). For example, and this remains rather crudely articulated, could the perturbation (conflict) influence the emotional state of the individuals as they are making decisions through hyperbolic discounting or affective forecasting (Bechara and Damasio 2000; Loewenstein 2003)? Is the perturbation having an impact on higher cognitive functions (their ability to track interactions and exchanges) and hence the estimate of the costs and benefits of adopting a strategy? We can more finely resolve these questions by exploring how the physiological or neural factors underlying affect are influenced by perturbations like conflict.

Once we have established that the perturbation of interest is in principle relevant (e.g., on the right scale) to the cognitive component deemed to underlie strategy use, we can then ask whether strategy adoption remains invariant when the cognitive mechanism underlying it is perturbed. This would be something like a game theoretic stability concept defined at the level of the mechanical instantiation of strategy.

Although not typically framed as robustness questions in the sense presented here, there is a large literature on decision making under risk and ambiguity, stress, and imperfect and erroneous information (Kahneman et al. 1982). For example, it is now well understood that individuals asked to make decisions under stress accelerate processing and consider a smaller space of alternatives or only a subset of alternatives compared to those not exposed to stress (Keinan 1987; Payne et al. 1993). Increasingly, the neural bases for these differences are being worked out (Rilling and Sanfey 2011). The social factors leading to robust decision making have also been the subject of study as illustrated by recent investigations into the wisdom of crowds phenomenon (Surowiecki 2004) and cloud sourcing, which show that errors at the individual level can sometimes (with the caveat of the Roger's paradox) be corrected using population information. These ideas are closely related to what has been called *inferential robustness*: the social consensus process by which we come to accept, for example, that a result in science is trustworthy (Wimsatt 2007). Within the evolution literature, a variety of game theoretic strategies has been identified that compensate for individual mistakes, and hence make the functional consequences of decision making less sensitive to errors in judgment, state estimation, or "trembles of the hand" (Selten 1975). Among these are strategies like *win–stay, lose–shift* (Sigmund and Nowak 1993) and *tit-for-two-tats* (Axelrod 2006). These large, mechanistic literatures can be used to design experiments that explicitly investigate the collective, behavioral, and neurophysiological factors underlying robust decision making.

Methods of Experiment and Analysis

Methods for Studying Robust Decision Making

As we have summarized, studies that explicitly aim to provide a mechanistic account of robust decision making are uncommon, but key insights and methods from biological robustness studies of phenotypic invariance can be generalized into the decision-making arena, since the cognitive mechanisms underlying decision making are also phenotypic traits. Thus many of the experimental procedures, like knockout and knockdown, common to the study of robustness in molecular and genetic systems, could prove to be valuable if carefully adapted for use in the decision-making research.

Methods for Studying Societal Robustness

Studies of robustness at the societal scale require creative and novel approaches, because integrated social systems are hard to manipulate systematically and human social systems present important ethical limitations. When such experiments can be conducted, they are often hard to replicate as many factors have to be controlled to make comparison meaningful. Nevertheless, with caution and rigor, it should be possible to proceed; we also wish to emphasize that some progress has already been made.

For example, when the robustness mechanism is localized in a single or small subset of individuals, it is possible to conduct "behavioral knockout" experiments, as illustrated by the policing study. In these studies, the focal mechanism is disabled by temporarily removing key individuals. There are obviously many potentially confounding issues, such as how to ensure that only a single function is disabled when individuals typically perform multiple roles. However, these should be able to be resolved through careful and creative experimental design.

Similarly, it is possible to ask whether a system is robust to some specified perturbation if the components (individuals or behavioral mechanisms) compromised by the perturbation can be determined. Typically, obtaining rigorous and trustworthy results will require the use of animal society model systems. These systems involve animals that live in large, captive social groups, which permit highly resolved time-series and network data to be collected during the experiment before and after the experiment.

Down the road it might also be possible to find creative ways to conduct knockout studies on social systems with distributed robustness mechanisms. This has also been a challenge in the genetics and molecular cases, but recent studies designed to disrupt the distributed microtubule networks of cells suggest that this is also achievable (e.g., Noel et al. 2009).

When a controlled experiment is not possible, but time-series and network data are available before and after a documented "natural perturbation," it is possible with the help of appropriate null models to draw preliminary conclusions about whether the system is robust to the observed event. Hence direct manipulation might be circumvented when presented with suitably time-resolved observational data.

Finally, if time-series data are available on social interaction patterns, the decision-making rules individuals use can be extracted from data using a variety of nonparametric and machine learning approaches (e.g., Dedeo et al. 2010; Krakauer et al. 2010). These decision-making rules can then be used to build an empirically grounded virtual environment that allows for controlled virtual experiments to explore societal robustness.

Conclusion

The study of robustness represents an attempt to explain the continued function of complex systems in the face of frequent errors and perturbations. Historically, questions related to robustness have emphasized either functional frameworks (economic and game theoretic) or mechanical frameworks (developmental and engineering). The challenge is to unite these perspectives so that the evolution of robust mechanisms can be explained. A critical element of this objective is to formalize the logic of robust systems and present systematic experimental approaches to their analysis. With the appropriate experimental logic, we can then go about isolating the many mechanisms that yield robust functions, such as error correction, conflict management, developmental canalization, and niche construction. Decision-making rules represent an exemplary study system for robustness, as errors in individual judgment will need to be compensated through mechanisms of consensus formation at the collective level. In highly coupled systems that are intractable to experiment, simulated experiments, building upon frameworks such as agent-based models and inductive game theory, will prove to be important.

Acknowledgments

We are grateful to Bill Wimsatt for helpful comments on a preliminary draft. JCF and DCK received support during the writing of the paper from NSF BCS-0904863. Many of the ideas presented in the chapter were developed by JCF and DCK while they were supported by the SFI Robustness Program funded by James S. McDonnell Foundation. PH received support from the Deutsche Forschungsgemeinschaft (SFB 618).

9

Robust Neural Decision Making

Peter Dayan

Abstract

Animals are extremely robust decision makers. They make seemingly good choices in a very wide range of circumstances, using neural hardware that is noisy, labile, and error prone. This chapter considers dimensions of robustness that go beyond fault tolerance, including the effects of outliers and various forms of uncertainty, and discusses the multiple scales of robustness afforded by the rich complexities of neural control.

Introduction

Robustness is omnipresent in the natural world. This is evident at a macroscopic scale, with individual sequoia trees living for many hundreds of years, surviving fires, droughts, earthquakes, and predators, all without being able to move; or indeed in species adopting the strategy of producing vast numbers of offspring with at least some beating the rigors of early life. It is also evident at a microscopic scale, with the essential architecture by which biological information is represented in DNA surviving for hundreds of millions of years. To take an example: one can extract a piece of DNA from a species of algae, insert it into a mouse, and then find that mouse cells synthesize a correctly functioning form of its "payload"—a light-sensitive ion channel—for which it codes, and distribute this channel to sensible places on those cells' membranes (Gradinaru et al. 2007). This cannot be seen as anything other than the most amazing degree of systemic robustness.

In this chapter, we aim to characterize some facets of the robustness of mammalian neural decision-making systems. Although not even nearly approaching the feats of the previous paragraph, individual animals generally do a rather good job at coping with complex, only partially known, environments which pose a variety of decision-making challenges. We seek to understand what forms of robustness they exhibit, and how this arises from which implementational, algorithmic, and computational mechanisms and strategies (Marr 1982). I do not attempt to be comprehensive. I will merely present a number of problems and, where appropriate, discuss potential solutions.

A large fraction of work in decision making has concentrated rather simply on optimality; for instance, maximizing reward and minimizing punishment (Dayan and Daw 2008; Sutton and Barto 1998). There have been some attempts to consider aspects of the variability or riskiness of outcomes (see, e.g., D'Acremont and Bossaerts 2008; Weber et al. 2004). However, this work has only scratched the surface of the problems posed by robustness in the domain of choice. Thus, after describing a set of relevant dimensions, I present general features of neural computation that underpin robust control and will interpret specific characteristics of the multiple mechanisms believed to be involved in selection (Daw et al. 2005; Dayan 2008; Daw and Doya 2006), in terms of both the vulnerabilities they suffer, which limit robustness, and, conversely, what they contribute to the superior overall quality of control.

This chapter should be read in conjunction with Gluck et al. (this volume), which provides fuller illustrations and analysis of the dimensions of robustness that can influence decision making, and discusses the relationship between robustness and its confusingly close relations such as optimality, adaptability, and flexibility.

Robustness

Marr (1982) provided a framework for analyzing information-processing systems, suggesting that computational, algorithmic, and implementational concerns be considered separately. This provides a helpful, if crude, taxonomy for key elements of robustness.

In Marr (1982)'s conception, the computational level plays the most critical role; it is here that both the underlying issues and the logic of their solution are abstractly posed. Of course, this level does not have priority for all forms of robustness, for instance to hardware issues. The central computational concern for robustness is uncertainty arising from ignorance, change, and misspecification, all of which imply the possibility of erroneous choices. Such choices can have devastating consequences, either in terms of the demise of the agent or substantial lost opportunities, particularly in the case of distributions of outcomes with heavy tails, and thus substantial outliers. Since one branch of the field of robustness has concentrated rather narrowly on negative outliers (notably economics, with its black swans; Taleb 2007), we consider them below separately from more general forms of uncertainty. However, it is of course true that robust methods for finding near optimal gains, such as forms of defeasible optimism (Brafman and Tennenholtz 2003; Auer 2003; Kearns and Singh 2002), are also critical.

The solutions to the computational problems of robustness broadly include many ways of enhancing, speeding, and taming learning. Crudely, one can distinguish two sorts of problems. One set lies with data, such as outliers and nonstationarity; these are solved by robust statistical methods. The other problems concern reward or cost functions, and are solved by forms of worst-case controllers. Along with an analysis of approaches for individual decision makers, there are various methods appropriate to groups of organisms, with the power to distribute exploration across individuals, and even to learn (or at least follow, by herding) good and bad strategies from observation.

Marr (1982)'s implementational level considers how information processing is actually realized. Here, the issues for robustness are noise, sloth, temporal variability, and other forms of unreliable neural hardware. How can high-quality choices be guaranteed to arise from mechanisms of such apparently low quality? Various other implementational issues also arise; for instance, there are many rather stark distinctions between the neural systems involved in rewards and punishments (Daw et al. 2002; Boureau and Dayan 2011), suggesting that we may need to consider some of the elements of robustness separately in each. Equally, there appear to be strong implementational constraints on mechanisms such as working memory (Miller 1956), which are normally critical for high-quality planning.

Finally, the algorithmic level interpolates between computational and implementational levels. In a modern conception (Dayan 1994),[1] this indicates ways of addressing computational demands using only the available implementational substrates. At a conceptual level, it broadly introduces solutions for robustness, largely centered on the systematic involvement of multiplicity. This is seen at a microscopic level, for instance in terms of structures such as radically overcomplete codes, representing single quantities in the simultaneous activity of large populations of neurons (thus being robust to failure). It is also seen at a macroscopic level, with brain systems based on radically different precepts aimed at solving the same computational problem. These systems may involve parameters that control robustness-influencing factors such as trade-offs between exploration and exploitation. Although there will often, formally, be an optimal way to set and change these parameters, such methods are typically radically intractable and/or depend on a wide range of (computational) assumptions about the environments. Thus, algorithmic robustness is often achieved by methods involving many different settings, within and between individuals.

In the remainder of this section, I describe the key implementational and computational problems faced by organisms (notably faults and noise, outliers, and uncertainty), along with some general algorithmic strategies animals adopt to overcome them. Thereafter, we will specialize the arguments for the separate decision-making systems that appear to jostle for influence in the brain.

---

[1] At least modern compared to Marr (1982), for whom all the levels were substantially more distinct.

Fault and Noise Tolerance

The functional characteristics of the brain as an architecture for computation are well rehearsed (e.g., Koch 1999; Dayan and Abbott 2001) and do not need extensive discussion here. By comparison with modern computers, brains have two implementational advantages: dense three-dimensional wiring and an enviable power efficiency. However, key implementational challenges stem from the fact that neural elements are painfully slow, are subject to many different forms of noise, as well as death, damage, turnover, and (in some places) birth (Gould 2007), albeit the last to a degree that is insufficient to avoid the competition between stability and plasticity for storage (Carpenter and Grossberg 1988).

It is notable that neural information processing is extremely robust to damage, leading Lashley (1929), for instance, to the near *reductio ad absurdum* of equipotentiality, with any bit of brain being able to substitute for any other bit given an insult. One general strategy the brain appears to adopt for preventing damage from having a devastating effect involves population codes (e.g., Pouget et al. 2000, 2003). That is, single entities, such as the subjective utility of a choice, are represented by the simultaneous activity of a very large number of neurons, each of which responds to value in a more or less ordered manner (albeit modulated by an apparent handful of other characteristics; (Kennerley et al. 2009; Padoa-Schioppa 2009; Roesch and Olson 2005).[2] Literal duplication of responsivity is actually rare—partial overlap offers much more efficient codes (Shannon 1948)—however, the substantial redundancy inherent in overcomplete population codes implies that losing a few neurons need have little effect. Of course, there are limits to this: as witnessed by the field of neuropsychology (Shallice 1988), Lashley was wrong—localized damage, either from neurological incident or targeted experiment, leads to a panoply of more or less precisely characterizable deficits.

Another version of this general strategy is multiplicity at the level of large-scale neural systems, the most gross example of which is the existence of two hemispheres (although the degree of bipotentiality this confers by the time of adulthood is less clear; Moses and Stiles 2002). Again, in the context of decision making, literal duplication is rarely evident. Rather, there appear, as discussed in greater depth below, to be different decision-making systems that cooperate and compete to control action. Eliminating one system, for instance by a focal lesion, need not affect the other systems. Indeed, disturbing rather than eliminating its output might actually be more deleterious, since noisy or confidently expressed but actually incorrect input from a system could be worse than no input at all.

---

[2] There is, however, an active debate about the extent to which the apparent order is real. An alternative idea is that substantial aspects of the selectivities of neurons are actually rather random (Rigotti et al. 2010), with the order emerging from the tails of the resulting distribution, rendered via the filter of experimental report.

Of course, the overall ramifications of damage in a complex system are hard to trace. For instance, the striatum is a key subcortical region that is involved in prewired and learned control. Parkinson's disease results in a progressive, focal lesion that eliminates one particular form of input to the striatum: its dopaminergic neuromodulatory innervation. This disturbs the overall output such that sufferers are increasingly unable to execute voluntary actions recommended by any of their decision-making systems. This is true even though at least some of these systems appear themselves not to depend on dopamine (Robinson et al. 2005; Berridge and Robinson 1998; Dickinson et al. 2000).

The robustness to stochasticity conferred by large populations of neurons is not clear, mostly because the nature and implications of noise are hotly debated. For instance, one suggestion is that apparent noise is really best seen as the sort of stochasticity necessary to realize Markov chain Monte Carlo forms of normative Bayesian inference (Fiser et al. 2010). In this case, from a computational (rather than an algorithmic or implementational) perspective, the randomness is not really noise at all. Certainly, animals can detect signals near the quantum limit (Lakshminarayanan 2005), suggesting that any recurrent effect of frank stochasticity in successive stages of processing can be substantially quenched, at least under some circumstances. In fact, the utility of averaging over large populations to eliminate "true" noise might in any case be limited by correlations in the activity of these populations (Shadlen et al. 1996; Kohn and Smith 2005; Smith and Kohn 2008), although the nature and magnitude of these has also been challenged (Ecker et al. 2010; Cohen and Kohn 2011).

Finally large populations could potentially confer robustness against the sloth of individual neural elements (Tsodyks and Sejnowski 1995). Whether the appropriate algorithms are actually employed is less clear—not only are there delays in the first responses of early sensory neurons to stimuli (Thorpe et al. 1996; Foxe and Simpson 2002), the times to production of subsequent actions are also subject to unconscionable further delays (Smith and Ratcliff 2004).

Outliers

The recent financial crisis has focused attention on events that lie on the distant tails of probability distributions that characterize possible outcomes (Taleb 2007). Such events can have a catastrophic effect on the performance of complex systems that are designed or tuned to work in more normal regimes. It is particularly hard to learn about such tails in a well-controlled manner from observations, since, by definition, very few samples are ever available. Distributions that fall off as power laws rather than exponentially with magnitude have this characteristic and are ubiquitous across natural and artificial phenomena (Mandelbrot 1983; Barabási and Albert 1999).

The issues of mitigating the potential effects of such outliers pose tricky computational-level problems and have attracted some solutions. One example is *maximin control*, which seeks to find the best ("max") option assuming the worst ("min") conceivable consequence of that choice (Wald 1945), as if nature was a malicious opponent rather than just a tease. The popular method called H∞ control can also be seen as embodying a similar precautionary principle (Doyle et al. 1989). All such methods can certainly offer some protection against the worst rainy day; however, there is a cost to obsessive conservatism, in terms of lost opportunities for possible, or even likely, gains.

These forms of control can be seen as enhancing the sensitivity of control strategies to outliers so that a possible catastrophe can be averted. However, in the context of statistics and machine learning, robustness, quantified by the same methods, has almost exactly the opposite implication. Robust statistical techniques (e.g., Huber 1972) are designed to protect inference against the untoward effects of outliers by diminishing rather than enhancing sensitivity to them. For instance, the arithmetic mean of a set of samples is strongly influenced by extreme values, since it is the quantity that has the minimum mean square difference to all the samples. By contrast, the median, which minimizes the absolute difference rather than the square difference, is much less affected by extremes. Even more robust error measures or loss functions than the absolute difference have been suggested, including ones that approach asymptotic values as the difference increases, rather than growing unboundedly.

Data on the way the brain handles outliers is rather mixed. Sensory processing appears beautifully adapted to the heavy-tailed nature of natural phenomena. This is evident in forms of adaptation (Zhaoping 2006) as well as in accounts of the unsupervised acquisition of cortical representations that are based on fitting heavy-tailed generative models to input statistics (Olshausen and Field 1996; Schwartz et al. 2006). However, an explicit attempt to measure the loss function in the context of motor estimation and control showed that it is robust in the statistical sense; that is, it suppresses rather than enhances the effect of outliers (Körding and Wolpert 2004).

Uncertainty

Outliers are perhaps best seen as a particular special case of uncertainty: not only must more than one outcome be conceivable, but also gaining (or perhaps missing) one of those conceivable outcomes, though extremely unlikely, could be devastating. However, as a computational-level issue, uncertainty is much more general, and itself demands various apparently less extreme forms of robustness.

First, consider expected uncertainty (Yu and Dayan 2005) created by known distributions of possible outcomes. The data on sensitivity to risk as defined by the variance or higher-order moments of such distributions such as skew and kurtosis are multifarious (with such consequences as the widespread purchase of costly insurance). Given explicit information about what might happen, human subjects have been described as being risk seeking for losses and risk averse for gains (Kahneman and Tversky 1981). Sensitivity to skew and kurtosis have also been observed (Hsu 2006), but there is ample room for further work on the provenance of these effects in implicit, nonlinear utility functions and/or explicit, moment-based penalties (D'Acremont and Bossaerts 2008). Further, when the statistics of possible outcomes have to be acquired from experience rather than description, a broader range of risk seeking and risk aversion is observed (Jessup et al. 2008; Hertwig et al. 2004).

Animals are inevitably uncertain about many aspects of their environments. One primary source is ignorance and change, two factors that both humans and animals model. For instance, Bayesian treatments such as the Kalman filter have been used to characterize various higher-order features of animal conditioning (Dayan et al. 2000; Pearce and Hall 1980; Mackintosh 1983), and humans are known to be sensitive to "meta-characteristics" associated with change, such as the degree of volatility (Behrens et al. 2007).

More generally, Bayesian and other optimal methods can be seen as minimizing the requirement for learning from experience, and doing this learning as efficiently as possible. These are important, since experience in the world is at least costly (i.e., a threat to homeostasis) and, given potential predators, actually dangerous. Such a need can also be decreased by generalizing knowledge between environments, although paradigms such as learned helplessness (Seligman and Maier 1967; Maier et al. 2006) show problems with this. Animals can also limit the need to learn for themselves by learning from others (see Hammerstein and Boyd, this volume). Social learning is evident, for instance, in the social transmission of food preference (Galef and Wigmore 1983; Galef 1996), in which observational information about the utility of outcomes is provided, and in imitation learning, in which observation is used to suggest appropriate policies directly (Schaal et al. 2003; Wolpert et al. 2003; Oztop et al. 2006).

Being robust to uncertainty associated with ignorance and change, at least in an appetitive domain, implies being able to take advantage of potential opportunities, which in turn requires exploration. From a normative perspective, uncertainty engenders exploration via forms of exploration bonus, the most famous of which is the Gittins index (Gittins 1989; Berry and Fristedt 1985). Exploration bonuses quantify the potential benefits of the possible outcomes, weighed against the opportunity cost demanded by any failure to exploit. Bayes optimal exploration is computationally catastrophic; there are thus many, more or less completely non-Bayesian approaches (Thrun 1992; Dayan and Sejnowski 1996; Manyika and Durrant-Whyte 1995; Auer 2003; Simmons et al. 2000; Krause and Guestrin 2005; Ng et al. 1999).

In the aversive domain, at least two forms of robustness are apparent. First, along with the neophilia associated with exploration bonuses, animals can also be neophobic and exhibit thigmotaxis (the tendency to remain close to the periphery of an open area) and negative phototropism (growth away from a light source), both of which help keep them out of trouble in the face of unknown potential threats. As we note below, there are mechanisms for adjusting the relative balance between these two as a function of context, and so perhaps prior experience. This is robustness to potential threat, akin to allowing for the possibility of outliers.

Rather more dramatic are the consequences of observing an actual threat or a cue that predicts a threat in a context. This is evident both in the short term, as in forms of behavioral sensitization that are seen in everything from the sea slug *Aplysia* upward (Pinsker et al. 1973), and the long term, as in single-trial fear conditioning (Fanselow 1980), and indeed the debilitating condition of posttraumatic stress disorder to which this might be related. That is, one overall strategy might be described as "be overoptimistic, but overreact and overremember."

A much more radical form of uncertainty stems from mis-specification of the model of the world. Indeed, one of the main problems that animals face is working out the structure of the aspects of the world that can potentially be predicted and particularly those that can potentially be controlled. This includes determining the relevance of potential cues and also estimating appropriate ways of re-representing those cues so that predictability and controllability are clear. Of course, this poses a particularly severe problem for robustness, with the very class of threats being protected against, or opportunities requiring suitably availing, being mysterious.

One generic approach to potential mis-specification is to use classes of models that are expandable in the light of ever more subtle features that can be revealed by increasing amounts of data. There are, for instance, various so-called non-parametric Bayesian methods that allow for this (Ferguson 1973; Rasmussen and Williams 2006). However, they have not been well coupled to appropriate methods of exploration. An alternative strategy embodied by certain of the non-Bayesian algorithms is, as for outliers, to assume a potentially malicious environment, but then to adopt a more benign measure of acceptable performance (a form of regret; Auer 2003). By making substantially weaker assumptions about the environment, these methods are at less risk from mis-specification.

Finally, uncertainty can be created by the computational demands of performing inferences to make predictions and of deciding upon appropriate controls, possibly in the light of those predictions. Many methods for control require more substantial amounts of online storage than our relatively anemic working memories offer (Miller 1956). This is particularly true in the central case when one is trying to optimize responses for long-term gain, and so must consider whole trajectories. Further, despite a common view of working memory as involving a set of rather independent "slots" that can be full, but otherwise offer the same sort of high-quality storage as computer memories (Zhang and Luck 2008), there is in fact substantial evidence that information instead degrades more or less gracefully as memory is taxed (Bays and Husain 2008). Degradation implies that calculations associated with those methods for control will become progressively more inaccurate or uncertain as the problem gets more difficult. Such inaccuracies inspire the use of alternative methods for estimation (Sutton et al. 2007) and control; these can then have their own trade-offs (Daw et al. 2005).

In fact, the complexity of control is not just a problem for brains, with their limited working memory capacities. Rather, many interesting control problems, particularly in the face of uncertainty, suffer a calamitous computational complexity—formally, being NP-hard or NP-complete (Littman et al. 1998; Allender et al. 2003). This has been a spur to the use of heuristics and approximations.

Neural Reinforcement Learning

Having discussed some dimensions of robustness, together with computational ideas about their implications and effects in biological decision making, let us turn to the specification of the neural architecture of control, and how its various parts contribute to, and detract from, robustness. At a coarse level, one can view many algorithmic features through the lens of robustness.

Current conceptions of the general architecture have recently been the subject of extensive investigation (Daw et al. 2005; Dayan and Daw 2008; Dayan 2008; Doya et al. 2002; Samejima and Doya 2007). Very briefly, omitting a wealth of detail (particularly about their neural realizations) and citations that can be found in those treatments, there is a critical distinction between instrumental and Pavlovian control, as well as a separation between two methods of instrumental control.[3] As we will see, the dividing line between instrumental and Pavlovian control is contingency: in instrumental control, actions are chosen because of the actual observed or inferred outcomes to which they lead in particular circumstances; in Pavlovian control, outcomes and predictions of those outcomes elicit preprogrammed responses automatically, whether or not those responses do in fact lead to those outcomes. Although such an absence of contingency may seem strange, we will interpret Pavlovian responses as being substantial contributors to robustness. These various influences over choice are supported by different neural systems and have distinguishable psychological properties. There remain, however, substantial and confusing interactions.

Pavlovian Control

Pavlovian control can perhaps best be seen as a form of evolutionary programming. Given the presence of a biologically significant stimulus, such as a morsel of food or a threatening predator, animals exhibit a set of stimulus-determined,

---

3 A third instrumental system, the episodic controller (Lengyel and Dayan 2008), has also been suggested but will not be discussed, because there are fewer data on its properties.

species-typical (Bolles 1970) reflexes, such as moving toward and eating the former, or fighting, fleeing, or freezing given the latter. The responses are sometimes called unconditioned responses (URs), since they occur without any need for learning or conditioning; the stimuli that elicit them are called unconditioned stimuli. Mechanisms exist to adapt these responses to circumstance. For instance, in food aversion learning, previously appetitive outcomes that subsequently make animals sick cease to elicit approach responses when subsequently encountered (Garcia et al. 1974). A more impressive form of evolutionary programming that we mentioned above is that certain species can engage in social forms of learning: rats and other species exchange information about the foods they have safely consumed using odors (Galef and Wigmore 1983; Galef 1996).

Even more pertinently for the present discussion, outwardly neutral stimuli, such as lights and tones that predict the possible future occurrence of the biologically significant stimuli, come to elicit responses, called conditioned responses (CRs). Some CRs are called *consummatory* and are closely related (though not necessarily identical) to URs for the unconditioned stimuli that are predicted. Other CRs are called *preparatory* and involve such behavior as approach and engagement for appetitive predictors (e.g., the SEEKing responses of Ikemoto and Panksepp 1999, which appear to depend critically on the neuromodulator dopamine) as well as withdrawal and behavioral inhibition for aversive ones; the latter have a more complicated relationship with the neuromodulator serotonin, which may be partly opponent to dopamine (Deakin 1983; Deakin and Graeff 1991; Graeff et al. 1998; Daw et al. 2002; Boureau and Dayan 2011; Cools et al. 2008, 2011).

Pavlovian control confers a certain critical robustness. In particular, it solves key problems that have to do with uncertainty. No matter how minimal the opportunity for learning, a subject will have a repertoire of essential actions shaped by evolution to be appropriate in the face of a range of stimuli. Indeed, one might expect hard-wired controllers for aspects of the environment that are stable, since there is no benefit to having them be learned.

The appropriateness of extending these responses to conditioned stimuli is a little more subtle. As an example, predators should associate rustles in the foliage with the prey they subsequently predict, and thus approach future rustles. The converse is true for prey-detecting predators. In both cases, the range of possible predictors is too great for these also to be preprogrammed. In computational terms, these predictive Pavlovian responses eliminate one of the two problems involved in learned control. That is, although subjects still have to learn the predictions themselves, they do not have to learn which action is appropriate given the prediction; this speeds up the acquisition of good performance. Exactly the same is true for single-trial fear conditioning, in which animals learn the relationship between a conditioned stimulus and an aversive outcome in a single shot.

Hard-wired influences to which we might extend the term Pavlovian are also apparent in robustness-relevant aspects of neophilia and neophobia. First, novel objects are known to generate activity in systems associated with the processing of rewards (notably the dopamine system that was mentioned above), albeit to a degree that decreases back to baseline as stimuli become familiar (Horvitz 2000; Ljungberg et al. 1992; Schultz 1998). It has been suggested that this activity is a form of exploration bonus, masquerading as a true reward, providing an automatic incentive for new entities to be approached and explored (Kakade and Dayan 2002).

Second, one of the key substrates of Pavlovian influence and control is a ventral region of the striatum called the nucleus accumbens. Chemical manipulation of activity in this region leads to Pavlovian responses such as approach and withdrawal, but in the absence of eliciting stimuli or predictors (Reynolds and Berridge 2001; Faure et al. 2008). A recent experiment showed that the relationship between location of stimulation on the accumbens and the affective valence of the response (i.e., whether it would be appropriate for appetitive or aversive unconditioned stimuli) depended on subjects' familiarity with their environment. In the familiar home cage, potentially neophilic, appetitive, responses dominated. In an unfamiliar environment, neophobic, aversive responses prevailed (Reynolds and Berridge 2008). This distinction can be seen as a form of adaptive robustness.

Third, humans and other animals appear to employ a variety of strategies such as forms of win–stay, lose–shift (Nowak and Sigmund 1993). These work well in many settings in which more sophisticated learning might appear to be necessary. The neural substrate underlying them is not yet clear.

Fourth, there are widespread signaling modalities such as the norepinephrine system which have been suggested to act as forms of neural interrupts (Bouret and Sara 2005; Dayan and Yu 2006) in the face of unexpected uncertainty (Yu and Dayan 2005). This form of uncertainty arises when what happens is grossly at variance with what is expected, and thereby engages automatic mechanisms associated with outliers, including a complex stress response (Elenkov and Chrousos 2006).

On the other hand, the sheer inflexibility of Pavlovian control is its Achilles heel. As noted, Pavlovian responses are automatic and programmed in the light of the evolutionary environment. They are not contingent on their success in gaining rewards and avoiding punishments, and the choices (rather than the predictions) are not dependent on the actual individual experience of the subject rather than the catastrophes of its hapless non-ancestors. This presents a key challenge to robustness: if Pavlovian control was all that there was, there would be circumstances under which subjects would experience unpleasant outcomes but nevertheless, apparently blithely, repeat the offending actions that were responsible. Indeed, some such cases will be described below.

Instrumental Control

In instrumental control, responses are determined not through evolutionary programming but by contingency, or the appropriateness of their consequences (Dickinson 1985, 1994). Very crudely, consider a lever that can be moved either up or down. If a hungry subject is always and only fed just after it has moved the lever down, then it will learn to move the lever down rather than up. If the subject moves the lever appropriately, there will be a Pavlovian relationship between the lever and the delivery of food. Thus, we can expect the subjects to emit Pavlovian responses such as approaching and engaging with the lever to perhaps move it up and down. However, by themselves, Pavlovian mechanisms are unable to support the crucial extra step of reiterating just one of those movements because of its contingent success.

Instead, there are instrumental mechanisms that can learn the correct action to take in this and other more general cases. There is good evidence for two separate systems: one goal directed or model based; the other habitual or model free (Daw et al. 2005; Doya et al. 2002; Samejima and Doya 2007).

Model-Based Control

Model-based control depends on knowing, or learning, the transition structure of the environment (i.e., how the states change consequent on possible actions) and the appetitive and aversive outcomes the environment affords. A simple example might be knowing the map of a maze, along with the location of the treasure it contains. Given such a model, it is formally straightforward, if computationally potentially rather challenging, to choose the action that maximizes long-run rewards or minimizes long-run punishments by inference, i.e., roughly by executing a form of dynamic programming (Howard 1960; Puterman 2005) or forward or backward search (Foster and Wilson 2006, 2007; Johnson and Redish 2007; van der Meer and Redish 2010).

Model-based control is highly flexible on at least two scales. First, if the transition contingencies or the nature or utilities of the outcomes change, then the choice of action can change immediately too, since the outcome of dynamic programming will be different (Dickinson and Balleine 1994, 2002). It is because of this characteristic that model-based control is referred to as being goal directed, since actions are chosen in the light of current goals. This constitutes a very important form of robustness: inference, which is relatively safe and cheap, can replace experience, which is potentially dangerous and expensive.

Second, on a longer timescale, when the model of the world has to be learned, there is an opportunity for prior information gleaned from related environments to be used to hasten learning, or indeed limit or structure exploration. This again can lead to robust control, since information that can be generalized does not need to be learned afresh. Limiting generalization appropriately is, however, rather hard. Paradigms such as *learned helplessness* (Seligman and

Maier 1967; Maier et al. 2006) show that when subjects discover that there are sharp limits to their ability to manipulate an aversive environment, they can generalize this information widely, and so typically fail even to try to control a new, and actually normal, environment (Huys and Dayan 2009).

The major problem with model-based control is that it is not robust to the size or scale of the environment. As the complexity grows, measured for instance by the number of possible states and trajectories, the demands on working memory for performing inference rapidly become unreasonable, rendering model-based control catastrophically noisy and inaccurate.

Model-Free Control

Model-free methods of control occupy exactly the opposing sweet spot in the balance between inference and experience (Daw et al. 2005). They set out to calculate the same quantity as model-based methods; namely, the long-run utility associated with each possible action in a state. It is these values that are required for performance. However, they acquire them through extensive learning from experience, without ever needing to acquire a model of the world, rather than intensive inference (Sutton 1988; Sutton and Barto 1998; Watkins 1989). This means that at the time of use, the values are immediately available, permitting control even in very large environments. This is the key form of robustness that was missing from model-based control.

However, model-free control requires substantial experience, since it typically depends on a bootstrapping process of eliminating inconsistencies between successive estimates of long-run values along trajectories. Early in learning, the inconsistencies do not provide a useful learning signal. Thus model-free control is not robust to changes in the environment (since substantial learning has to be carried out with the new transitions or outcomes). Further, since it lacks a model, model-free control cannot directly incorporate prior information about the nature or structure of the environment and thereby reduce or improve learning or exploration. That is not to say that there is no way for general facets of an environment can affect model-free control. Rather, its parameters, such as the rate of learning from inconsistencies, can reflect expectations about ignorance and ongoing change (Dayan et al. 2000). Moreover, there are generic ways to influence exploration, such as the strategy of being overoptimistic, so that states and action which have not been visited or tried always start off looking more attractive than those that have, and so should be attempted (Brafman and Tennenholtz 2003; Auer 2003).

Finally, model-free methods swap time for space in a different sense. In a complex environment, the model-free system faces the daunting task of representing the mapping from a potentially huge number of states and action to the long-run utilities. This can tax the representational capacity even of the large, distributed, population codes that we discussed above, thereby inducing a further form of noise.

Interactions

We have discussed three rather different schemes for control, which exhibit different, and sometimes complementary, forms of robustness. Let us now consider how they interact and whether this achieves the intersection or the union of the available robustness.

Under normal circumstances, model-based and model-free instrumental control appear to interact with each other in a generally appropriate manner. That is, the initial choices in a new environment appear to be under the influence of model-based control, for instance being immediately sensitive to changes in contingencies or outcome values (Adams and Dickinson 1981; Dickinson and Balleine 2002; Holland 2004). After so much learning that model-free values should be accurate, the model-free system can take over control, in a process known as habit formation. This can be interpreted as a battle between two sources of non-robustness: the noise inherent in the complex calculations of the model-based controller and the noise associated with incomplete and inefficient learning in the model-free system. The locus of behavioral control is transferred as these noise levels change and is indeed appropriately sensitive to various factors, such as the complexity of model-based inference and the amount of learning (Killcross and Coutureau 2003).

Of course, the signature of habits is choice that is insensitive to change in the world. That this happens after substantial learning is a clear form of non-robustness. One contributor to this non-robustness may be that the model-free system suffers from a poor model of its uncertainty, and so is unaware that changes in the world should have made it less certain (Daw et al. 2005).

There are other issues for robustness in the interaction between model-based and model-free control. For instance, robustness can be enhanced if the model-based system can train the model-free system offline, during periods of quiet rest or sleep, allowing model-free control to exhibit the sensitivities to change of the model-based system, but again without the need for complex online inference (Sutton 1990). There is indeed some evidence for this form of offline training (Ji and Wilson 2007). Conversely, if model-based control is responsible for the initial choices in an environment, but fails to explore (e.g., owing to an overgeneralized form of learned helplessness), then the experience the model-free system will be able to use to learn its own estimates will be incomplete, so that, even if it is itself undamaged, it will never seize control.

There may also be more intimate interactions, with the model-based system influencing the prediction error based upon which the model-free system learns (Daw et al. 2011). Furthermore, model-free values may provide estimates which the model-based system employs when forced by computational complexity to prune its online evaluation of options (Samuel 1959).

By contrast, the interplay between Pavlovian and instrumental control is rather richer and less well understood. Under very many circumstances, and particularly when experimenters want to train their subjects quickly, the outputs of the systems are designed to align. This allows the robustness of Pavlovian control, in terms of its lesser requirement for learning, to be combined with the robustness of instrumental control, in terms of its ultimate flexibility. However, Pavlovian and instrumental control can be put into conflict, with highly entertaining and deleterious consequences (e.g., Hershberger 1986; Breland and Breland 1961; Dayan et al. 2006; Williams and Williams 1969).

Hershberger (1986), for example, engineered an environment in which a chicken interacted with a chicken feeder in a mirror-like world. If the chicken ran toward the feeder, the feeder retreated at twice the speed; if the chicken ran away from it, then the feeder approached at twice the speed. Thus, to get fed the chicken had to run away from the feeder. The chickens could not learn this behavior. We can analyze this in terms of Pavlovian/instrumental interactions as follows. The instrumental requirement is straightforward: run in one direction with respect to a relatively arbitrary object to acquire a food reward. The Pavlovian requirement is harder: having at one time been fed by it, the feeder is an appetitive cue; it therefore inspires approach and engagement. Crucially, the Pavlovian system requires approaching the feeder because of the prediction of food, not because approaching is the action that leads to the food; indeed, this fact underlies the fiendish design of the experiment. This outfoxes the chicken. In another, conceptually similar paradigm called negative automaintenance, one can observe cycling: extinction of an appetitive association leads to the equivalent of successful retreat (under instrumental control) but is followed rapidly by renewed approach, leading once more to extinction.

It is noteworthy in such cases that Pavlovian responses transcend instrumental ones. This is perhaps reasonable, given the comparison between millions of years of evolution and just a few sessions in a laboratory. It does, however, invite consideration of different timescales of robustness.

Discussion

In this chapter, I have considered a number of dimensions of robustness in decision making, notably noise in the processes involved, outliers, and various forms of uncertainty. There is a distinction between robustness in appetitive and aversive contexts: the former leads to considerations of exploration whereas the latter results in a variety of issues about potential and actual threats. I have outlined the architecture of neural control and highlighted how the existence of multiple mechanisms influences the robustness of the ultimate behavior both positively and negatively.

Robustness occurs at various levels of computational analysis (Marr 1982). At the implementational level, I presented various population coding strategies that have been suggested to ameliorate the relatively poor-quality (albeit extremely power-efficient) neuroelectronic components of the brain. At the algorithmic level, I discussed a wealth of strategies that different systems adopt which can lead to robust control, along with various ways in which their interaction is suboptimal. Finally, at the computational level, I discussed a number of approaches to outliers and trade-offs between exploration and exploitation. These provide the foundational underpinnings for the overall degree of robustness that is apparent.

One of the most attractive facets of the field of decision making is that computational, psychological, and neural findings are closely linked; indeed, predictions and observations at one level are directly used to suggest methods or experiments at others (Dayan and Daw 2008). The issue of robustness, however, has yet to become a focus for integrated investigations of this sort. Thus, for example, the implications of robustness for the adaptive balance between exploration (neophilia) and exploitation (neophobia) have not been well examined.

Perhaps the primary issue for robustness concerns learning. Many mechanisms are devoted to obviating or minimizing the amount of experience necessary, including evolutionary priors over values or actions, generalized priors from past experience, and carefully targeted exploration in the light of uncertainty. The work on social dimensions, such as the transmission of food preferences and imitation learning, can also be seen as ways of reducing the requirement for experience. Along with this, the baseline strategy of being generally optimistic, but highly reactive in the face of a discovered threat provides a relatively safe "sandpit" in which exploration can robustly occur.

Various other social factors have also been explored. For instance, there has been substantial work on robust algorithms in the face of game theoretic competitive interactions between individuals. Methods such as "tit-for-tat" and other variants of "win–stay, lose–shift" are robust enough to perform well against many opponents in iterated versions of the prisoner's dilemma game (Axelrod and Hamilton 1981; Nowak and Sigmund 1993). The neural realization of these, and indeed of more sophisticated interactive controllers that build models of the opponent as well as the environment, are only just starting to be explored (King-Casas et al. 2005; Dorris and Glimcher 2004; Barraclough et al. 2004; Hampton et al. 2008; Yoshida et al. 2008, 2010; Ray et al. 2009). Note, though, that uncertainty and noise of the forms we have discussed above, may limit the robustness of some of these strategies (Stevens et al. 2011).

Another social dimension to robustness is the possibility of adaptive allelic variation across the whole population of decision makers. We know that there are relatively common polymorphisms in genes coding for mechanisms involved in decision making. These might endow their carriers with systematically different biases, for instance by influencing parameters such as the bias toward optimism that underlies exploration in model-free control. If some individuals are more ready to explore, allowing others, who are themselves more conservative, to learn from their successes and failures, then the population as a whole can benefit (Suomi 2006; Williams and Taylor 2006).

In sum, robustness in biological decision making depends on the detailed interaction between many mechanisms operating across a wealth of scales. We should not be surprised at this complexity (whose surface we have doubtless barely scratched), since decision making is a mandatory core competence underpinning the survival of organisms. We have rather powerful approaches to understand instrumental control, whose sensitivity to contingency places learning, rather than hard-wiring, at its core. Perhaps inevitably, we struggle more with Pavlovian control to avoid telling a collection of disparate, plausible, just-so evolutionary stories about robustness that might possibly be nothing more than entertaining fiction.

Acknowledgments

I am very grateful to David Silver for discussion and very helpful comments on an earlier version of this chapter. I would also like to thank the members of my group for their engagement with and discussion of the issues, and Peter Hammerstein, Julia Lupp, Jeffrey Stevens and an anonymous reviewer for helpful comments and edits. Thanks also to Julia Lupp, Marina Turner and the ESF team for their excellent and efficient hospitality. My work is funded by the Gatsby Charitable Foundation.

10

Advantages of Cognitive Limitations

Yaakov Kareev

Abstract

Being a product of evolutionary pressures, it would not be surprising to find that what seems to be a limitation of the cognitive system is actually a fine-tuned compromise between a set of competing needs. This thesis is demonstrated using the case of the limited capacity of short-term memory, which is often regarded as the prime example of a cognitive limitation.

Short-term memory (STM) can hold only a small number of items; originally estimated at 7±2 items, it is now believed to be closer to 4 or 5. Because STM is the part of the cognitive system that holds the information available for conscious processing, its capacity sets an upper limit on the size of the sample that may be considered to determine characteristics of the environment. Obviously, the smaller that size, the higher the risk of obtaining inaccurate estimates of important parameters. However, at the same time, it can be shown that the very same limitation also carries with it a number of advantages.

First, small samples lead to systematic, and arguably beneficial, biases in the estimation of parameters that bear on the perception of regularity: Correlations are likely to appear stronger than they actually are and variances (i.e., risks) smaller than they actually are. People have been found not to correct for these statistical biases, further implying that it is beneficial to perceive the environment as more regular than it actually is. Second, because decisions based on small samples are bound to result in occasional judgment errors and inconsistent behavior, the somewhat erratic behavior that inevitably results from the use of small-sample, error-prone data can benefit the organism in situations in which it is to the organism's advantage to behave in an unpredictable manner (e.g., prey-predator interactions, repeated choices between service providers, other situations in which a mixed strategy is called for, and auto-correlation between successive decisions is undesirable). Third, by relying on only a limited number of recent experiences, the organism is better attuned to changes in its environment.

It is therefore maintained that what might look upon first sight as a major cognitive limitation, actually represents a fine-tuned compromise between a set of competing needs.

Introduction

Defining a characteristic of an information-processing system to be a limitation is a tricky business. On one hand, one could easily propose how essentially any characteristic of the system could be improved: having three or four eyes would give humans a more complete picture of their immediate surroundings, faster neuronal transmission would speed up reactions, etc. In this respect, the current value of any characteristic of an information-processing system might be regarded, almost by definition, as a limitation. On the other, increasing the amount of information available to the system—whether by way of the amount of input transmitted by a sensory organ, the amount of output resulting from internal calculations, or just the speed with which the same amount of information is made available—may require system-wide changes, such as increased storage capacity or increased processing speed, which may require compromising some other aspects of performance. As a result of such considerations, many characteristics of the human information-processing system are not considered limitations: neither having only two eyes nor the speed of transmission through the nervous system is regarded as a limitation.

Here I will focus on what it takes for a characteristic of an organism's information-processing system to be considered a cognitive limitation. No rules for this enterprise exist, of course, but I would like to propose a few. First, for a characteristic to qualify as a cognitive limitation, its value should impose a clear cost in terms of functioning relative to the level of functioning that could be achieved with other, more desirable (and potentially reachable) values of the same characteristic. Second, there should be some variability—inter-species, intra-species, or both—in the values of the characteristic, and there should be at least one way of moving from the undesirable range of values to a more desirable range, such as through actions by the individual (exertion of effort, learning), situational factors (education, changes in the environment), or easily envisioned evolutionary pressures.

Of the characteristics that are commonly considered limitations of the cognitive system (e.g., processing speed, computational complexity or lack thereof, and memory capacity), none looms larger than the limited capacity of short-term memory (STM). This limit, reflecting a structural characteristic of the cognitive system, was first pointed out by G. A. Miller (1956) in his classical study "The Magic Number Seven, Plus or Minus Two." There are numerous sophisticated measures of STM capacity (e.g., Engle et al. 1999; Just and Carpenter 1992), but the uninitiated can easily get a feel for this limit by considering the digit-span task, which is a component of the Wechsler Intelligence Scale for Children (Wechsler 1949). In this task, a tester reads a list of single digits at a rate of one digit per second and instructs the test taker to repeat the list in exactly the order it was read. The first list is short and easy to repeat (e.g., "3, 8, 6"). Another, longer list (e.g., "4, 5, 9, 2") follows a correct repetition. Increasingly longer lists are read until two successive failures occur. The number of digits in the longest list correctly remembered serves as the estimate of the test taker's STM capacity. Analyzing people's performance in a number of converging tasks, Miller estimated the STM capacity of normal adults to be around seven items, ranging from slightly less to slightly more depending on the characteristics of the individual and the specific task. Miller's original observation generated enormous interest and follow-up research, and has become a cornerstone of cognitive psychology (Simon 1990). Although estimates of STM capacity vary—in fact, modern estimates place its value at four or five items (Cowan 2001)—it is uncontested that STM capacity is limited to a small number of items.

Before getting further into a discussion of whether or not a limited STM capacity constitutes a cognitive limitation, it should be pointed out that, by imposing a relatively rigid constraint on information processing, the limited capacity of STM renders information processing quite robust: An important aspect, the amount of information likely to be considered when a judgment, decision, or choice is being made, is unlikely to be much affected by extraneous factors such as context or (to some extent) even the ability or mood of the decision maker.

Is STM Capacity a Cognitive Limitation?

The limited capacity of STM has obvious implications when one considers that STM is regarded to be the workspace of the cognitive system. Only items in it—whether selected from sensory input or retrieved from long-term memory—are actively available for processing. Once STM is full to capacity, any new information added necessarily pushes out some information. Therefore, a clear implication of this capacity limit is that one can simultaneously consider only a small number of items—whether they are observed in the outside world, retrieved from memory, or both. This, in turn, implies that the cognitive system can act on only a small sample of the information available at any time. Because the variance of the sampling distribution of any statistic is larger the smaller the sample, small-sample data are likely to result in inaccurate estimates, and give rise to disagreements between different people observing the same phenomena. The built-in inaccuracy that comes with small samples is orthogonal to the lack of bias in estimates of certain population parameters (more on that later).

If we assume that having accurate estimates of our environment is essential for efficient functioning, the limited capacity of STM appears to be a major handicap, one that increases the chances of erroneous estimates, which would lead to sub-optimal decisions and actions. Furthermore, it is also a limitation by the other criterion I suggested earlier: First, there is natural variability in built-in STM capacity—more intelligent people have, on average, larger STM capacity than less intelligent people (e.g., de Jong and Das-Small 1995; Engle

2002; Jurden 1995), adults have larger capacity than children (e.g., Case et al. 1982; Huttenlocher and Burke 1976)—and STM capacity also varies with situational factors, such as cognitive load (Gilbert et al. 1988) or emotional stress (e.g., Diamond et al. 1996). Second, there are ways to reduce the impact of the limitation: Although the number of items in STM is limited, the items themselves are chunks—units that have a reference in long-term memory. For example, the sequence G O L D B A R, read a letter at a time, may constitute 7 one-letter chunks for a person whose knowledge of English is limited to the names of the letters, but only two chunks (or just one) for people familiar with English words. Thus, whereas the number of chunks is limited, the amount of information per chunk can vary, depending on a person's knowledge. This point was demonstrated in studies comparing laypersons' and chess masters' memory for briefly presented board positions (Chase and Simon 1973; De Groot 1965).

To sum up, by the criteria suggested earlier, the limited capacity of STM is definitely a cognitive limitation: Its effects are costly, it has a range of values that vary in the severity of their implications for functioning, and there are well-defined processes that affect the value of this characteristic.

However, although it is clear and uncontestable that the limited capacity of STM has some detrimental effects on performance, I will argue that the very limited capacity of STM, and the use of small-sample data that it imposes, has some effects that not only are not detrimental, but may in fact be outright beneficial. Some of these effects are obvious, and a brief discussion of them may suffice, while others are more subtle and require a more elaborate discussion.

Saving Search Cost and Time

It is a truism that the expected accuracy of an estimate increases with an increase in sample size. It is therefore generally believed that better decisions are reached on the basis of a larger sample. However, making do with small samples saves search time and effort. Saving on search time translates into earlier choice and more time for enjoying the benefits of that choice. Such savings are even more important when a decision is taken in the context of a race for the possession or right of use of a unique, indivisible good (e.g., buying or renting a house, buying a used car, choosing a mate). In such cases, competitors who make do with less information, who spend less time on search and accept the risk of having made a nonoptimal decision, may win possession or right of use of what would have been one's eventual choice after collecting all the data necessary to reach a well-justified decision. Thus, shorter search and decision times are obvious advantages of the constraints imposed by the limited capacity of STM. Although the benefits in this respect are difficult to assess, it is clear that they mitigate the costs incurred by the inaccuracy of the estimates themselves.

Furthermore, whereas search time and costs usually go up linearly with sample size, the expected accuracy of estimates increases linearly only with the square root of the size of the sample. As a result, the benefits of extra items diminish quickly. Indeed, analyses have shown that in judgment and choice tasks, accuracy quickly approaches an asymptote with a surprisingly small number of sample items (Hertwig and Pleskac 2010; Hertwig and Todd 2003; Johnson et al. 2001). Other indications that good performance can be built on relatively little data come from work on the effect of observing the behavior of a small number of neighbors and imitating the behavior of a successful neighbor (Aktipis 2004; Morales 2002; Nowak and May 1992; Schlag 1999). Work on the implications of adopting the "Win—Stay, Lose—Shift" decision rule, a rule that bases decisions on the most recent outcome only (e.g., Nowak and Sigmund 1993; Posch 1999; Posch et al. 1999), also demonstrates that high levels of performance may be achieved while relying on little data. Other analyses of performance with relatively little data (Kandori et al. 1993; Sandholm 1998) also point to the fact that using small amounts of data is not very costly, if indeed it is costly at all.[1]

Beneficial Effects of Biased Estimates

Although sample statistics provide unbiased estimates of a population mean (and hence, also the difference between two means), they provide biased estimates of other population parameters—most notably variance and correlation. Moreover, for these latter parameters, the degree of bias is negatively related to sample size: The smaller the sample, the larger the bias. These biases are a statistical fact. Unlike the biases so much studied in the behavioral sciences (e.g., Kahneman et al. 1982; Gilovich et al. 2002), which are assumed to result from the biased processing of unbiased input, the biases that I am referring to apply when the data are well sampled and the processing unbiased: By using sample data, one ends up seeing a systematically distorted picture of the world.

The two parameters for which sample statistics provide a biased estimate are the variance of a distribution and the correlation between variables. Both involve aspects of the regularity of the environment, and if their bias is not

---

1 In this respect, it is also appropriate to mention work that demonstrates, theoretically and empirically, that simple decision rules can lead to good performance. These studies include the initial work within the heuristics and biases framework, particularly work on the availability heuristic (for summaries of early studies, see Kahneman et al. 1982; Tversky and Kahneman 1974), its later emphasis on biases notwithstanding; work in the tradition of the fast and frugal heuristics (Gigerenzer et al. 1999); work on simple learning rules in complex environments (Houston et al. 1982; McNamara and Houston 1985a, 1987); and work on a simple heuristic that leads to regret minimization (Hart 2005; Hart and Mas-Colell 2000). Mention of this work has been relegated to a footnote because many of these simple rules or heuristics still rely on the accumulation of much data.

corrected for, the world is perceived as more regular than it really is.[2] I argue that the biases these statistics introduce to one's picture of the world are likely to be beneficial.

Assessment of Variance

The variability, or heterogeneity, of a distribution—measured by the second moment around the mean[3]—is of great import for the organism: It provides an indication, for example, of the error (squared deviation) that is expected when predicting the mean, the risk that is assumed when taking an action, and the variety of strategies that may be called upon when having to deal individually with all items in a distribution (e.g., an instructor of a class of students).

It is a well-known statistical fact, although one whose behavioral implications have hardly been considered, that sample variance is an attenuated estimator of the variance of the population from which that sample has been drawn. The attenuation, very familiar to practitioners of the t-test, is by a factor of $(N-1)/N$, where N denotes the sample size. To obtain an unbiased estimate of the population variance, σ_x^2, one needs to apply the following correction: $\text{est}(\sigma_x^2) = (N/(N-1))\, S_x^2$, where S_x^2 is the sample variance. The degree of the bias is small when N is large, but substantial for the size of the sample imposed by the limited capacity of STM: If one literally accepts the classical estimate of STM capacity, then adults observe, on average, a world that is only 6/7, or 86% as variable as it really is. Another way to consider this effect is to note that, for samples drawn from a normal distribution and consisting of 7 items, the estimated variance will be lower than the population variance 65% of the time. Thus, if uncorrected, small-sample-based estimates of variability are likely to be lower than the true variability—to project a picture of a world that is more homogeneous, less risky, than it actually is.

Of course, subjective perception need not reflect the (distorted) picture observed. The cognitive system could have evolved, or could learn, to apply some correction to estimates of the variability observed in samples, much as statisticians correct their estimates of variability. However, empirical studies addressing this question (Kareev et al. 2002) indicate that this is not the case. These studies compared performance in consequential decision-making tasks between people who had, or were likely to have at their disposal, samples differing in their size: In one study, people with small STM capacity were compared with people with large STM capacity; in another people presented with

[2] Interestingly, as long ago as 1620, Francis Bacon, in describing the "idols"—the bad habits of mind that cause people to fall into error—noted that people expect more order in natural phenomena than actually exists: "The human understanding is of its own nature prone to suppose the existence of more order and regularity than it finds" (Bacon 1620/1905:265).

[3] In using this definition, I assume that variables are measured on interval or ratio scales, and that variance is the measure of variability. My arguments, however, also apply for variables measured on ordinal and nominal scales, or for measures of variability other than variance.

small samples were compared to people presented with large samples; in a third study, people who had a whole population in full view were compared to people who saw only a sample of the population. In all cases, people who saw, or were likely to rely on smaller samples behaved as if the variability of the population in question was smaller. Such results would not have been observed had people corrected for the bias introduced by the use of small samples (or if they had been altogether inattentive to variability).

One implication of these findings is that people tend to underestimate risk and errors of prediction, and consequently feel more in control of their environment than they actually are. The bias, which affects everyone, will be more pronounced among the less intelligent, children, and people under stress than among the more intelligent, adults, and people who are relaxed. The bias also implies that individuals will perceive others as more homogeneous than themselves, and out-groups as more homogeneous than in-groups (Linville et al. 1989).

Would such a bias benefit or harm the individual? It is difficult to answer this question. On the one hand, it is possible to claim that any bias is dysfunctional. On the other hand, consider the fact that feelings of control contribute to organisms' physical and psychological well-being (Alloy and Tabachnik 1984; Langer 1975; Seligman 1975) or the claim that "cautious optimism" is a key to the risk-taking behavior and successful entrepreneurship that are so important for society (Selten 2001). It may be the case that the distorted view resulting from the use of small-sample data to assess variability provides just the right amount of distortion.

Detection of Correlations

The detection of correlations is one of the more important undertakings of the cognitive system. It underlies classical and operant conditioning, it promotes the emergence of concepts, and provides the basis for learning about causal relationships. When correlations exist and are detected, they help understand the past, control the present, and predict the future (Alloy and Tabachnik 1984; Shanks 1995). Organisms could hardly be expected to negotiate the vast complexity of tasks facing them if they did not detect and utilize the relationships existing in their environment. What relationship could one expect to observe when drawing a sample of bivariate observations and calculating the correlation between them? As is the case with other statistics, the sampling distribution is more variable the smaller the size of the sample on which it is based. Thus, when variables are actually unrelated, the use of small samples could well result in a false alarm—the "detection" of a correlation that does not exist. However, a very different picture emerges when one considers the sampling distributions of non-zero correlations and tries to infer relationships between variables that are in fact related. It is a well-established, though little known, fact that the sampling distribution of non-zero correlations is skewed, and increasingly so the smaller the samples on which they are based (e.g., Hays 1963). Figure 10.1, based on the tables compiled by David (1954), presents the sampling distributions for a population correlation, ρ, of .50 for sample sizes of 4, 7, and 10. The skew is striking: The medians—at .59, .54, and .52, for the three sample sizes, respectively—and even more so the modes of the sampling distributions are more extreme than the true values, and the bias is stronger the smaller the sample size. To be sure, the means of the sampling distributions are all slightly less extreme than the population values, and the more so the smaller the sample. However, the following is of great import: initial impressions not only carry greater weight than later impressions (Hogarth and Einhorn 1992), but also often bias the interpretation of further information (Nickerson 1998); in addition, small-sample data, more often than not, produce estimates that are more extreme than the population parameters, but deviate in the right direction. This points to the possibility that the limited capacity of STM acts as an amplifier of correlations (Kareev 1995b).[4]

Empirical research (Kareev et al. 1997) lent support to this hypothesis, and further theoretical analyses (Kareev 2000) showed samples in the range of 7 ± 2 items—the very range originally identified by Miller (1956)—to be of particular import in the detection of useful relationships between binary variables. These results point to the intriguing possibility that the limited capacity of STM in fact facilitates the detection of important relationships. The premise that organisms are intent on detecting effects when they exist (i.e., avoiding misses), even at the cost of making false alarms, sits well with recent findings that show people invest misses with greater value than false alarms (Kareev and Trope 2011; Wallsten et al. 1999; Wallsten and González-Vallejo 1994) and "design" their sampling procedures in ways that would increase the power of a statistical test, at the cost of observing inaccurate, inflated sample values that they do not correct (Kareev and Fiedler 2006; Soffer and Kareev 2011).[5] Theoretical analyses as well as experimental findings also support the notion that in reaching decisions, people rely on a small number of instances retrieved from memory (e.g., Selten and Chmura 2008; Stewart et al. 2006). The claims about the beneficial effects of small samples with respect to the detection of correlations also sit well with earlier accounts of the benefits of small capacity for perceptual (Turkewitz and Kenny 1982) and language (Elman 1993; Newport 1988, 1990) development.[6]

4 The bias discussed is larger, the more extreme the true correlation.

5 With respect to the analysis put forward by error management theory (Haselton and Buss 2000; Haselton and Nettle 2006; Nettle, this volume), an implication of the limited capacity of STM is that over the whole range of decisions, misses (false negatives) are more costly than false alarms (false positives). Interestingly, in all of the examples used by Nettle (this volume), which apparently capture natural and important decision-making scenarios, this is indeed the case.

6 It should be noted that the claims about the correlation-amplifying effect of small samples have been challenged (Anderson et al. 2005; Gaissmaier et al. 2006; Juslin and Olsson 2005), but also that these challenges have been responded to (Kareev 2005).

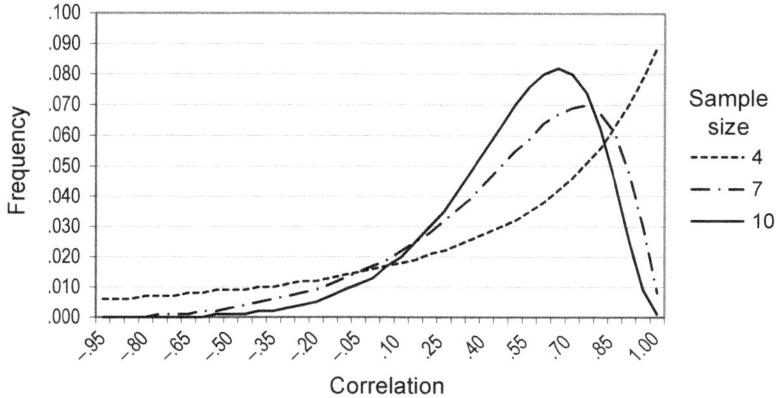

Figure 10.1 Sampling distributions of the correlation coefficient, for ρ = .50, and for sample sizes of 4, 7, and 10.

Being Less Predictable

As has been pointed out time and again, when sample size is small, sample statistics—hence, estimates of population parameters—are likely to vary. To an outside observer, behavior based on repeated small samples may look erratic and unpredictable. In previous sections of this paper, I assumed that it is to the organism's benefit to obtain accurate estimates; in the present section, I point out that, in many situations—situations that involve interactions between agents whose interests do not completely match—being less than perfectly predictable may in fact be in the organism's best interest. In such situations, the interests of the individual may be well served by the unavoidable reliance on small, error-prone estimates.

Many interactions are of the pursuit-avoidance type, in which one agent wishes to catch another agent, whereas the latter wishes to avoid the former. Predators and prey, police and criminals, players of matching pennies (and of other games modeled by it) all engage in such interactions. The prescription in such situations is to behave in a random, unpredictable manner—to employ, so to speak, a mixed strategy, or to engage in Protean defense (Driver and Humphries 1988; Humphries and Driver 1967, 1970). Use of small-sample data renders one's behavior noisier, and thus more difficult to characterize and predict. In this respect, the limited capacity of STM endows behavior automatically with a desirable characteristic.

Erratic, unpredictable behavior is of even greater value in another important class of situations—those involving choices between adaptive agents who are eager to be selected by the decision maker. Examples of such agents include service providers, who are selected on the basis of the quality of their service, and employees who are competing for a bonus to be awarded on the basis of their productivity, as well as flowers, which are selected by pollinators on the basis of the quality and quantity of their nectar. In all cases, the decision maker's main interest is to maximize his or her utility—to obtain the best service, the highest production, or the most nectar.

Which monitoring regime should one employ to achieve that goal? A straightforward suggestion would be to employ full scrutiny: to observe each option numerous times, in order to obtain an accurate estimate of its value, and then choose the option with the highest value. However, when it comes to choices between adaptive agents that are eager to be selected by the decision maker, it can be argued that larger samples do not necessarily lead to better decisions. To see why, consider the owner of a small shoe factory, who has two employees producing shoes. Wishing to increase production, the owner decides to offer a monthly bonus, to be awarded to that employee who produced more shoes during the month.

How should production be assessed? Should the owner take stock of each day's production and award the bonus on the basis of the monthly sum total? What effect would such a monitoring scheme have on the workers' motivation? Given that the workers are likely familiar with each other's productivity, it could turn out that neither of them would exert greater effort: The less productive worker would see no chance of outperforming the more productive one and would continue to produce at the previous level; the more productive worker would be likely to realize that and therefore not exert higher effort either. With such a scenario, the owner would indeed award the bonus to the truly deserving worker, but might end up paying the extra amount of the bonus without reaping any benefit.

Now consider an alternative monitoring regime—minimal scrutiny. Suppose the owner announces that production will be assessed on the basis of one day's production, with the crucial day chosen at random, separately for each worker, at the end of the month. With such a regime, if the distributions of daily production of the two workers overlap, the less productive worker does stand a chance of occasionally winning the bonus. Realizing that, he or she may work harder, to increase the overlap between the two distributions. Sensing the danger, the more productive worker may also try harder, to maintain or even reduce the overlap between the two distributions. With such a monitoring scheme, the principal will occasionally (but infrequently) commit an error, awarding the bonus to the less deserving worker, but may also bring about a more desirable situation.

A formal analysis of similar situations has been advanced within the framework of labor markets and tournament theory (Cowen and Glazer 1996; Dubey and Haimanko 2003; Dubey and Wu 2001). In all these studies it has been concluded that less, or even minimal, scrutiny can result in an outcome that is more desirable from the viewpoint of the decision maker because of the very uncertainty that a lower level of scrutiny introduces into the resolution process. An experimental test of the viability of this line of reasoning (Kareev and Avrahami 2007) demonstrated that in a two-person competition involving the performance of a routine, simple task, minimal scrutiny indeed resulted in higher performance; the effect depended on the existence of a bridgeable initial gap between competitors and showed up in competitive situations only.

Increased productivity is not the only possible benefit of minimal scrutiny. As pointed out by Kareev and Avrahami (2007), by leading to the occasional reward of a weaker competitor, minimal scrutiny also helps maintain competition and sustain diversity. In contrast, the consistent choice of a currently superior option may eliminate weaker competitors altogether, eventually leaving the decision maker at the mercy of a monopoly or a provider with a limited pool of characteristics.

Thus, by necessarily using small-sample data and therefore being susceptible to error, decision makers signal that they are error prone, or even "irrational"—that knowledge of the objective, prevailing conditions does not necessarily lead to successful predictions of their behavior. It is as if the decision makers announce: "I am not, I cannot be, fully rational. Take note of that and now act as you please." The notion that irrationality can breed rational behavior and desirable outcomes sits well with work in game theory on equilibrium refinements (Myerson 1978; Selten 1975; see also the related work of Kreps et al. 1982). That work "indicates that rationality in games depends critically on irrationality. In one way or another, all refinements work by assuming that irrationality cannot be ruled out, that the players ascribe irrationality to each other with a small probability. True rationality needs a 'noisy,' irrational environment; it cannot grow in sterile soil" (Aumann and Sorin 1989:37–38). Game-theoretic work on the role of bounded recall in fostering cooperation (Aumann and Sorin 1989) demonstrates another benefit of limited capacity. Thus, a cognitive limitation that introduces a grain of inconsistency or irrationality into behavior may act to bring about an environment that is more beneficial to the constrained decision maker.

Detection of Change

Thus far, I have assumed a static environment. In such an environment, the inaccuracy that necessarily results from limited capacity may still be viewed as a drawback, even in view of the benefits I have discussed. However, environments are hardly ever static; with the advent of time, many characteristics change in value: Whether constituting abrupt changes in quality—an inferior service provider becomes superior, a foe becomes a friend, a cooperator starts defecting—or gradual, periodic shifts—the yield of one type of flowers diminishes while that of another increases—changes abound. To function properly, to make sound decisions that respond to current conditions, organisms must continuously monitor the environment and act in a way that is responsive to changes in absolute and relative values (Cohen et al. 2007; Gross et al. 2008;

Speekenbrink and Shanks 2010). It is obvious that using numerous observations and averaging over a large number of items may provide an inaccurate estimate of the current situation. As a result, the detection of change would be slow. Indeed, it has long been recognized by students of behavior that in changing environments, remembering too many past events is counterproductive (McNamara and Houston 1985a, 1987; Rakow and Miler 2009; Shafir and Roughgarden 1994). Furthermore, both human (e.g., Massey and Wu 2005) and animal (Gallistel et al. 2001; Shettleworth et al. 1988) studies of behavior in changing environments reveal fast adaptation to change. The adaptation is so fast, in fact, that Gallistel et al. (2001) explicitly state that it poses a challenge to any learning theory. In contrast to learning theories, which postulate that behavior reflects response propensities accumulated over all past experience, models of behavior that rely on small samples, observing the environment through a narrow window of the most recent events, monitor the environment continuously and adjust to change automatically. In fact, the detection of change is faster the narrower the window. To illustrate that last point, Figure 10.2a–d presents the predictions derived from two models of choice behavior. Both models—win–stay, lose–shift (e.g., Nowak and Sigmund 1993; Posch 1999; Posch et al. 1999) and no regret–stay, regret–shift (Kareev, Avrahami, and Fiedler, in preparation)—take into account the last trial only. The graphs show how choice probabilities for two or three options change over time, both before and after a change in value is introduced. The important point is that although the two models employ different dynamics, and neither involves learning, both predict smooth changes with time and quick adaptation to change. Here, too, it is the apparent limitation which ensures efficient and robust performance.

Thus, with respect to the detection of change, relying on a small sample of recent events confers a real, uncontestable advantage. One may even speculate that a species' STM capacity has been shaped, at least in part, in response to the amount of change that the species faces in its environment, with more change resulting in smaller capacity.

Conclusion

The argument put forth here is that, far from the being the cognitive limitation it may seem at first, the limited capacity of short-term memory is in fact the result of multiple trade-offs, and represents a compromise between the values that would be optimal for competing needs.

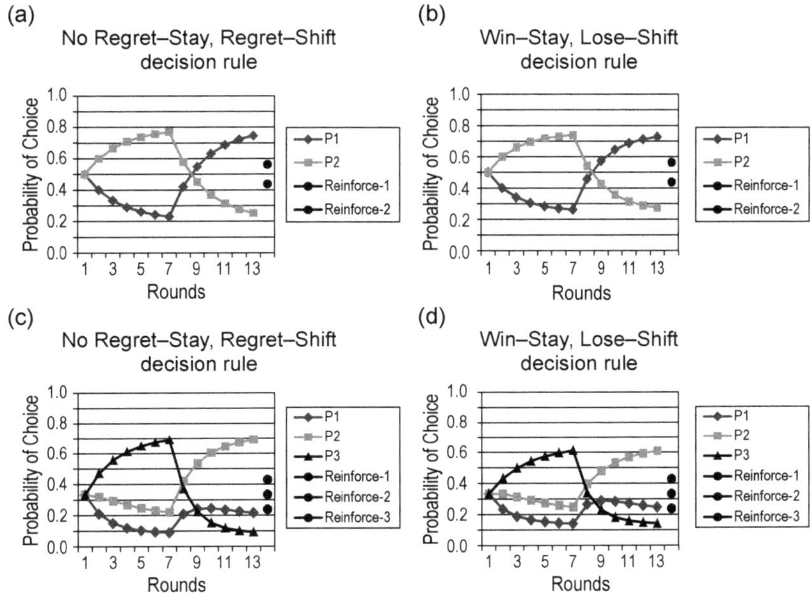

Figure 10.2 Choice probabilities over time, as predicted by the "no regret–stay, regret–shift," and the "win–stay, lose–shift" models. Choices are either between two options, whose probabilities of providing a reward are .7 and .9 (a and b), or between three options, whose probabilities of reward are .5, .7, and .9 (c and d). In all cases, initial choices are assumed random (no prior knowledge). The probabilities of reward change on trial 8. The reinforcement values, represented by black dots on the right, are asymptotic values following the matching law. Results are analytical, not simulations.

11

Modularity and Decision Making

Robert Kurzban

Abstract

Mechanisms that are useful are often specialized because of the efficiency gains that derive from specialization. This principle is in evidence in the domain of tools, artificial computational devices, and across the natural biological world. Some have argued that human decision making is similarly the result of a substantial number of functionally specialized or "modular" systems brought to bear on particular decision-making tasks. Which system is recruited for a given decision-making task depends on the cues available to the decision maker. A number of research programs have advanced using these ideas, but the approach remains controversial.

> Adaptive specialization of mechanisms is so ubiquitous and so obvious in biology, at every level of analysis, and for every kind of function, that no one thinks it necessary to call attention to it as a general principle about biological mechanisms.—C. R. Gallistel (2000:1179)

Specialization Yields Efficiency

Tools crafted with some goal in mind are fashioned to have a particular shape that will efficiently allow the goal to be accomplished efficiently. A hammer has a large, heavy part at one end and a long, graspable handle—a shape that allows the force of arm muscles to be focused on a small area, such as the head of a nail, which makes hammers efficient at driving nails into wood. The shape that a tool must have to turn a screw is quite different, embodying quite different principles, which is why hammer-shaped objects are not good at driving screws.

Because different shapes are required to solve different problems efficiently, as objects take on one form, they necessarily become worse at solving a host of other problems. This is why a carpenter owns a large number of shapes. Using an effective tool for a range of jobs requires having a large number of options from which to choose so that an efficient tool can be found for the job at hand.

Having a collection of hammer-shaped objects, from ball peen to sledge, allows many different jobs to be completed efficiently because of the advantages that accrue to using shapes specialized for particular tasks.

This is not to say that greater specialization is always better than less. Having multiple tools carries costs, which trades off against the efficiency gains of specialization. At some point, having an additional hammer to use for similar nails carries little marginal efficiency gain, and one is better off using the same hammer for slightly different nails.

In many cases, natural systems show trade-offs in the direction of narrow specialization, proliferating functional units. White adipose cells' structure—round, expandable, with a small amount of cytoplasm and a vacuole containing fat—reflects its energy storage function. Per unit weight, twice as much energy is stored in fat compared to protein or carbohydrate. Similarly, the structure of neurons—small cell bodies with lengthy axons and dendrites capable of conducting electrical impulses which can influence other cells—reflects their function of transmitting energy. One could imagine that the same cell might be able to both store energy and transmit energy, but in order for a single cell's structure to accommodate both functions, there would be substantial compromises regarding at least one, if not both, functions.

Specialization yields efficiency in economic production, as Adam Smith (1776/2005) pointed out in *The Wealth of Nations*. Workers and firms, by focusing on one area or task, become better in those areas, gaining comparative advantage over competing workers or firms.

Computer scientists take advantage of the efficiency gains that specialization produces. The notion of "shape" in the context of tools has an analog in information processing. Different computational tasks can be solved efficiently with different computations. That is, the computations that a subroutine must perform depend on the desired input-output relationships in a manner that mirrors the way that different tasks require different shapes. Different desired relationships require different computations.

This idea, referred to as *separation of concerns*—broadly, the idea that one should produce specialized subroutines to accomplish different information-processing goals—is seen as "a key guiding principle of software engineering" (Ossher and Tarr 2001:43). Subroutines that perform a narrow, well-specified task can be built to be efficient at performing just that task.

Nonhuman animal minds are collections of specialized information-processing devices. Specialization is evident in the very different computational capacities required to meet the adaptive challenges that animals face in their particular ecological niche. Web-spinning mechanisms require different computations from those required by navigational systems for migration or memory systems for storing and recalling the location of cached seeds. In the same way that the nonhuman animal (and plant) world consists of species with different specialized physical parts to solve their specific problems, nonhumans similarly have specialized computational mechanisms to guide their behavior (Alcock 2001).

Certain aspects of the human nervous system show exquisite specialization. Photoreceptors in the retina have the narrow function of transforming electromagnetic radiation of particular wavelengths into information to be used to recover information about the external world. These cells are divided more finely still, with the structure of the cells reflecting their slightly different functions (responding to different parts of the spectrum, high vs. low light levels, etc.).

Further, large collections of cells in the central nervous system are organized around particular functions. The visual system efficiently builds a representation of the external world. Specialized memory systems store semantic information or episodic information (Sherry and Schacter 1987). The observed specialization of these systems reflects the ideas above, the gains from specialization and functional incompatibilities, such that the computational desiderata of one function are inconsistent with the desiderata of another, leading to multiple systems.

One possibility is that the efficiency gains for problem solving are reflected in cells, organ systems, economies, computer programs, animal minds, and aspects of evolved human cognition ranging from perception to memory systems, but that this principle of specialization systems ends there, and does not apply to "central systems," the computational procedures responsible for human decision making and social behavior (Fodor 1983). Alternatively, it could be that these systems, the mechanisms that underpin human decision making, are also functionally specialized, or, to use the more common term in cognitive science, "modular" (Fodor 1983; Barrett and Kurzban 2006).

The Problem of Deployment

Drawing on the above logic, some have argued that the advantages to specialization gave rise to a collection of evolved decision-making systems tailored to the array of adaptive problems, including social problems, faced by our ancestors (Tooby and Cosmides 1992). Exhaustive accounts of this idea are available elsewhere (Pinker 1997; Sperber 2005).

It is important to note that the semantics of "modularity" has been richly debated (Barrett and Kurzban 2006). Many modern modularity theorists follow Pinker, who suggested that modules "are defined by the special things they do with the information available to them" (Pinker 1997:31). That is, modules are not defined by Fodor's (1983) list of criteria; instead, modularity is a property of an information-processing device, and the device's specific properties—including the special things it does with the information it receives—will depend on the function of the mechanism in question. This idea grounds empirical work because different putative functions will require different design desiderata, including the specialized computations the mechanism performs, the inputs it takes to execute its function, whether its processes are "interruptible" by other processes, and so forth. Further, this view clarifies the limited utility of asking whether a given system is modular or nonmodular, instead shifting the debate to asking questions about the correct description of its proper domain (Sperber 1994), the details of the computations it performs, the inputs it takes, and so on.

Systems that consist of large numbers of modular subsystems gain the advantages that specialization of the subsystems afford, but give rise to another problem, that of deployment. With a large collection of computational devices, the use of which will generate very different input-output relationships, selecting the appropriate device is a problem of obvious importance (Gigerenzer 2008; Marewski and Schooler 2011).

In human-made devices with a large number of specialized functional devices, such as a smart phone loaded with applications (Kurzban 2010), the choice of application is done by the user. Natural computational devices such as the human mind must make this decision by themselves. One way this problem is solved is by virtue of the input structures. For instance, in the context of the human sensory apparatus, eyes and ears are struck by both electromagnetic radiation and by changes in air pressure but they act on only the appropriate form of energy because of the design of photoreceptors and hair cells (Barrett 2005). In the context of other tasks, this option is not available; so, to determine which systems to deploy, the mind must use information in the sensory array to determine which adaptive challenge or opportunity is currently at issue, and recruit the mechanisms designed around this problem.

One area in which the mechanism-selection process has been discussed at some length is emotions. Tooby and Cosmides (2008) have argued that the emotions can be thought of as one way of solving the problem of selecting among candidate computational mechanisms.

This view suggests that there are particular cues in the environment that will reliably activate certain suites of systems, deactivating others. The selection of the cue/system mapping is possible because of the reliable value of deploying particular systems contingent on the presence of particular environmental cues. (No specific commitment is made about the ontogenesis of these cues or the level of abstraction of the cues. The presence of a sexual rival might reliably give rise to jealousy, but which people count as rivals must be acquired ontogenetically.)

Emotions can be thought of, then, as being activated by a specific set of cues in the environment, which activate some (and deactivate other) computational systems whose function is to solve the adaptive problem associated with the cues. The activation of these systems leads to the characteristic phenomenology associated with emotions, as well as the physiological correlates that prepare the organism for appropriate action.

To illustrate, consider fear. This emotion might be activated in a situation in which there is immediate threat of attack, such as when one is walking in an unfamiliar environment with predators or enemies possibly about. This situation activates a suite of mechanisms, including those relating to vigilance to increase the chance of detecting threats, while simultaneously suppressing less urgent priorities, such as the need to sleep or eat. At the same time, there is the phenomenological experience of fear, as well as physiological changes (e.g., increases in heart rate, production of adrenaline, etc.) that prepare for the action that might be required in response to the current priority, the possibility of physical attack.

If, as in the case of emotions, decision-making systems for different domains exist side by side in the human mind—including systems for choosing and attracting mates, finding food and foraging efficiently for it, building a network of allies, and so on—then several empirical issues immediately come to the fore: What are the systems, and what computations do they perform? How is the selection among candidate systems made?

Empirical Investigations

If there are specialized, distinct computational devices with evolved functions, then it should be possible to specify the functions of these devices. In turn, these functions should make commitments in terms of their computational properties, including the circumstances that recruit their operation, the inputs that these systems take, the specific computations that the systems perform, and the outputs of these systems, which are potentially visible through decision making. There is substantial research driven by these sets of ideas. Here, I will discuss a very small number of them to give examples of the overall approach.

Detecting Cheaters

Probably the best known example of work along these lines is Cosmides and collaborators' work on the putative *cheater-detection module* (for a review, see Cosmides and Tooby 2008). As Cosmides et al. (2010:9007–9008) recently put it, modular systems "succeed by deploying procedures that produce adaptive inferences in a specific domain, even if these operations are invalid, useless, or harmful if activated outside that domain...by exploiting regularities—content-specific relationships—that hold true within the problem domain, but not outside of it."

It follows that for the putative computational mechanism they posit—the cheater-detection algorithm—to implement its function, the cognitive system must be able to delineate the conditions under which the algorithm is to be deployed. Their "social contract theory" points to the cues to such situations, namely cases in which there is an allocation of costs and benefits, along with costs that must be paid or obligations that must be met. (For a more detailed specification of the computational model they have in mind, see Cosmides and

Tooby 1992.) The cheater-detection algorithm is recruited when an individual could have intentionally taken a benefit without having paid the cost or met the requirement.

Cosmides et al. (2010) used a standard method in this area, the Wason Selection Task (Wason 1983), in which subjects are given a conditional rule (if *P* then *Q*), and asked to see if the rule has been violated. This is typically done with cards, or pictures of cards, with writing on both sides. So, if the rule is *if you drink beer, then you are over 21*, then a card with *P*: *drinking beer* on one side and ~*Q*: *17 years old* on the other would be a violation of the rule.

Cosmides et al. (2010) varied the content of the narrative surrounding the task, adding or removing elements which, by hypothesis, would be differentially likely to evoke the cheater-detection algorithm, but which other models (e.g., permission schema) predicted would have no effect. Because of the adaptive logic surrounding social exchange theory, this model commits to the view that particular contents will affect performance, allowing the hypothesis to be put at risk by varying the relevant contents.

Specifically, holding the rest of the content of a conditional rule constant, they varied whether the person in the conditional rule was taking a benefit, paying a cost, or neither. Under social contract theory, the presence of someone taking a benefit should recruit the cheater-detection algorithm. As predicted, when the content of the rule involved someone taking a benefit, performance was very good (82%), significantly better than a nearly identical rule, in which all was constant except the benefit was transformed to a cost or to something neutral. Similar manipulations, which varied other aspects of the rule contents relevant to recruiting the cheater-detection algorithm, showed similar results, suggesting that the algorithm is differentially likely to be activated depending on the details of the contents of the conditional rule.

Predicting Events

While cheater detection is among the best known applications of the ideas surrounding modularity to decision making, it is not the only one. Recently, Wilke and Barrett (2009) suggested a novel explanation for the *hot-hand phenomenon* that draws in large part on the "ecological rationality" approach (Todd et al. 2011b) as well as the sorts of ideas sketched here. The hot-hand phenomenon refers to the observation that people's predictions about events suggest that they expect these events to come in "streaks," with the probability of an event occurring being higher after one has just occurred. The classic finding in this literature is in the domain of basketball, where Gilovich et al. (1985) found that observers thought a player was more likely to make a shot after having just made one.

While this is often referred to as a "fallacy"—and indeed, people do make systematic errors—Wilke and Barrett (2009) argue that this phenomenon is the manifestation of a well-designed modular system built around foraging.

Specifically, because items in the world for which people search (e.g., prey items, sources of water, people) are generally clumped spatially and temporally, a well-designed prediction mechanism should take this into account. This produces an implicit assumption that the world is autocorrelated.

Most centrally to the point here, they delineated the context in which the hot-hand phenomenon should be observed, specifically in cases of sequential search (in space or in time) in which there is a binary result of each search, a "hit" or a "miss." (This does not characterize all searches; sampling from a fixed pool, for instance, they would predict, should not elicit this pattern.) As an ancillary prediction, they suggested a content effect, such that the phenomenon should be evoked to a greater degree for some types of objects—naturally occurring resources—than for others (evolutionarily novel ones).

Investigating these ideas in both American undergraduates and Shuar hunter-horticulturalists in Ecuador, Wilke and Barrett (2009) presented subjects a large number of observations in sequence of the presence or absence of an array of items (e.g., fruits, nests, bus stops). The task in each case was for the subject to predict whether or not the next event in the sequence would be a hit or a miss.

As predicted, the hot-hand phenomenon was observed in both cultures, suggesting that the implicit assumption that items for which one is sequentially searching in the world is autocorrelated is something of a default. Further, the Shuar showed a greater effect for coins than the American undergrads, raising the possibility that the default assumption can be "unlearned" over time with cultural experience with certain types of objects. Together, the pattern of data suggest that the implicit assumption that the world is clumpy is a component of an aspect of the foraging system, applied to new contents during certain types of sequential search (see also Scheibehenne et al. 2011).

Choosing Levels of Altruism

Research in behavioral economics has shown systematic deviations from predictions of standard economic models (Camerer 2003). Nonetheless, in some settings, people's decisions conform to the predictions from these models with surprising precision. More specifically, in some cases, people look as though they are motivated purely by self-interest, maximizing their earnings from the experimental context, whereas in others people seem to be generous, delivering benefits to others, showing so-called social preferences. What explains these patterns?

The modular view here suggests that one source of difference derives from the bringing to bear of different computations depending on the cues of the decision environment. According to this view, properties of the task recruit different algorithms, which in turn guide choice.

Consider the canonical double auction, in which subjects are given the role of either buyers or sellers for some abstract commodity. Values for items are assigned to buyers and sellers and trades are allowed; subjects earn money by selling above their value or buying below their value.

In these environments, the behavior of subjects as a group is well predicted by standard models, and subjects look, more or less, as though they are perfectly selfish, or money maximizing. One possibility for explaining this is Smith's (1998) notion of "impersonal exchange." There are no cues in these settings that one is part of a group; rather, the framing is competitive, with people assuming roles as in stock exchanges, where gains are made at the expense of other agents.

Compare this to public good games, in which subjects must make decisions about allocating resources (selfishly) to themselves or (generously) to an account that will grow the pie that members of their group share. In these settings, people contribute to the group—though rates of cooperation declines over rounds of play—suggesting social preferences of some sort.

Although a number of proposals have been made to explain the results of the large number of public goods experiments (Camerer 2003), one possibility is that the structure and terminology of the game activates mechanisms designed around reciprocity (Nowak and Sigmund 2005), alliance building (DeScioli and Kurzban 2009), or other social systems. Indeed, manipulations of "identity," for instance, lead to increased contributions (Brewer and Kramer 1986).

More generally, there is evidence that players in these games search for and use a framing with which they are familiar. Henrich et al. (2001) conducted a series of studies among 15 small-scale societies, using a number of tasks from behavioral economics, including the public goods game. Members of one society, the Orma, dubbed the public goods game the *harambee* game, mapping it onto a local means of producing things such as roads and schools (Gintis et al. 2003). Gintis et al. (2003) argue that many local social norms are reflected in game play across cultures, suggesting that subjects in these studies search for mappings between the game and some aspect of social life.

The self-interested play in auction games set against the prosociality in public goods games suggest that people can, and often do, compute where their economic self-interest lies and pursue it; but they certainly do not always do so, choosing instead to deliver benefits to others. This is consistent with the idea that the game framing recruits different modular structures, though the details of the mapping might depend on the details of local institutional norm clustering.

Summary

These examples are obviously not intended to be exhaustive, but rather a very small sampling of the types of areas to which a modular perspective can be applied.

Indeed, while research informed by these ideas is continuing, a particularly interesting line of work places subjects into relatively unstructured decision environments. Consider the recent work of DeScioli and Wilson (2011), who placed people into a foraging environment in which subjects' avatars could move around a virtual world, exploiting patchy resources. Given considerable freedom of action, subjects spontaneously developed property right norms in some ecologies, in particular, those with patchy rather than uniform resources. This suggests that properties of the ecology are recruiting different systems, but in similar fashion across subjects in the experimental setting. Such work holds considerable promise for illuminating how different psychological systems are recruited as a function of parameters of the decision environment.

Controversies

The modular approach sketched here is not free from controversy, though the details of the controversy depend on what one takes "modularity" to mean. In contrast to Fodor's (1983) initial formulation, modern views take modularity to turn on functional specialization (see above; Barrett and Kurzban 2006; Pinker 1997). Consequently, a system that is modular in the sense of functionally specialized can simultaneously fail to fulfill, as an empirical matter, criteria that Fodor (1983) assigned to modules (e.g., localized, shallow outputs).

Here I present two of the most prominent issues raised in the context of this more recent construal of modularity. This is not intended as an exhaustive treatment of critiques of modularity, but should provide a sense of the larger issues and entry points into this literature.

Information Integration

Chiappe (2000) and, more recently, Chiappe and Gardner (2011) have suggested that the human ability to integrate information across different domains, along with the human capacity to reason analogically and use metaphors, undermine the view that the mind's architecture is modular.

Defenders of the modularity thesis, however, would reply that these abilities, while they support the notion that particular systems can take informational inputs from other systems, do not carry as a logical entailment that the systems doing this integration do not have a (narrow, or, at least, specifiable) function. That is, to show that a given input influences a particular process demonstrates that the process in question is not encapsulated with respect to that particular computational input, not that the system does not have an evolved function (Barrett and Kurzban 2006, 2012).

To return to the reasoning task discussed above, consider Gigerenzer and Hug's (1992) demonstration that performance depends on whether the subject is given the perspective of an employer or employee. For precisely the same problem content, one's role interacts such that those primed with the employer perspective identify cases in which someone received an underserved pension as violations of the rule; from the employer's perspective, such individuals stand out because they got a benefit without having met the requirement, "cheating" from the standpoint of the employer. Symmetrically, those primed with the employee perspective identified cases in which a deserved pension was not received as violations of the rule; from the employee's perspective, such individuals were denied a benefit even though they met the requirement, being "cheated" from the standpoint of the employee. The content of the problem was integrated with one's knowledge about one's role to produce the solution; however, in both cases, a solution predicted by the hypothesized function detected instances of cheating.

Information integration, then, is relevant to arbitrating claims that a given mechanism is (or is not) encapsulated from another, which in turn can be relevant to distinguishing among hypotheses regarding function.

Novelty and Flexibility

A second frequent critique of modularity revolves around novelty. Chiappe and Gardner (2011:2) write that modularity "has limited usefulness in explaining the existence of mechanisms designed to deal with novel challenges and with the development of novel solutions to longstanding adaptive problems."

There is, however, a sense in which modules only address novel challenges. Any particular given object is one that was never seen during evolutionary history, yet our visual systems build representations of them. This is because modules are designed to process inputs in a systematic, useful way, even if the inputs are novel tokens of the types over which they operate (Barrett and Kurzban 2006).

Further, an advantage of modular systems is that they potentially allow vast new combinations. The adaptive immune system, for instance, is modular, allowing surface immunoglobulins to take a vast array of new shapes, which in turn contributes to the system's function. Modularity allows novelty through combinations, just as in generative natural language grammar.

Broadly, novelty is a problem for any evolved system, since the system at any given moment is the result of the interaction between the organism's developmental environment and the genes that have been brought through the process of selection to that point, tested against previous, historical adaptive challenges. Successful architectures are ones that led to adaptive behavior given tokens never previously encountered. Modular architectures, such as the adaptive immune system, allow the combination of many different types of parts, allowing for flexibility and dealing with novelty (on this point, see Sperber 2005).

Two final points surrounding novelty are worth brief mention. First, while humans are surpassingly good at navigating certain types of novel contexts, as cultural adaptation illustrates, some sorts of novelty (e.g., the availability of inexpensive high energy foods) has presented significant challenges (Burnham and Phelan 2000). Second, some human evolved systems seem specifically designed around novel tokens, as in the case of social learning systems, which, some have suggested, are designed to use features of other individuals to pick out which (novel) ideas to attend to and differentially acquire (Henrich and Gil-White 2001).

Summary and Conclusion

Specialization is common across the biological world. In the domain of human cognition, the idea that the computational mechanism of the mind might be specialized, including those mechanisms designed to solve adaptive problems associated with navigating the social world, has fallen under the rubric of "modularity." While some empirical evidence has been gathered to distinguish between claims that the systems which underlie behavior are more or less specialized, considerable debate remains to sort out the details.

First column (top to bottom): Jeff Stevens, Rob Kurzban, Henry Brighton, Rob Kurzban, Bill Wimsatt, Bernhard Voelkl
Second column: John McNamara, Group discussion, Yaakov Kareev, Reinhard Selten, Yaakov Kareev, Kevin Gluck, Jeff Stevens
Third column: Kevin Gluck, Jens Krause, Reinhard Selten, Jens Krause, Bill Wimsatt, Henry Brighton, John McNamara

12

Robustness in a Variable Environment

Kevin A. Gluck, John M. McNamara, Henry Brighton, Peter Dayan, Yaakov Kareev, Jens Krause, Robert Kurzban, Reinhard Selten, Jeffrey R. Stevens, Bernhard Voelkl, and William C. Wimsatt

Abstract

Robustness is a prominent concept in technical sciences and engineering. It has also been recognized as an important principle in evolutionary biology. In this chapter, it is proposed that the term "robustness" be used to characterize the extent to which a natural or artificial system can maintain its function when facing perturbation and that this concept is relevant in considerations of Darwinian decision theory. Situations in which the action of natural selection is liable to lead to the evolution of robust behavioral strategies are highlighted along with some psychological mechanisms that might lead to robust decision-making processes. Robustness describes a property of a system varying on a continuous scale rather than existing as a dichotomous feature. Degree of robustness depends on the details of the interaction of system characteristics and environmental contingencies, as well as the specific types and extents of perturbations to which the system may be subjected. A system can be robust in one domain while remaining highly vulnerable to perturbations in others. As defined here, robustness is related to, yet distinct from, flexibility and optimality. The sorts of environmental variation, and hence perturbations, that an organism or technology is liable to face are described, as is the cost-benefit trade-off of robustness. Finally, the robustness of decision making at the level of social groups is considered.

Introduction

Soldiers fighting in the trenches during World War I reportedly developed a "live and let live" mentality in which mutual truces spontaneously arose, but a violation of the truce resulted in quick retaliation. Axelrod (1984) proposed this as an example of tit-for-tat, a strategy that can maintain cooperation by copying a partner's previous action. In the trench warfare example, ceasefires led to ceasefires and unprovoked attacks led to retribution. Though tit-for-tat can generate cooperation, a simple mistake can spawn a spiral of escalating aggression. An itchy trigger finger can start a cascade of violent attacks in the trenches. A forgotten or misplaced thank you note can explode into a dramatic family feud. As a decision strategy, tit-for-tat may work well in a perfect world with no mistakes or accidents, yet it is not *robust* to error in the environment (Selten and Hammerstein 1984).

The topic of robustness is integral to the evolution of biological organisms (Hammerstein et al. 2006; Kitano 2004; Wagner 2005). The same is true for the evolution of nonbiological systems, such as scientific communities and technologies. Because decision makers' environments are and have been subject to perturbation, robustness is important to consider in the context of a general theory of decision making, and especially so in a decision theory built on the principles of Darwinian evolution.

Examples of robust solutions to problems posed by perturbations include, but are not limited to, the existence of multiple (often redundant) mechanisms, tolerance buffering, and systemic stochasticity. The competition and cooperation among these and other mechanisms in complex and uncertain environments give rise to the observed complexity of decision-making behavior (Simon 1996).

Our main argument is that robustness has been a key driver for the evolution of different decision-making and control mechanisms. Choice reflects the competition and cooperation of these mechanisms and therefore does not fit nicely into categories which arise from a simple set of axioms. Much hinges on the necessity of learning, which is a central battle between flexibility and speed of response.

Understanding the evolutionary origins of decision mechanisms requires us to address the nature of robustness, the nature of environments, and the costs and potential benefits associated with robustness in these environments. Acknowledging that robustness can result from social interactions, we also consider robustness in group decision making, especially via emerging results from the literature on swarm intelligence.

Robustness

The concept of *robustness* is important to many fields of science. In structural engineering, structures must meet a fundamental requirement of robustness. For example, as described in structural design codes EN 1990 and EN 1991-1-7, "a structure shall be designed and executed in such a way that it will not be damaged by events such as explosion, impact, and/or the consequences of human errors, to an extent disproportionate to the original cause." In biology, robustness addresses the probability of survival and reproduction for members of a given species. From economies to ecologies, from *Homo sapiens* to

Drosophila, from the physical landscape in which animals forage to the conceptual landscape in which scientists forage, and across all spatial and temporal levels of analysis, robustness matters. The broad relevance and complexity of the concept of robustness prompt us, however, to ask: Is there a level of description of robustness that can aid our understanding of decision making?

Defining Robustness

To begin, we must clarify our perspective by defining robustness and some of its features. Robustness can be conceptualized as a function between some dimension of the environment and performance. For example, for a given automobile engine, there is a relationship between temperature and performance such that performance deteriorates at low and high temperatures. Different engines will exhibit different curves illustrating this relationship. Such curves can be summarized in various ways (e.g., mean, standard deviation, area under the curve) to characterize how "robust" the engine is *with respect to* temperature. Greater performance across the environmental dimension equates to greater robustness. A selling point of the original air-cooled Volkswagen engine, for example, was that it would actually start (at least more reliably) at colder temperatures than its water-cooled competitors. Within the Volkswagen engine, a key performance characteristic (startability) was more robust to temperature variation than the other engines on the market at the time.

In the context of biological systems, different phenotypes can be understood to have different performance robustness curves—how well the system in question executes its evolved function—under a range of different conditions. Over evolutionary time, mechanisms will be selected whose properties maximize reproductive success. The specific mechanisms selected by Darwinian evolutionary processes depend strongly on the frequency and magnitude with which different environmental contingencies are encountered and on the implications of the interaction of biological mechanism and environmental contingency for reproductive success.

In the context of computations, one can construe performance as a relationship: the production of a particular output (e.g., representation, behavior) given a particular state of the world. The same analysis applies. Different mechanisms will maintain performance—this adaptive, systematic input-output relationship—to a greater or lesser extent, and we can try to quantify how "robust" each mechanism is: How well does the mechanism maintain its input-output relationship as a function of some environmental parameter? We define robustness as:

> *the extent to which a system is able to maintain its function when some aspect of the system is subject to perturbation.*

Robustness is not a binary category but rather a continuously varying *n*-dimensional state space. This conceptualization of robustness is consistent with

Levins's (1966) emphasis on *invariance*, Campbell's (1958, 1966) emphasis on *multiple determination*, and Wimsatt's (1981) methodological prescription for *robustness analyses*. High dimensionality makes quantifying robustness in a general way more challenging and less tractable but not necessarily impossible, provided one can formally specify the dimensions and possible perturbations. Systems, then, will exhibit a "degree of robustness" as opposed to being dichotomously robust or nonrobust.

We are using the term *system* here in the most general sense, and an aspect of that system is any variable, component, characteristic, or policy that could affect its function. We adopt a similarly broad interpretation of *function*: it encompasses a system's purpose, process, or level of performance or outcome. A *perturbation* is any pressure or stressor or environmental change that could in principle affect the system's function.

Flexibility, Optimality, and Robustness

Robustness is associated with *flexibility* and *optimality*. Though related, these concepts are not identical to robustness. Confusion surrounding the relationships between and distinctions among these terms seems to originate in the fact that flexibility, optimality, and robustness are all potential capacities or characteristics of systems. The difference is that robustness depends on, as defined earlier, *the maintenance of function when some aspect of the system is subject to perturbation*. Thus, robustness is only identifiable within the context of environmental variation and perturbation, and is only present when the system's function is preserved. In other words, in a static environment it may be possible to demonstrate that a particular system is flexible or even optimal (when the environment is both static and certain) within whatever fixed dimensions define the operating environment, but this tells us nothing about the system's robustness. Robustness is only demonstrable when the function of the system is assessed across perturbations in environmental contingencies.

Figure 12.1 provides an abstract illustration of some of the key issues for robustness and allows us to distinguish it from optimality (defined according to some performance criterion) and flexibility (in terms of adapting to prevailing conditions). The figure is intentionally abstract to make general points about systems and environments and the influence of interactions of their characteristics on functional performance.

For expository purposes, let us make consideration of the abstractions in Figure 12.1 more concrete by assuming that the figure represents the performance of a hypothetical agent foraging in an environment, which can potentially exist in different states, modeled as a scalar value $e \in [0, 10]$. This could represent some characteristic of the environment, such as the rate at which prey arrive. Similarly, the agent's decision policy is parameterized by a scalar parameter $p \in [0, 10]$ (say a measure of its preference for exploitation rather than exploration). Figure 12.1a shows the performance of the agent as a function of

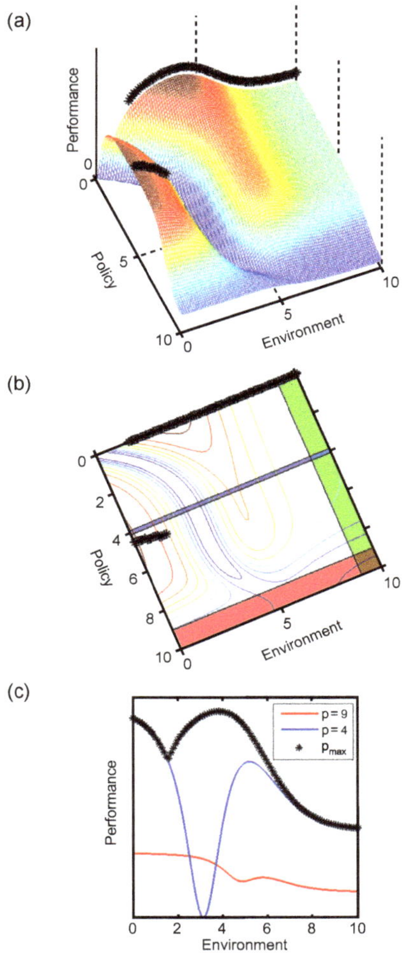

Figure 12.1 Issues surrounding robustness: (a) and (b) illustrate a hypothetical case in which the performance of a system depends on a single scalar quantity representing the external environment (designated *e* in the text) and a single scalar policy (designated *p* in the text) representing an aspect of the system. A 3D plot of the performance surface that results from interactions of the policy-environment interactions is shown in (a); a contour plot of the same surface is provided in (b). The black points show the optimum value of the surface for each value of the environment for the parameter range shown. The colored rectangles on (b) refer to ranges of values of the environment or the policy whose performance characteristics we highlight in the text. In (c), performance is shown as a function of the value of the environment (*e*) for various policy values in (b), as described in the text.

e and *p*, which, for the moment, we will consider to be measured over a short time period. Figure 12.1b shows the same surface, but now as a contour plot.

Let us further assume that in this context the function of the agent is to maximize its foraging performance. We can now consider how the performance of the system varies over the interaction of environment and agent characteristics:

1. If the environment only occupies a limited range of values of e, say $e \cong 2$, then an agent with a fixed policy (here with $p \cong 4$) will optimize performance.
2. However, if the environment changes a little, so that $e \cong 3$, now performance with $e \cong 4$ is catastrophically bad. Thus, we might say that performance with a fixed parameter p has low robustness to environmental perturbation. The blue curve in Figure 12.1c shows the performance as a function of e for $p = 4$; the catastrophe at $e \cong 3$ is apparent.
3. If, on the other hand, we choose $p = 9$ (the red curve in Figure 12.1c), then performance is more stable as a function of e. Maintenance of performance, in terms of minimizing performance variability, is one popular measure of robustness; note, however, that this comes at the cost of rather poor performance in environments with low values of e compared with what could possibly be achieved for those environments. Indeed, a credible strategy for explaining apparent sub-optimalities in the behavior of an agent in some environment is to declare that this provides robustness in other possible, but not current, environments.
4. Conversely, if the environment happens to take the value $e \cong 10$ (the green rectangle in Figure 12.1b), then performance is very insensitive to the agent's policy. This is another form of robustness; that is, if the policy is subject to internal perturbation arising from damage or decay.
5. A further possibility is that the agent might be able to measure the state of the environment and adapt its parameter p accordingly. The black asterisks in all of the subplots in Figure 12.1 show what happens if the agent can choose p to optimize performance as a function of e. Performance is attractively good. This shows how flexibility can aid robustness. However, this flexibility comes with three potential costs: two statistical and one computational. The first statistical cost stems from the need to measure the value of e. This requires samples, which could be expensive to acquire (if the values of p are inappropriate during this period). These costs can be mitigated by knowledge of the likely environments e, which will help constrain the estimation problem. The second statistical cost is that of estimation error/uncertainty. Even if estimation is performed as well as possible, there is a chance of unlucky samples that could skew the value of e that is inferred, giving rise to poor performance. The computational cost comes from the extra machinery necessary to make the agent exhibit this flexibility.
6. For the case of the blue curve, even a small change in e suffices to ruin performance if that small change results in $e \cong 3$. Another popular notion of robustness concerns outliers (see Dayan, this volume). What

happens if the environment can sometimes take the value $e \cong 20$, which is outside any previous experience? Human scientists and engineers can use theories about the world to extrapolate surfaces, such as the one in Figure 12.1, beyond any extant observation, and so protect against imaginable but not experienced environments. Natural selection does not have this luxury. Indeed, as we discuss below, natural selection is condemned to respond to the history of recent environments, and thus can exhibit a form of maladaptive and non-robust overfitting.

7. Finally, we face the necessary task of integrating the net performance curves, such as those in Figure 12.1c, in order to assess the total performance of a particular policy. For instance, it might be catastrophic for performance ever to go to 0, or maybe only the average performance over the distribution of possible environments matters. Unfortunately, this step is completely dependent on the range and distribution over possible environments, and also the problem. Even for the case of natural selection, it will be different if we measure the expected number of offspring at different times in the future, or the probability of not having been eliminated. Thus, whether we consider the red curve in Figure 12.1c to be better or worse than the blue curve is formally unanswerable in any absolute sense. The answer is contingent on the specific circumstances.

In summary, there are clear circumstances under which we can distinguish robustness from at least a local notion of optimality and from flexibility. It is less possible to distinguish robustness from a more holistic notion of optimality, since the latter can always be structured so as to include robustness.

It is sometimes supposed that there are trade-offs between the plasticity or flexibility of behavior and its robustness. In such circumstances one must first ask whether the correct parameter is being evaluated. Thus flexibility of behavior may contribute to the robustness of a fitness outcome. With the robustness of a fitness outcome, would there then be no pressure for evolutionary change, restoring the opposition between robustness and flexibility?

At other times, robustness is pitted against optimality in discussions of decision making (e.g., Rosenhead et al. 1972). Arguments are usually about whether the optimal solution for a model is robust in a wider modeling setting or in the real world. Of course, in the real world, it is not possible to identify with certainty what is optimal (or what is robust) because it is not possible to specify fully the real world in modeling terms. Nevertheless, assuming that what does evolve is adaptive, this argument can be loosely reformulated as: Is it adaptive to be robust, or is it optimal to be robust?

Here we focus on the implications of robustness across evolutionary environments for the case of decision making. In some cases (such as inbuilt fight, flight, or freezing responses), the agent's behavior appears to be fixed and genetically determined, perhaps as a response to the dangerous costs of learning. In other cases agents achieve good performance across a range of environments (which we consider a form of robustness) by learning (Dayan, this volume). However, there are multiple mechanisms for learning, which, again, may putatively be a response to the demands of performance across variable environments.

Variable Environments

If everything about the environment in which a system exists is static, then robustness is irrelevant. In the real world, of course, environmental variability is ubiquitous. It is not a question of *whether* there is environmental variability, but rather *what, when, where*, and *how much* variability exists. In our definition of robustness, we refer to these environmental variations generically as perturbations.

The notion of robustness has broad application because operating conditions are rarely stable. Assessing a system's robustness requires specifying a range of environmental perturbations over which the system will to a greater or lesser extent maintain its function. Yet what forms of variation do organisms commonly encounter? In particular, by categorizing common forms of variation, is it possible to shed light on robust design patterns? Before considering some key dimensions of environmental variation, it is worth stressing that the list of potential sources of variation is unbounded. Furthermore, the relevance of a particular form of variation will depend both on the organism or technology in question and the level at which we seek to understand it. A cognitive scientist, for example, will likely find irrelevant the fact that all biological systems operating at temperatures above 0 Kelvin face the perturbations arising from thermal noise, even though such perturbations place constraints on functional design (Wagner 2005). For the purposes of this discussion, we wish to describe a generally useful categorization of dimensions of environmental variability relevant to robustness:

1. *Variation within and between environments*: Variation occurs both within and between environments. For instance, within a specific environment, the relevant operating conditions faced by an organism could be relatively stable, such as the temperature and light conditions of deep-sea dwelling creatures. However, the range of potential environments that organisms of the same species may face could be highly variable and uncertain. Consequently, the level at which we examine variation (e.g., the population vs. the individual) will determine which forms of variation are relevant to robustness (Wagner 2005). Indeed, the following forms of variation have the potential to occur both within and between environments.
2. *Internal versus external variation*: Variation can refer to, say, changes in the state of an organism's endocrine system or its energy levels.

Should such factors be considered part of the environment or part of the organism? Intuitively, the body defines a boundary separating the internal and external environment, both of which present sources of variation which could pose a threat to functioning. Faced with the problem of regulating its own temperature, for instance, an organism will be subject to both internal and external determinants of temperature change. Robust responses to these factors may require different designs, due to internal and external variation occurring over different timescales (Wagner 2005). More generally, fatigue, aging, and a developing immune system are all examples of variation in the internal environment which place constraints on functional design.

3. *Degree of predictability*: Uncertainty exists when an action, such as a movement, a decision, or an interpretation, has more than one potential outcome. Variability in these outcomes can be more or less predictable; that is, cues available to an organism vary in how well they correlate with the true state of the world. For example, distinguishing predator from prey, moving from one food patch to another, or interpreting a visual scene all require inferring latent properties of the environment from environment cues. In each case, the available cues provide more or less predictive indicators of events. At one extreme, all mechanisms are equally robust in a completely random, maximally uncertain environment. The presence of regularities, however, will mean that mechanisms will be robust over certain ranges of variation at the expense of others (Geman et al. 1992). Thus, a key dimension of all forms of variation, and one which strongly influences robust design, is the degree of predictability.
4. *Degree of stationarity*: Some forms of variation, such as the caloric content of alternative food items, will likely remain constant over time. They are *stationary*, in that the statistical properties (e.g., mean, variance) of the caloric content of specific foods remain constant over time. Other forms of variation, such as the where the food is located, will likely change over time. They are *nonstationary* in some combination of time and space. On the one hand, a temporal or spatial dependency can be seen as another cue and simply an additional source of uncertainty which renders events predictable to a greater or lesser extent. On the other hand, both the prevalence and significance of nonstationary properties of environments suggests that they are worthy of study in their own right. For example, the balance struck by an organism when facing the trade-off between exploration and exploitation will likely depend on how, and to what extent, the environment changes over time and place (e.g., McNamara and Houston 1985b). In this way, it is not only environmental variability but also variability (nonstationarity) in environmental variability that is a dimension on which it is relevant to consider robustness.

5. *Social determinants of variation*: Social variation arises from the actions of other agents, rather than, say, nonsocial variation (e.g., the caloric content of food items or seasonal variation). Although the dividing line between social and nonsocial environments can often be unclear, the social environment nevertheless introduces forms of variation which would otherwise be absent (Hertwig et al. 2012), and therefore it is also germane to considerations of robustness. Strategic, game-theoretic settings represent one class of variation in which other individuals' behaviors depend on the agent's own actions. This contingency can result in complex and dynamic variation in the social environment. The human linguistic environment is a striking example, as the generative nature of human language continually leads to the creation of novel utterances and the evolution of interpretation of meaning (Brighton et al. 2005). Just as the consumption of food and water resources by others will influence the hunting and gathering decisions of organisms, the communicative acts of others can influence interpretation, understanding, and decision making.

The existence of multiple types of environment, affording different opportunities, allows for the possibility of modeling the environment. This can be done by setting hyperparameters in parameterized algorithms (i.e., parameters that describe the possible distribution of other lower-level model parameters which may, in turn, describe distributions of resources within the environment) or by employing preprogrammed modules when these have been hard-wired. The trouble with the latter is choosing between modules and organizing learning within the modules. Modules may or may not compete; an example of the latter is when modules process different information about the same underlying quantities. Generic strategies may exist to cope with these opportunities; for instance, apparent over-optimism (exploration bonuses; Kakade and Dayan 2002) in the face of ignorance, followed by over-caution in the case of disaster. Robust exploration methods, such as regret bounds for bandit problems (Auer 2003), are another possibility, with attractive, though sometimes suboptimal, characteristics. Representational learning, such as prolonging development so that the representations of the world on which decisions hang can be adapted to the prevailing statistical properties (see Dayan and Abbott 2001, chapter 10), is a further critical form of robustness.

Evolutionary Selection Pressures

There would be no evolutionary selection pressures to produce robustness unless population members experienced environmental perturbations. Thus we might expect the robustness of current decision mechanisms to be positively correlated with the degree and dimensionality of ancestral perturbations.

However, even when there are perturbations, whether we expect robustness to evolve depends on the particular perturbations and the performance measure considered, as the following examples illustrate.

Case 1: Amount of Food Obtained Each Day during the Winter

Consider a small bird or mammal that is trying to survive the winter. This animal has a very limited ability to store food as fat, and so must get enough food each day or it will die. Suppose that food availability varies from day to day. Then we might measure the performance of a foraging rule that the animal might use as the amount of food found for each level of food availability when using this rule. Some rules may do well on average, some may be especially good when food is plentiful, and some may be very efficient at finding food when food is scarce. Unless a rule has reasonable performance on each day during the winter, the animal will not survive the winter. Thus we would expect natural selection to produce foraging rules that maintain the capacity to find sufficient food each day, despite adverse environmental perturbations (i.e., we expect the evolution of rules that are robust given our performance measure).

Case 2: Number of Surviving Offspring Produced in a Breeding Attempt (Demographic Stochasticity)

Consider a hypothetical female bird that breeds once a year. During breeding she decides whether to lay one or two eggs. After the eggs hatch, the chicks tend to attract predators through their call. Thus, the more eggs the female lays, the greater the probability that a predator will find the nest and kill all of the offspring. Suppose that a predator may be present in the local area or not; each event will occur with a fixed probability. If no predator is present, then each egg laid results in a mature offspring. If a predator is present and the female lays one egg, then the predator finds the nest with a low probability. If she lays two eggs, the predator finds the nest with a higher probability. We assume that whether this female's eggs survive is independent of what happens to other breeding individuals (demographic stochasticity). Here our performance measure is the number of surviving eggs. The environmental conditions are whether a predator is present. Table 12.1 gives a numerical example.

In this setting we expect natural selection to favor the egg-laying strategy that maximizes the expected (average) number of surviving offspring. Suppose that the predator is present with probability 0.5. Then laying one egg will result in $(0.9 + 1)/2 = 0.95$ surviving offspring on average; laying two eggs will result in $(0.2 + 2)/2 = 1.1$ surviving offspring on average. Thus the laying of two eggs will be favored by selection. Note, however, that this strategy is not robust in that few offspring survive when the predator is present.

Table 12.1 Number of surviving offspring per year in nests with one and two eggs depending on whether a predator is present. The probability a predator detects a nest with one egg is 0.1; the probability that it detects a nest with two eggs is 0.9. The probability that a predator is present is 0.5.

	One egg	Two eggs
Predator present	0.90	0.20
Predator absent	1.00	2.00
Average over conditions	0.95	1.10

Case 3: Numbers of Surviving Offspring (Environmental Stochasticity)

Consider an annual plant species. Every year in the spring, a seed will either grow or remain dormant in the ground for another year. The decision of whether to grow or not is made before environmental conditions are known for that year. Conditions during growth can be "Good" or "Bad"; most years are Good. If it is a Good year, a germinating seed grows to a mature plant during the summer, produces many seeds, and then dies. If it is a Bad year, a germinating plant dies. Seeds that remain dormant are not affected by the type of year. A strategy for a plant specifies the proportion of its seeds that germinate each year.

We can take the performance measure to be number of seeds in the ground next winter that are descendants of a seed that is in the ground this winter. Consider first the performance of the strategy of having all seeds germinate. This strategy does very well in Good years, leaving many descendants, but leaves no descendants in Bad years. Since most years are Good and a mature plant produces many seeds, this strategy maximizes the mean number of descendant seeds left next year per current seed. However, any genotype that coded for this strategy would be wiped out in the first Bad year—so this strategy would not evolve. Instead, we expect a bet-hedging strategy to evolve: some seeds germinate, others remain dormant as insurance against Bad years (Cohen 1966). This strategy would be favored over immediate germination because it is more robust against seasonal variation.

General Discussion

In both Cases 1 and 3, there will be selection for robust strategies because of multiplicative effects. In Case 1, the probability of overwinter survival is the product of the probabilities of survival on each day. In Case 3, the number of descendants left far into the future is the product of the numbers left from year to year. In general, whether effects are multiplicative or additive depends both on the performance measure and on the spatial and temporal structure of the environment (McNamara et al. 2011).

At the other extreme, there can be selection to take risks. For example, in elephant seals a few males are able to monopolize most of the breeding females; thus, it may be worth it for a male to take big risks to become dominant.

In situations where the current habitat properties are unknown, there may be selection to be "optimistic" and take a risk that it is a good habitat (McNamara et al. 2011). In these cases there is no selection for robustness. In fact, adaptive strategies are far from robust.

In general more modeling work is needed on the question of whether it is adaptive to be robust. In the construction of models, it is important to broaden the worlds that are considered. It is impossible to build models of the real world. Nevertheless, more complexity is needed than has been used in most previous work. For example, consider the foraging strategy of a small bird that must gain enough energy to survive the night. If we model a world in which the bird "knows" the availability of food, and knows it will not change during the day, we get one prediction about risk-sensitive foraging. If we allow for uncertainty and change, we get another that is more robust (McNamara 1996).

Overfitting

Natural selection shapes the strategies of population members so that they are roughly adapted to the local environmental conditions experienced by ancestors. Mutation tends to work against this process, so we would expect the more recent past to have a greater influence than the more distant past on current population strategies. There is a tendency to "overfit" in that current individuals are adapted to the recent past and may not be adapted to future environmental conditions.

As an example, consider an environment that is stable apart from the occasional El Niño year. (El Niño is a warming of Pacific Ocean surface temperatures that occurs approximately every five years and influences global weather patterns.) We would expect a different distribution of strategies immediately after an El Niño event than after a run of normal years. After an El Niño event the population is more adapted to another such event, although it is likely to be a normal year. Even if the environment is stable over evolutionary time, the experience of an individual over its lifetime is liable to be different in detail to that experienced by any of its ancestors; so there will even be a tendency for short-term overfitting in evolutionarily stable environments.

In the El Niño example, robust mechanisms might be able to deal well with both normal years and El Niño years. But what are the opportunity costs of robust decision mechanisms? In periods of more stable or predictable environments, individuals using a robust mechanism would show a gradual proportional decrease in a population because there is no selection pressure for the robustness. Therefore, the timing of unpredictable events and the difference between the performance of other mechanisms compared to robust mechanisms is critical.

In terms of moment-to-moment decision making, a strategy specifies the rule by which an individual integrates its past experience in deciding how to respond to current stimuli. In a complex world, past experience is liable to be unique in detail. Thus, the decision-making rule is dealing with a situation never experienced by any of its ancestors. How well it copes with this situation depends on the robustness of its rule. Overfitting will tend to produce rules that are not robust. This is analogous to statistical overfitting of data, in which over-parameterized models fit noise rather than the underlying pattern (cf. Goldstein and Gigerenzer 2009). Whether overfitting occurs may depend on the statistical properties of the environment, the learning capacities of the agent or organism, and the classes of rules on which selection acts. To the extent overfitting is likely to result in fragility (non-robustness), more general processes will be adaptive and likely selected, if they are able to evolve before catastrophe strikes.

Psychological Mechanisms and Robustness

Optimality modeling can tell us about broad selection pressures acting on the robustness of behavioral strategies, but to understand the fine detail we need to consider underlying psychological mechanisms (McNamara and Houston 2009).

There are both external and internal factors associated with robustness. For the former, from a computational or functional perspective, it can be useful to consider separating the chance of substantially negative outliers from the opportunities associated with variable, but typically beneficent, environments. Learning ontogenetically in the face of looming catastrophe would be highly maladaptive; therefore we have hard-wired, evolutionarily programmed mechanisms that tell us what to do. These can also be considered to be heuristics, rules of thumb (Hutchinson and Gigerenzer 2005), or modules (Kurzban, this volume). It is not clear how mutable they are in the light of experience (e.g., can we learn to climb some particular tree to flee a bear) or how generic policies (approach/withdrawal) interact with specific ones. We can use notions from control theory as ways of formalizing the effects of outliers (Doyle et al. 1989; Wald 1945). Finding ways to study these modules and getting insights into their historical appropriateness is an important task, lest this critical part of the architecture be just arbitrary. One route might be to compare defense mechanisms across species (Bolles 1970).

Some have distinguished at least two mechanisms of the internal environment that pertain to robustness: model-based and model-free control mechanisms (see Dayan, this volume). These controllers have different abilities to work in the face of computational and statistical uncertainty. Each embodies prior information in different ways (model-based control in a much richer manner than model-free control) and so can adapt to environments differently. A challenge is that the controllers interact richly, making it hard to tease apart their individual contributions.

The Cost-Benefit Trade-offs of Robustness

From an engineering standpoint, building robustness with respect to one dimension generally carries some sort of cost elsewhere for the system. For example, robustness to damage can be gained from adding redundant backup systems that mirror the functionality of primary systems. This approach is commonly used when primary system failure can lead to very costly outcomes, as in the case of hospital power generation systems and expensive and sophisticated machines, such as the space shuttle.

Thus, robustness generally entails a cost-benefit trade-off: the cost carried by the means used to gain robustness trades off against the benefit of the robustness gain. In the case of redundant systems, the benefit is gaining robustness to failure of the primary system, but at the additional cost of the backup system. Similarly, robustness to different dimensions of environmental perturbation trades off against one another. For instance, in designing an airplane, one might choose to include a redundant hydraulic system so that if the primary system is disabled (e.g., by a bullet or by a material fatigue-induced structural failure), the backup system can be relied upon to maintain the airplane's function. This increases survivability, which is a benefit. However, this benefit is achieved at the cost of maneuverability, due to the weight increase that results from the inclusion of the extra equipment.

Similarly, achieving robustness in biological systems generally carries some cost, often in the allocation of energy for the tissue required to increase robustness. The human ability to recognize faces is robust with respect to things such as viewing angle, lighting conditions, and so on. This, presumably, requires expensive nervous tissue to implement this functionality (for a discussion of the costs and benefits of larger brains, see Chittka and Niven 2009). Adding robustness, in this and other senses, can be understood as imposing some cost that must be made up either by sacrificing energy in some other domain or with additional collection of energy.

Costs can come in any number of forms. Consider a system that samples the environment in order to estimate some parameter, such as the mean of the distribution. Sampling from a large number of instances reduces the error (i.e., increasing the robustness of the estimate). However, reducing the error this way increases the time and energy spent sampling (Kareev, this volume).

The trade-offs inherent in achieving robustness help to explain why systems cannot be universally robust across all possible dimensions. Though there might be other reasons (e.g., phylogenetic constraints, local minima), selection will generally favor robustness mechanisms for which marginal robustness benefits outweigh their marginal costs, whatever those costs must be. Because many different robustness trade-offs must be addressed by any given design, selection can be understood to resolve a range of robustness trade-offs.

Note that robustness cannot be universally optimized because selection pressures change and there is uncertainty regarding the future environments that organisms face. However, the same uncertainties that make optimization impossible can and do result in both artificial and natural forms of "tolerance buffering," which serves to increase the robustness of the system to variation. For example, when designing and constructing a bridge, engineers build in a degree of safety margin assuming certain standard and extreme stresses. This is an intentional buffering for robustness that is possible in domains characterized by decision makers that are goal-directed, creative, and purposeful. Natural selection is characterized by fitness to the environment rather than creative intention, but we also see tolerance buffering in naturally selected biological systems. The liver, for instance, has substantially greater capacity than necessary for maintenance as well as the ability to regenerate or replace lost tissue. Robustness through excess capacity (for normal maintenance) is likely to prevail throughout biological design. While tuned to past environments in evolution and ontogeny, the capacity will often and at most times be in excess of that typically required to handle less common stress. There is an evolutionary advantage for mechanisms that buffer the system against higher levels of environmental variability, independent of whether the system is a natural or artificial one.

Robustness in Multi-Agent and Group-Level Decision Processes

The social environment offers a unique set of issues for robust decision making. Uncertainty associated with interaction with social agents may require robust mechanisms to cope with this variation. The strategic components of social interactions provide a particularly challenging environment for robustness because mechanisms must be robust not to an independent environment but one that responds directly to the organism.

In addition to providing obstacles to robustness, social agents can also enhance robustness by using group decision making. Group living enables animals to address many problems that would be difficult or impossible for single individuals. For example, individuals in groups can catch bigger prey or better protect themselves against predators (Krause and Ruxton 2002). Groups can afford forms of robustness that are not available to individuals. They permit things like learning from observation, which is another way of alleviating the requirement for individual acquisition (Oztop et al. 2006; see also Hammerstein and Boyd, this volume), herding, and also group rather than individual exploration of parameter spaces.

The recent introduction of self-organization theory into the behavioral sciences has led to many case studies which show that group living can also facilitate better decision making. The solution of cognitive problems that go beyond the capability of single animals is known as *collective intelligence* or *swarm intelligence* (Krause et al. 2010). A process definition is that two or more individuals independently, or at least partially independently, acquire information, and these different packages of information are combined and processed through social interaction, which provides a solution to a cognitive problem.

There are many examples of swarm intelligence in invertebrates (particularly in social insects) and more recently in vertebrates. The perception in most case studies (particularly in invertebrates) has been that the individual animal is cognitively relatively simple and restricted in what it can achieve, whereas the group collectively is capable of solving difficult problems.

One of the first scientists to point out that individuals can benefit from collective decision making was the French mathematician and political scientist Nicolas de Condorcet. He assumed that each individual can either be right or wrong with a certain probability p. Provided that $p(right) > 0.50$, the probability of a correct collective decision of the group will increase as a function of group size, provided that the individuals that have the correct information are in the majority in the population.

Subsequent research has demonstrated, quite counterintuitively, that swarm intelligence does not require a majority of individuals who know the correct answer or, in fact, any individual to know the correct answer. This was first shown by Galton (1907) in an empirical study on humans based on what is now known as the *many-wrongs principle* (Bergmann and Donner 1964). The many-wrongs principle is often mentioned in the context of navigational problems where navigational accuracy is predicted to increase as a function of group size (Simons 2004). The assumption underlying the many-wrongs principle is that all individuals have a common target destination, but that each individual navigates with some error. If group members average over their directional preferences (through social interaction), then the error with which the group moves toward the target decreases as a nonlinear function of group size. In the example by Simons (2004), individual errors followed a normal distribution. However, the principle of the many wrongs producing a good overall decision is not restricted to a particular type of distribution. As long as the mean of the individual vectors approximates the target direction, there are different types of distribution that could produce a similar outcome (i.e., reducing navigational error with group size). Thus, group decision making provides a robust response to individual errors in the estimates.

Collective decision making is usually robust to changes in the environment, outliers, and loss of group members, but it also has costs and is based on a number of prerequisites, one of which is independence of individuals. For instance, if all individuals in a group have the same bias, then no degree of redundancy (i.e., no increase in group size) is likely to make the collective decision any better. In contrast, the greater the independence of individuals in a system, the greater the probability will be that (a) a solution will emerge and (b) decision-making quality will increase with group size.

The self-organized nature of collective behavior is a form of redundancy that protects against failure of system components. Self-organization means that the decision-making process is decentralized and therefore less vulnerable to localized damage than central-control systems, where the loss of only a few individuals can have a strong adverse effect if they are the leaders or central decision makers.

Robust Group Decision Making in Animals

In honeybees, *Apis melifera*, workers perform different tasks which correlate with their age and development. Younger bees usually work inside the hive building cells and feeding larvae, whereas older workers collect nectar and pollen outside the hive. The colony functions best when there is a certain ratio of bees to carry out the indoor and outdoor services. If, however, a large number of bees get killed while performing outdoor duties, then the colony can respond adaptively with younger workers developing more rapidly to take on outdoor tasks (Robinson 1992; Schulz et al. 1998). Likewise, if for some reason the number of young bees is drastically reduced, then older bees take up nest building tasks again. This process of worker allocation is not centrally controlled. It is achieved through the contact frequencies between the workers of different ages and task groups. During each contact between two workers, different hormones are exchanged; these hormones control worker development and thereby regulate the ratio of indoor and outdoor workers (Huang et al. 1998). This example clearly shows how a self-organized process can allow a robust response to the loss of workers, not only through redundancy but also through reorganization of task allocation.

Many examples of swarm intelligence come from the social insect literature, where quorum sensing has been identified as an important mechanism for decision making (some cases of quorum decision have been identified in vertebrates as well; Ward et al. 2008). Insect colonies sometimes need to find a new nest location because their old nest has either become too small or has been damaged. The problem that the colony needs to solve frequently takes the form of a complex trade-off between speed and accuracy of decisions. A proportion of scouts (individuals that explore the surroundings for suitable nest locations) leave the nest and, if successful, these individuals then try to recruit others to the new potential nest location they encountered. After the number of individuals in support of a particular nest location reaches a threshold (i.e., a quorum), the entire colony will favor this location (Franks et al. 2009). If speed is important, then the quorum threshold can be low, resulting in lower accuracy and reducing the probability of deciding in favor of the best new location among the available options. If speed is not a constraint, then the quorum threshold can be high and accuracy is increased (Franks et al. 2009). A similar type of quorum-based decision making has also been described in fish shoals that have to make a decision about which path to take to avoid danger or which leader to follow (Ward et al. 2008).

Group Decisions and Robustness in Science and Technology

One of the richest areas for research and application in group decision making is in science and technology. These communities have sophisticated methodologies for combining inputs from different sources to construct solutions or answers that are more robust, more reliable, and less prone to error. When software is sent out for beta-testing, a substantial population of unpaid testers explore it under diverse conditions, reflecting their diverse interests, and report failures that are used to produce corrections incorporated in the released version. This could be seen as a richer and more sensitive extension of the methodologies discussed above for decision making in groups.

In scientific investigations, it is standard to assume that a measurement or result derived using two or more different modes of instrumentation is more reliable, and that cross-checking in this way is commonly expected and accepted as a test for the "reality" or "non-artifactuality" of the measurement, result, or detected property or entity (Soler et al. 2011; Wimsatt 1981). Levins (1966) talks about "robust theorems"—results derivable in diverse models of a phenomenon using different assumptions, which thereby do not depend upon the details of the various specific models. Campbell (1958, 1966) talks about the importance of "triangulation" using different methods, and Campbell and Fiske (1959) expound the use of a "multi-trait, multi-method matrix" to correct for biases in methods of measurement or bad choice of indices for a trait. Physicist Richard Feynman (1967) contrasts "Babylonian" vs. "Euclidean" formal methodologies and argues the advantages of the multiply connected inferences in the former over the minimalist serial inferences of the latter. These all result in methods of scientific inquiry that provide robustness to the vagaries of individual biases in model creation and data collection.

In the larger social structure of science, demands for repeatability, public disclosure of methods, and the peer review system are all attempts to secure greater reliability through robustness. Are these methods foolproof? Of course not. If the various methods fail to be independent in relevant respects, this can compromise results. Wade (1978) reviewed twelve different models of group selection by various authors, representing advocates and opponents of group selection, including some of the most distinguished evolutionary biologists. His study demonstrates that their near-universal claims to show (robustly) that group selection is not a significant force is undercut by the fact that the models shared five simplifying assumptions that were (a) false and (b) each biased the case against group selection. Wimsatt (1980) analyzed these cases and found that the biased assumptions are products of simplifications resulting from heuristics used in reductionistic methods of formulating and solving problems, paralleling earlier work by Tversky and Kahneman (1974) on heuristics and biases in probabilistic inference. Any purported case of robustness is corrigible by demonstrations that the methods are not independent, or as Levins (1966) would say, "not a representative sample from the space of possible models."

This domain suggests that case studies of scientific inference, technology testing, the organization of laboratories, and the composition of peer review panels all provide rich possible sources for understanding methodologies for generating and testing robustness in group decision making. We would like to know particularly how such methods may fail, and how induced biases might be detected and corrected.

Concluding Remarks

Robustness is an integral concept in the evolution of biological organisms and nonbiological systems, and a principal driver in the evolution of different decision-making and control mechanisms. To understand the evolutionary origins of decision mechanisms, therefore, we must address the nature of robustness, the nature of environments, and the costs and potential benefits associated with robustness in these environments.

In this chapter we have highlighted a number of issues that are central to robustness. For example, we have argued that in organisms, redundancy in the mechanisms of decision making is often not achieved by duplicating mechanisms but rather through the cooperation and competition of somewhat different mechanisms. Our primary emphasis has been on the extent and type of variation that individual organisms or technologies face. This variation provides a source of perturbation and is therefore central to whether we expect natural selection to produce optimal strategies, flexible strategies, and/or robust strategies. Selection acts on the underlying psychological mechanisms which must implement the behavioral strategy, and different mechanisms are robust to different sorts of environmental perturbation and may have different costs. We have highlighted how robustness can be achieved via multi-agent decision making, in the hopes that this discussion will stimulate further study into these important issues.

Variation in Decision Making

13

Biological Analogs of Personality

Niels J. Dingemanse and Max Wolf

Abstract

Individual differences in behavior that are stable over time and correlated across different contexts can be found in a wide range of species across the animal kingdom. Such structured behavioral differences have been termed "animal personalities" (behavioral syndromes). This chapter provides a brief introduction into this research area and discusses the two most common examples of behavioral variation associated with animal personalities (boldness-aggression syndrome and responsiveness to environmental stimuli); key genetic and physiological correlates of animal personalities; and fitness consequences of animal personalities in natural populations. The discussion makes clear that animal personality is a ubiquitous characteristic of animal populations, that personality variation is heritable and underpinned by variation in neuroendocrine and metabolic profiles, and that fluctuating selection pressures act on this variation in a wide variety of taxa. The widespread existence of personality variation implies that the study of decision making (i.e., the cognitive process that helps individuals select among alternative behavioral actions) should explicitly incorporate between-individual variation.

Introduction

Individual differences in behavior are a ubiquitous phenomenon in animal populations. Great tits (*Parus major*), for example, differ in the speed with which they explore a novel object, three-spined sticklebacks (*Gasterosteus aculeatus*) differ in their aggressiveness toward territorial intruders, and mice (*Mus musculus*) and rats (*Rattus norvegicus*) differ in how quickly they solve a maze problem. The existence of quantitative behavioral differences such as these, however, is hardly surprising. After all, for most quantitative traits, phenotypic differences are to be expected due to, for example, noise in the development of the phenotype or stochastic state differences among individuals (e.g., in energy reserves). At first sight, these differences do not seem to call for a deeper explanation.

Two basic observations suggest, however, that there may be more to behavioral variation than meets the eye (Gosling 2001; Sih et al. 2004; Bell et al. 2009). First, behavioral differences are often stable for some period of time.

A great tit that explores a novel object faster than a conspecific will tend to be faster also several weeks later (Verbeek et al. 1994). Similarly, the rank order in the level of aggression that individuals show toward territorial intruders tends to remain stable throughout the breeding cycle in sticklebacks (Huntingford 1976).

Second, behavioral differences often extend to a range of different situations and contexts in a systematic way. The faster a great tit approaches a novel object, for example, the faster it explores a novel environment, the more aggressively it behaves toward conspecifics, the quicker it forms rigid foraging habits, and the more willing it is to take risks (Groothuis and Carere 2005). Similarly, the more aggressively a stickleback behaves toward a conspecific territorial intruder during the breeding cycle, the more aggressive it is also toward a hetero-specific intruder and the bolder it is when approaching a predator outside the breeding season (Huntingford 1976).

Such findings are surprising. If behavioral differences were solely due to random factors, such as noise in the development of the phenotype, these differences should be uncorrelated, both through ontogeny and across different situations and contexts. The above findings, however, indicate that behavioral differences are often much more structured. Behavioral differences that are maintained through time and across contexts are termed *personalities* in humans (Pervin and John 1999) and, analogously, the term *animal personalities* (or coping styles, temperament, behavioral syndromes) has been adopted in the literature. Throughout, we will follow this usage, using the term animal personalities to refer to behavioral differences that are (a) stable through part of the ontogeny of individuals and (b) correlated across a range of situations and contexts.

Human Personalities

Compared to the study of animal behavior, human psychology has a long history of personality research. Others have discussed this huge literature extensively (Buss and Hawley 2010); thus we will limit ourselves here to a summary of a few basic facts on human personalities.

Personality research is, almost by definition, concerned with those traits that show some stability through ontogeny, since the term personality refers to "those characteristics of individuals that describe and account for consistent patterns of feeling, thinking, and behaving" (Gosling 2001). This is not to say, however, that personality characteristics are thought to be absolutely stable (Fleeson 2004). In fact, human personality traits are known to differ in their stability and to undergo systematic changes with age, maturational events, and social-contextual transitions (Caspi et al. 2005).

What are the basic dimensions of human personality differences? The "currently most widely accepted and complete map of personality structure"

(Gosling and John 1999) is the Five Factor Model of human personalities. This descriptive model is based on the following approach: Subjects are asked to rate themselves, or people they know, according to a long questionnaire describing behavioral dispositions. These ratings are then distilled into a smaller set of variables using factor analysis. The consensus from a large number of studies that have followed this approach appears to be that behavioral variation can be described along five independent axes: extraversion, agreeableness, conscientiousness, neuroticism, and openness (Pervin and John 1999). These axes reflect distinct dispositions, each subsuming a large number of phenotypic attributes that tend to be intercorrelated. Individuals that score high on the extraversion axis, for example, tend to be talkative, assertive, active, energetic, outgoing, outspoken, dominant, forceful, enthusiastic, sociable, adventurous, noisy, and bossy. A number of studies have shown that an individual's personality score predicts outcomes in his or her life, such as juvenile delinquency, performance in school and at work, psychopathology, and longevity (Ozer and Benet-Martinez 2006; Roberts et al. 2007). Twin studies have fairly consistently reported, for each of the five axes, broad-sense heritabilities in the range of 0.4 to 0.6 (Bouchard and Loehlin 2001), implying that roughly 50% of variation in personality is of genetic origin.

Animal Personalities

The study of personalities in nonhuman animals has a much shorter history. An early landmark was a set of studies by Nobel Laureate Ivan Pavlov, in which he identified three temperamental types in dogs (Pavlov 1928:363–364):

> Finally, we have been able to distinguish several definite types of nervous systems. To one of these types, then, I take the liberty to direct your attention. This type of dog is one which judging by his behavior (especially under new circumstances) everyone would call a timid and cowardly animal. He moves cautiously, with tail tucked in, and legs half bent. When we make a sudden movement or slightly raise the voice, the animal draws back his whole body and crouches on the floor.... As I gradually analyzed the types of nervous systems of various dogs, it seemed to me that they all fitted in well with the classical description of temperaments, in particular with their extreme groups, the sanguine and the melancholic.... Between these extremes stand the variations of the balanced or equilibrated type, where both the process of excitation and the process of inhibition are of equal and adequate strength, and they interchange promptly and exactly.

Despite several other seminal contributions (Hebb 1946; Huntingford 1976; Yerkes 1939), research on animal personalities was almost nonexistent during most of the 20th century. Several explanations are offered for this in the literature. Clark and Ehlinger (1987) attribute the neglect of individual differences to the focus of early ethologists on stereotyped, species-typical behaviors and the assumption of comparative psychologists (behaviorists) in search of general laws of learning and cognition that variation in their subjects' responses results from uncontrolled factors in the environment. Wilson (1998a) emphasizes the historical trend in the use of natural selection to explain differences between organisms at increasingly finer scales, starting with adaptive explanations for differences between genera and higher-level taxa (1940s) to differences between closely related species (1960s) to adaptive explanations for differences among subpopulations (1980s). He argued that differences between individuals are the next in line to be considered from an adaptive perspective. Several other authors point to a rigid refusal to "anthropomorphize" (e.g., Réale et al. 2007).

Times have changed, and during the last decade animal personalities have been the subject of considerable scientific interest. A survey by Gosling (2001, 2008) counted personality studies from a wide range of vertebrate and invertebrate taxa, including mammals, fish, birds, reptiles, amphibians, arthropods, and mollusks. The research questions in these studies are diverse, addressing methodological issues (e.g., assessment of the reliability and validity of particular behavioral tests), the structure of personalities (e.g., the correlation of traits with each other and the stability of these correlations), and the causes and consequences of personalities (e.g., the physiological correlates of personalities, the effect of experience on personalities, the issues of whether personalities are favored by selection). Reviews of this huge and growing literature are provided elsewhere (Dingemanse and Réale 2012; Dingemanse and Wolf 2010; Réale et al. 2007; Wolf and Weissing 2012; Sih et al. 2012). Here, we limit ourselves to a brief introduction of two common behavioral characteristics of animal personalities which figure prominently in the literature. Thereafter we discuss the genetic and physiological correlates of animal personalities and provide an overview of selection pressures which act on animal personality traits. We hope that this discussion increases awareness that humans and other animals vary consistently in suites of correlated behavioral traits, and that such variation warrants inclusion in a Darwinian framework of decision making.

Two Common Behavioral Characteristics of Animal Personalities

The Boldness-Aggression Syndrome

The term "boldness-aggression syndrome" refers to a positive correlation between individual differences in boldness (e.g., in the response to a predator or in exploring a novel environment) and level of aggression toward conspecifics. To our knowledge, such differences were first described in three-spined sticklebacks by Huntingford (1976), who found that (a) within the breeding season, male sticklebacks differed significantly in their aggressiveness toward territorial intruders, (b) the rank order of aggressiveness remained stable both when confronted with different types of intruders throughout the breeding cycle, and

(c) individuals that were more aggressive toward territorial intruders during the breeding season tended to be bolder in response to a predator outside the breeding season.

Since Huntingford's (1976) seminal study, the boldness-aggression syndrome has been described for a variety of other taxa (Sih et al. 2004), including birds and rodents. At present, the boldness-aggression syndrome is one of the most reported findings in the animal personality literature. Having said this, it is far from universal. For example, even within a single species, population comparisons have shown that the boldness-aggression syndrome is present in certain types of populations (those with high predation pressure) but not in others (Dingemanse et al. 2007).

Responsiveness to Environmental Stimuli

Individual differences in responsiveness to environmental stimuli are a second behavioral characteristic of animal personalities that appears to have some universality. While some individuals appear to be very responsive to all kinds of stimuli and readily adjust their behavior to the prevailing conditions, others show more rigid, routine-like behavior. Consistent differences in responsiveness (also referred to as coping style, reactivity, flexibility, and plasticity) have been documented in several taxa including birds, rodents, pigs, and humans (Dingemanse et al. 2010; Koolhaas et al. 1999). In both mice and rats, for example, individuals differ substantially in their responsiveness to environmental changes in a maze task. Some individuals quickly form a routine, are not influenced by minor environmental changes, and perform relatively badly when confronted with a changing maze configuration. Others do not form a routine, are strongly influenced by minor changes, and perform relatively well when confronted with changing maze configurations (Koolhaas et al. 1999). Similarly, some great tits readily adjust their foraging behavior to a change in the feeding situation while others stick to formerly successful habits (Verbeek et al. 1994). Individual variation in responsiveness is also increasingly studied in the wild (Dingemanse et al. 2010). In great tits, for example, individuals in the wild differ in how quickly they habituate behaviorally to repeated exposure of novel environment rooms in each of four West European populations (Dingemanse et al. 2012).

Genetic and Physiological Correlates of Animal Personalities

At the proximate level, the behavioral phenotype of an individual is shaped by its genetic, physiological, neurobiological, and cognitive systems. It is therefore not surprising that differences in personalities are often associated with differences in such systems.

Genetic Influences

The question of whether and to what extent animal personality variation has a heritable basis has received considerable attention in the recent literature (Dochtermann 2011; Réale et al. 2007; van Oers et al. 2005). The genetic basis of personality has been addressed in quite some detail using both quantitative and molecular genetics tools in a wide diversity of taxa, including birds (Drent et al. 2003), fish (Dingemanse et al. 2009), and primates (Weiss et al. 2000). Both approaches suggest that animal personalities are heritable but that their genetic underpinning might be complex.

Quantitative genetics. Animal personality variation, addressed from a quantitative genetics perspective, has primarily focused on estimating how much of the variation in behavior can be explained by additive effects of genes. Such studies have largely confirmed that (a) behavioral traits are moderately repeatable (i.e., they differ consistently between individuals, and repeatability values of behavior are roughly 20–40% according to a recent meta-analysis; Bell et al. 2009) and (b) behavioral traits are also moderately heritable (i.e., the consistent differences between individuals have a heritable basis; narrow-sense heritability estimates range roughly between 10–40% for behavioral traits; Réale et al. 2007; van Oers et al. 2005). These estimates of behavioral heritability are based either on classic artificial selection experiments or variation in relatedness within pedigreed natural populations.

Quantitative genetics studies have given rise to a number of further key insights concerning the genetic underpinning of personality. First, genetic correlations between the same behavior expressed in different contexts (e.g., activity in the presence vs. absence of predators, aggressiveness in juvenile vs. adult life-history phases, or exploratory behavior of novel environments vs. objects) are often extremely tightly correlated; genetic correlation values are close to 1 (Dingemanse et al. 2009). In other words, it appears that the same genes are often expressed when the same behavior is expressed in different contexts, giving rise to the cross-context correlations that characterize animal personalities. Second, genetic correlations between functionally distinct behaviors are, in contrast, relatively labile, as exemplified by work on stickleback which shows that both phenotypic and genetic correlations between components of the boldness-aggression syndrome vary between populations on relatively small spatial scales (Bell 2005). In other words, genetic correlations between different types of behavior might typically occur not because of genes with pleiotropic effects[1] but rather because of linkage disequilibrium[2] induced by natural selection (Dochtermann 2011).

[1] Pleiotropic effects occur when one gene influences multiple phenotyphic traits.

[2] In population genetics, linkage disequilibrium is the nonrandom association of alleles at two or more loci.

Candidate genes. Molecular geneticists studying animal personalities have focused primarily on candidate genes—genes known to affect personality in humans or husbandry animals (Munafò et al. 2008). In recent years, much research has centered on a single, though prominent, human personality gene: the dopamine receptor gene *DRD*4. This gene is part of the dopaminergic system that affects the motivation of behavior and influences human novelty seeking. Recent work shows that genetic variants (i.e., single nucleotide polymorphisms or SNPs)[3] exist for this gene and predict exploration behavior in animals of a wide range of taxa (Munafò et al. 2008). For example, SNP830 in the *DRD*4 gene predicts exploration of novel environments in a Dutch nest box population of great tits (Fidler et al. 2007). However, the same polymorphism in the *DRD*4 gene does not predict exploration behavior in three other nest box populations of this species (Korsten et al. 2010). This implies that links between genes and personality may be relatively complex and influenced, for example, either by gene–gene or gene–environment interactions. Such a complex genetic underpinning has recently been shown experimentally in stickleback, where the expression of heritable variation in personality traits is influenced by perceived predation risk during ontogeny (Dingemanse et al. 2009).

Physiological Influences

Evidence is accumulating to indicate that personality differences are often systematically associated with the physiological setup of individuals.

Neuroendocrine underpinning. Much research in this area has focused on the notion that individual animals might vary consistently in their physiological responsiveness to mild stressors, indicated by the term "coping style" (Koolhaas et al. 1999). "Proactive copers" are characterized by a relatively (a) low hypothalamic pituitary adrenal (HPA) axis reactivity, (b) high sympathetic reactivity, and (c) low parasympathetic reactivity. In contrast, "reactive copers" are characterized by a relatively (a) high HPA axis reactivity, (b) low sympathetic reactivity, and (c) high parasympathetic reactivity (Koolhaas et al. 1999). Over the last few years, such variation in physiological setup has been reported for a wide range of taxa, including birds, domestic pigs, rodents, fish, and primates (Carere et al. 2010).

Variation in stress physiology is systematically associated with the boldness-aggression syndrome and individual differences in responsiveness (Koolhaas et al. 1999; Koolhaas et al. 2010). Proactive copers—whose behavior is thought to be driven by internal mechanisms—are typically active, aggressive, and bold, and relatively unresponsive to environment stimuli.

3 Single nucleotide polymorphisms reflect DNA sequence variation and occur when a single nucleotide differs either between chromosomes of the same individual or different individuals.

Reactive copers, on the other hand, behave in a nonaggressive and cautious manner and are much more responsive to environmental stimuli (Koolhaas et al. 1999). Detailed experimental work in behavioral stress physiology suggests that variation in physiological and behavioral stress responsiveness is underpinned by two independent personality axes: one represents the quality of the response to challenging situations (coping style), the other the quantity of that response (stress reactivity) (Koolhaas et al. 2010). Finally, there is growing awareness that behavior and neuroendocrinology interact dynamically, and that one (endocrinology) should not necessarily be regarded as the proximate underpinning of the other (behavior) (Koolhaas et al. 2010).

Metabolism. Other recent research on the proximate underpinning of animal personality has focused on whether individuals differ consistently in their energy metabolism (e.g., resting metabolic rate) and whether such differences are systematically associated with personality differences. A recent meta-analysis shows that this is indeed the case (Biro and Stamps 2010). Individuals within several animal species differ consistently in resting metabolic rate, and these differences are associated with differences in activity levels, levels of aggression, and dominance status. Typically, active, aggressive, or dominant individuals have higher values of resting metabolic rate. These links appear plausible at first sight: highly aggressive and active individuals need the physiological machinery to fuel such costly activities. Phrased differently, individuals that have high metabolic rates might need to behave in such a way that they acquire resources needed to meet their daily energy requirements.

Natural Selection and Animal Personality

Why has natural selection not given rise to a single "optimal" behavioral type but rather to a mixture of behavioral types? From a theoretical perspective, several different lines of explanations have been proposed (Dingemanse and Wolf 2010; Wolf and Weissing 2010). Empiricists, however, have focused primarily on spatiotemporal variation in selection pressures as an explanation for coexistence.

Spatiotemporal Variation in Selection Pressures

Different behavioral types may be favored in different local environments (spatial variation) or at different times (temporal variation) and such variation in selection pressure may give rise to the coexistence of behavioral types. A recent review reported ten (out of eleven) studies reporting some form of spatiotemporal variation in selection pressures acting on behavior within the same population (Dingemanse et al. 2012). These studies comprised a large diversity of taxa (birds, insects, mammals, reptiles), implying that this form of selection may be widespread in nature. Temporal variation, for example, has been documented in birds (great tits), mammals (bighorn sheep, *Ovis canadensis,* and North American red squirrels, *Tamiasciurus hudsonicus*). In red squirrels, aggressive females were favored in certain (but not in other) years, based on growth and survival of their offspring. Spatial variation in selection pressures has been documented, for example, in birds (great tits), where fast explorers had relatively high breeding success in high density areas but slow explorers in low density ones.

Missing Pieces of Information

Despite great interest in the fitness consequences of animal personality, two key questions have been largely neglected in the empirical literature (Dingemanse et al. 2012). First, how does natural selection act on behavioral consistency? Do consistent individuals enjoy higher fitness than individuals that are less inconsistent in their behavior? Second, how does natural selection act on behavioral correlations? Do bold animals, for example, have higher fitness when they are also relatively aggressive? Addressing such questions would enable us to better understand why individuals differ consistently in suites of correlated traits, and whether such differences are perhaps favored by natural selection.

Over the past few years, researchers have started to expand their horizons and restrict their research not solely to understanding the proximate and evolutionary mechanisms maintaining personality variation. Specifically, there is growing interest in the question of whether personality variation, in itself, has ecological and evolutionary consequences. A recent modeling study (Wolf and Weissing 2012; Sih et al. 2012), for example, shows that populations consisting of a mix of social and asocial individuals spread faster than populations consisting of either one (Fogarty et al. 2011).

Summary

Animal personality refers to individual differences in behavior that are stable over time and correlated across different contexts. Such differences can be found in a wide range of taxa across the animal kingdom. Here we have introduced the empirical research on animal personalities. At present, there seem to be two common (i.e., to some extent universal) behavioral characteristics of animal personalities: the boldness-aggression syndrome and individual differences in responsiveness to environmental stimuli. Work on the proximate underpinning of animal personality shows that animal personality is heritable, that the expression of the same behavior across different contexts is often influenced by the same genes, and that genetic correlations between different behaviors (e.g., aggression and boldness) may result from linkage disequilibrium rather than gene pleiotropy. Animal personality is often associated with differences in stress physiology and metabolism, though behavior and stress physiology might interact dynamically and cause-effect relationships might therefore be complex. Selection on the behavioral phenotype often fluctuates in space and time, while selection pressures favoring personality per se (i.e., consistency, correlations between behaviors) await quantification. Finally, the ecological and evolutionary consequences of such variation await further study.

Acknowledgments

Both authors contributed in equal part to this manuscript. The introductory sections of this paper are modified versions from the introductory sections of M.W.'s Ph.D. thesis. N.J.D. was supported by the Max Planck Foundation.

14

Sources of Variation within the Individual

Gordon D. A. Brown, Alex M. Wood, and Nick Chater

Abstract

This chapter reviews sources of variation in decision making and choice that arise within the individual, and discusses the possible significance of such variation for the development of an evolutionarily plausible account of decision making. Different sources of within-individual variation are reviewed, including noise, general cognitive ability and life span changes, mood, and personality.

The approaches taken by economics, cognitive psychology, and the psychology of personality and individual differences are compared to and contrasted with ecological and evolutionary approaches. It is argued that sampling models may provide a unified cognitive approach to within- and between-individual variation in human judgment and decision making, albeit one that has yet to be linked sufficiently to ecological and evolutionary approaches.

Introduction

Our focus in this chapter is on sources of variation *within* the individual and on the possible significance of such variation for an evolutionarily plausible account of decision making. We therefore enumerate different sources of within-individual variation, focusing mainly on the human literature, in each case noting adaptive approaches where available. Our discussion contrasts with much work on human personality and individual differences, and in some parts of economics, which have focused on (assumed stable) *between*-individual sources of behavioral variation.

First we map out the different approaches to within-individual behavioral variation, and its relation to between-individual variation, which have developed largely independently in literatures that have concerned themselves with (a) the psychology of human personality and individual differences, (b) the cognitive psychology of judgment and decision making, (c) neoclassical economics, and

(d) evolutionary and ecological approaches in nonhuman animals. Much of the research is described in more detail under separate headings below.

The study of human personality has focused for half a century on identifying stable characteristics of individuals that capture systematic differences in people's "average" behavior (e.g., whether they are extraverted, conscientious, etc.) across different social situations. Thus the focus has generally been on between-individual rather than within-individual variation, with the long-standing person-situation debate concerning whether behavior is better predicted by personality or by the environment (Epstein 1979; Funder and Ozer 1983; Mischel 1968). Only relatively recently has research into human personality begun to examine within-individual variation in behavioral reactivity (Fleeson 2001, 2004; Sheldon et al. 1997).

Evolutionary and ecological approaches to studying human personality have differed greatly to how psychology has treated the subject. Integration of the across-individual human and animal literature is therefore difficult, although a start is being made (e.g., Nettle and Penke 2010). The concept of personality as applied to nonhuman animals is relatively recent (Dall et al. 2004; Nettle and Penke 2010) and more grounded in behavior (for a review, see Dingemanse and Wolf, this volume). In contrast to the human personality literature, ecological approaches have moved rapidly toward developing adaptive accounts of between-individual variation in behavior (e.g., Dingemanse and Wolf 2010a; McNamara et al. 2009). Moreover, evolutionary approaches have begun to emphasize the importance of providing an integrated and adaptive account of both within- and between-individual behavioral variation ("personality and plasticity"), emphasizing the fact that the two sources of variation can be considered as complementary aspects of a phenotype (Dingemanse et al. 2010b; Nussey et al. 2007) and demonstrating that behavioral patterns evolve in population environments in which natural selection favors them (e.g., Dingemanse et al. 2007).

Until relatively recently, economics and cognitive psychology have taken different approaches to variation in decision making. Within one tradition of neoclassical microeconomics, normative models of decision making—and many of the descriptive accounts derived from them—focus typically on between-individual differences in risk aversion, temporal discounting, etc. Such characteristics and preferences have traditionally been assumed to be stable, although subject to noise. Normative accounts, however, do allow for within-individual variation over time and as a function both of the environment and in response to the strategic choices—and unpredictability—of other players. Furthermore, much of behavioral economics and consumer psychology has focused on the construction of preferences (e.g., Lichtenstein and Slovic 2006; Plott 1996; Prelec et al. 1997; Simonson 2008), while experimental economists have extensively examined variation that arises from interactive decision making and its development over time (e.g., Duffy and Nagel 1997; Selten and Stoecker 1986).

Cognitive psychological models of judgment and decision making have, in contrast, taken a central explanandum to be the effects of context on decision making; they have asked, for example, how the introduction of additional options to a choice set can influence the option that is preferred or chosen, and how apparent anomalies and inconsistencies arise. Thus a key focus in psychology has been the effects of situational (and hence within-individual) variation, rather than across-individual variation or variation arising from strategic interactions, on decision making. Many similar issues, especially concerning preference reversals, are addressed by behavioral economics.

Below we review sources of within-individual variation, consider the implications of such variation for adaptive accounts, and suggest that sampling-based accounts may have the potential to offer a more ecologically plausible account of individual variation and decision making across contexts than current models.

Sources of Within-Individual Variation: Importance to Evolutionary Approaches

Why is within-individual variation interesting for an evolutionary plausible model of decision making? Any such account must explain the systematic ways in which choices and decisions are determined by the context of choice options available. A notable weakness of narrow neoclassical accounts, which assume that individual preferences are (within limits) stable and consistent across different times and contexts, is their inability to account for such apparent inconsistencies. We argue below that sampling-based accounts, based on very simple processing principles, may offer better models. It is at least plausible that an evolutionary foundation may be provided for such accounts, although this remains a neglected area.

There are, however, more general considerations. Following the development of evolutionary game theory (Maynard Smith and Price 1973; Maynard Smith 1982), it has been clear that, due to frequency-dependent selection, an evolutionarily stable strategy (ESS) may involve the simultaneous presence of different decision-making personalities within a population. Recent work has examined how the existence of variation in decision making might foster the evolution of cooperation within social networks (e.g., Lotem et al. 1999; McNamara et al. 2004). However, the coexistence of different decision-making styles or personalities at a given time could involve a mixed strategy (e.g., an individual could behave in an extraverted fashion for 65% of the time, perhaps to facilitate exploration and development of social relationships, and behave in a non-extraverted fashion the remaining 35% of the time, to reduce risk), such that stable individual differences did not exist. Alternatively, the ESS could involve 65% consistent extraverts and 35% consistent non-extraverts. Consistency can be selected for when reputation matters and personality is publicly observable (Nowak and Sigmund 1998). Thus, at a global level, it may be better to be consistently rather than inconsistently aggressive, effectively because of reputational considerations: individuals who are consistently hawkish in their behavior may avoid conflicts if their "personality" is publicly known. A different tradition of research within experimental economics, as noted above, focuses on how players of strategic games learn how to defect or cooperate in different contexts (i.e., to respond strategically to produce best responses).

Stability could arise for other reasons. Lukaszewski and Roney (2011) suggest that extraverted behavioral strategies are more likely to be successful when associated with physical strength and attractiveness. Consistent with their expectations, they found that physical strength and attractiveness accounted for much of the variance in extraversion. Such considerations, therefore, lead to an expectation of within-individual stability in general decision-making style. Nonetheless, as will be shown below, within-individual variation in both personality and decision making can arise from a number of sources, and such variation remains a key target of models.

Sources of Within-Individual Variation

We now turn to a review of within-individual variation that arises as a result of noise, general cognitive ability and life span changes, mood, and personality. We then argue that sampling models may provide a unified cognitive approach, albeit one that has yet to be linked sufficiently to the ecological and evolutionary approaches noted above. To make the review manageable, we confine ourselves predominantly to internal sources of variability. Of course, variability may also result from noisy environments and in response to the behavior and payoffs of other agents, but we largely pass over these exogenous sources of variability.

Within-Individual Variation Due to Noise

Let us first consider noise and inconsistency in risky choice. An individual may choose option *A* over option *B* at time 1, yet choose *B* over *A* at time 2 because of the noise within the decision-making system (Hey and Orme 1994). Choice reversals may occur on around 25% of choices, depending on context. Insofar as higher general cognitive ability is associated with greater consistency in risky choice and discounting (Burks et al. 2009), it seems unlikely that this particular type of within-individual noise-related inconsistency serves a direct adaptive function. Other types of noise and probabalistic strategy may well, however, be adaptive, as reviewed below.

The operation of a stochastic component has been examined extensively within models of economic choice between risky prospects (e.g., Blavatskyy and Pogrebna 2010; Blavatskyy 2011; Loomes 2005; Loomes and Sugden

1995). Models that are derived from an economic framework can incorporate a stochastic element in a number of ways. One possibility is that the utility of the two choice outcomes is calculated according to a standard framework, and that noise is then added to the resulting utilities (or perhaps to the difference between them). Such noise can result in violations of normative axioms. An alternative idea is that noise is added to model parameters (e.g., the curvature of the utility function) on each occasion that utility is calculated. In this case, the calculated utilities (and hence choices) may differ from one occasion to the next, but violation of axioms will not result. The choice of stochastic decision rule is important: Blavatskyy and Pogrebna (2010) distinguish seven different possible probabilistic accounts and, by combining these with various decision-making models, find that the rank ordering of model architectures in terms of their fit to the data can change as a result. Thus models of decision making cannot be evaluated independently of the account of noise they assume.

Sampling models of decision making provides another approach to within-individual variation in judgment and decision making more generally. For example, Vul and Pashler (2008) find that successive estimates of a quantity may have uncorrelated errors, such that the average of the two estimates is more accurate than either of them individually (the "crowd within"). Such effects fit naturally with an account in which a new mental sample is drawn from a remembered distribution on each occasion that an estimate is made, with successive estimates differing because they are made on the basis of different mental samples on each occasion (see also Stewart et al. 2006).

To what extent is the existence of noise in decision making relevant to the development of evolutionary-based models of decision making? We have already noted that variability may play a role in the evolution of systems, including cooperation. Some mathematical models of evolution accord a role to randomness as a part of development of successful strategies. Thus, in the prisoner's dilemma and other games, probabilistic versions of *win–stay, lose–shift* strategies can outperform other versions by taking advantage of unconditional cooperators (Nowak and Highfield 2011). More generally, unpredictability is crucial to the successful play of many predator-prey models. Noise is central to the concept of *trembling hand equilibrium* (Selten 1975) and noisy behavior (e.g., based on small-sample judgments), with the consequence that others may occasionally behave irrationally; this can provide an essential underpinning to the development of rational strategies (Aumann and Sorin 1989). Kareev (this volume) ably reviews a number of relevant strands of evidence, which will not be repeated here.

Within-Individual Variation Due to General Cognitive Ability and Life Span Changes

Overall cognitive capacity (as indexed, e.g., by measures of working memory capacity or general intelligence) has a substantial impact on individual decision making. To the extent that cognitive capacity changes within an individual, either because of distraction (divided attention), lack of energy (e.g., due to sleep deprivation or hunger), or during cognitive development, we would expect consequent within-individual variation in decision making. Such variation is indeed observed empirically in all these cases. For example, changing glucose levels lead to changes in human decision making—higher glucose levels lead to reduced temporal discounting (Wang and Dvorak 2010)—and mental fatigue has been linked to a "jumping to conclusions" style of reasoning (Webster et al. 1996).

Research on the effects on decision making of within-individual variation in general cognitive ability relies on the much larger body of between-individual variation, so we note the latter first. Across individuals, higher levels of general intelligence are associated with reduced use of heuristic strategies in reasoning (Stanovich and West 2000). In terms of specific decision making, Burks et al. (2009) find higher cognitive ability to be associated with reduced temporal discounting, higher social awareness, better prediction of others' first choices in a prisoner's dilemma game, superior planning, and greater risk seeking for gambles where the expected gain is positive. Reduced working memory capacity is associated with specific types of reasoning. For example, within the economics literature, memory limitations have been invoked to explain why agents might rationally select apparently dominated strategies (Bernheim and Thomadsen 2005), to understand the effects of price competition (Chen et al. 2010), the forgetting of past costs (Smith 2009), and the optimal way for rational agents to release information (Sarafidis 2007). Stevens et al. (2011) shows how imperfect memory for other players' actions can lead to biased strategy selection in repeated games, and Kareev (1995a and this volume) reviews evidence for adaptive roles of cognitive limitations.

The importance of working memory and sampling in judgment and decision making is addressed by various models (Fiedler and Juslin 2005). Judgments of likelihood have been explained within the framework of the MINERVA-DM memory model (Dougherty et al. 1999), and working memory capacity limitations have been linked to biased probability judgments (Dougherty and Sprenger 2006) and hypothesis testing (Sprenger and Dougherty 2006).

We would expect, therefore, that within-individual variation in general cognitive capacity would result in changes in decision making. Such variation is indeed seen as a result of cognitive aging and the concomitant reduction in cognitive ability. Thus aging is generally associated with increased risk aversion (cf. Deakin et al. 2004; Sanfey and Hastie 1999) and larger offers in dictator games (Bosch-Domenech et al. 2010). Temporal discounting is reduced in older adults (Read and Read 2004; Reimers et al. 2009), whereas loss aversion is increased (Johnson et al. 2006). As yet there is relatively little research on how reduction of cognitive capacity by distraction or fatigue influences such parameters as risk aversion, although this is a growing area (e.g., Hockey et al. 2000).

In summary: There is clear evidence of within-individual variation in decision making over the life span, and there is well-documented within-individual change in memory capacity and general intelligence, both of which impact on decision making. Mata et al. (2007) found that older adults used simpler decision-making strategies with reduced cognitive requirements, but argued that aging did not reduce the adaptiveness of decision making. However, there is little conclusive research on evolutionary accounts of this source of within-individual variation.

Within-Individual Variation Due to Mood

In contrast to general cognitive ability, and in partial contrast to personality (discussed below), strong within-individual variations in mood are part of common experience. Here, we confine ourselves to the human literature, while noting behavioral research on nonhuman animals that adopt a similar perspective. For example, behavior consistent with a "pessimistic cognitive bias" appears even in invertebrates: stimulus categorization by honeybees appears more pessimistic (i.e., an ambiguous stimulus is more likely to be perceived as predicting punishment) when the bees are, literally, agitated (Bateson et al. 2011).

Are within-individual mood variations related to variations in decision making, and can such variations inform evolutionary models of decision making? As Hermalin and Isen (2008) note, effects of mood have not yet generally been incorporated into economic models of human decision making, and there is a need to do so. There are strong effects of mood on decision making (Werner et al. 2009). For example, a considerable body of research conducted by Isen and colleagues has examined the effect of positive mood on reasoning and decision-making style. Positive affect generally enhances decision making (Isen 2001; Isen and Means 1983; Isen and Patrick 1983), perhaps because increased dopamine levels enhance cognitive flexibility (Ashby et al. 1999). In evaluations of gambles, positive affect has been associated with a shift in attention toward outcomes rather than probabilities (Nygren et al. 1996), and with increased confirmation bias in a selection task (Oaksford et al. 1996).

There has been little in the way of evolutionary models of mood shifts within human individuals, although Nettle (2009) develops a model according to which low mood states can lead to either risk seeking or risk avoidance according to the severity of the situation, such that mood states may adaptively adjust risk-taking propensity in response to environmental circumstances (for nonhuman animals, see Dingemanse et al. 2010b).

Given that mood appears to show substantial within-individual variation, with consequences for decision-making style, we identify a need for more research to develop theoretical accounts of the adaptive significance of mood shifts.

Within-Individual Variation Due to Personality

To what extent does within-individual variation in personality exist, and are the data of relevance to evolutionary models of decision making? We begin with research on between-individual variation. The idea that a small number of basic traits characterize individuals and their decision making has a long history, going back at least to Galen (129–200 AD) and the description of humans as phlegmatic, choleric, etc. Modern theories of human personality have made use of statistical advances to go beyond intuitive classification, converging on the "Big 5" personality factors (extraversion, neuroticism, agreeableness, openness, and conscientiousness), each with their own subcomponents (facets) as a stable, culturally generalizable, and reliable descriptive characterization of human personality (e.g., Goldberg 1993; McCrae and Costa 1987). The Big 5 has recently begun to be used as behavioral predictors and linked to specific components of decision making. Thus, Roberts et al. (2007) argue that individual differences in human personality predict important life outcomes (e.g., mortality, divorce, and career success) as well as or better than do socioeconomic and cognitive variables.

However, if the study of within-individual variation in human personality is to inform models of decision making, we need to know how personality characteristics relate to decision-making style. Behavioral economists have recently begun to use the techniques and methods of experimental economics to examine decision making and human personality. Of particular interest is the interaction between personality variables and economic variables (Borghans et al. 2008). Early work within economics focused on questions such as the relationship between personality and economic success as measured, for example, by income (e.g., Bowles et al. 2001). More recently, alongside examinations of the role of general intellectual capacity in economic behavior (Burks et al. 2009), economists have examined the effects of personality traits, particularly agreeableness and conscientiousness, on choices in economic games (such as prisoner's dilemma, ultimatum and dictator games, and cooperation games). It has been found that general intelligence and extraversion together predict attitude to risk (Anderson et al. 2011) and that individuals scoring highly for conscientiousness gain greater additional utility from increases in income (Boyce and Wood 2011). Extraversion and agreeableness influence cooperation in resource dilemmas (Koole et al. 2001), and gratitude prompts prosocial behavior (Tsang 2006).

Overall there is now ample evidence which indicates that between-individual variation in the human personality characteristics studied by psychologists is associated with different decision-making and fitness-related life outcomes. Furthermore, as reviewed below, there are numerous evolutionary accounts of why personality variation within the population might exist, with game-theoretic influences evident since Trivers (1971), although the bulk of this research has been carried out on nonhuman animals (e.g., Dall 2004; Dall et al.

2004; Gross 1996; Wolf and Weissing 2010). More recently, there have been the beginnings of parallel evolutionary approaches to human personality (Buss 2009; Nettle 2005, 2006; Penke et al. 2007a, b), along with consideration of the extent to which models of human personality generalize across species (e.g., Gosling and John 1999).

What of within-individual variation in personality, and the consequences of such variation for decision making? We consider a personality trait to be a consistent pattern of behaving, thinking, and feeling that can be used to predict behavior. Such definitions appear to exclude the possibility of within-individual variation in decision making. However, personality-related behaviors do vary within individuals over time and across situations, and it is possible at least to speculate on the fitness-relevance of such variation.

Specifically, recent work within personality theory has begun to move beyond the debate about whether situations or stable individual characteristics govern behavior and decision making to a focus on differences within individuals over time and across situations (Fleeson 2001, 2004). The idea is that, even though individuals can be readily identified by their "average" levels of personality, they routinely experience the full range of personality and can also be identified by the extent to which their personalities vary across situations. There are, of course, likely to be limits on such variation.

Changes in an individual psychological profile might be situation-dependent rather than time-dependent; consider, for instance, when people effectively adopt different personalities as they engage with different social worlds (e.g., with parents or with friends) or perhaps state-dependent changes as in the case of mood variation. There is clear evidence that individuals' personalities do indeed vary across the different roles they experience in life, and that there is individual variation in the extent of this differentiation (Sheldon et al. 1997). Thus the same individual may behave in an extraverted manner when they are in the company of their friends but be much more introverted in the presence of their parents, whereas another individual may exhibit much the same personality across each of these roles.

In terms of fitness, is high variation in personality and hence decision making across social roles good or bad? Although we have no direct evidence, higher levels of variation in personality across social roles has been related to lower levels of well-being (Roberts and Donahue 1994). Could we use measures of subjective well-being as a proxy for fitness? Consideration of such an assumption may be worthwhile, because human personality research has tended to look at the relation of personality (including within-individual variation in personality) to subjective measures such as well-being and life satisfaction. For example, it is striking that about 45% of the variance in overall life satisfaction can be predicted from conventional measures of personality—around ten times as much as can be predicted from measures of economic status (Howell and Howell 2008; Wood et al. 2008, 2009).

It is well established that subjective measures of well-being are reliable and valid in the sense that they are predictable (i.e., not just noise) and correlated with others' ratings, with "objective measures" (e.g., frequency of laughing and smiling), and with nonsubjective economic measures of quality of life (Layard 2005; Oswald and Wu 2010). Such findings, however, do not in themselves legitimate the approach of using well-being measures as a proxy for fitness.

Standing in favor of such an approach is the observation that measures of subjective well-being predict a large number of fitness-related outcomes. Subjective well-being is related to factors such as longevity (Diener and Chan 2011), heart disease, and strokes (Sales and House 1971) as well as wound healing, susceptibility to infection, and recovery time (Cohen et al. 2003). There is evidence that happiness leads to marriage as well as the reverse (Stutzer and Frey 2006). The correlation between an individual's income and their life satisfaction that is reliably observed within a country at a given time appears to be due not to the income per se but rather to the ranked position of that income within society (Boyce et al. 2010a). Overall, then, there is some plausibility to the suggestion that subjective well-being might in some circumstances be usable as a proxy for fitness.

However, there are also considerable difficulties with the idea that high subjective well-being is necessarily associated with fitness-maximizing decision-making personalities. The fact that particular personality traits are associated with systematically higher self-rated happiness (DeNeve and Cooper 1998) argues against the idea that different personalities and their associated different decision-making strategies represent different ways to achieve similar fitness outcomes. Furthermore, affective forecasting errors, such that people are wrong about what will make them happy, abound. For example, people think that having children will make them happier but, in fact, the reverse is the case. A plausible ecological account of affective forecasting errors, in general, may be possible—we overestimate how unhappy disablement will make us, or how happy we will be if we win the lottery. Given that we like to be happy, our erroneous forecasts will motivate us to avoid disablement and strive for large amounts of money—reasonably interpretable as fitness-directed motivations. When such states are imposed or attained, however, we adapt, perhaps because there is no longer any evolutionary advantage to feeling happy or unhappy about things that cannot be changed. It is clear, then (and even limited experience of the world surely suffices to confirm), that our happiness is not in itself something that will necessarily be maximized through evolutionary processes. Thus the use of well-being measures as a proxy for fitness seems risky at best.

A final consideration concerns the intuition from biology that, if different decision-making personalities exist and are heritable, there are likely to be both advantages and disadvantages associated with any one of them (Nettle 2006; cf. Sheldon et al. 2007). Such an assumption has been central to approaches within evolutionary biology, but it is only recently that human personality psychologists have begun to examine the downside of personality traits, such as conscientiousness, that have generally been viewed as unconditionally positive. Thus, for example, the "Big 5" personality trait of conscientiousness was for a long time seen as entirely positive and associated with higher well-being (DeNeve and Cooper 1998), motivation, and achievement (Judge and Ilies 2002; McGregor and Little 1998). There was hence a view that conscientiousness is always positive for well-being. However, Boyce, Wood, and Brown (2010b) hypothesized that experience of failure might result in greater loss of well-being for highly conscientious individuals. This is indeed what was found: After three years of unemployment, individuals high in conscientiousness (i.e., one standard deviation above the mean) experience a 120% higher decrease in life satisfaction than those at low levels. Similarly, secure attachment might seem an unalloyed good for fitness-related outcomes. However Ein-Dor et al. (2010) suggest that a mixture of social attachment styles may be beneficial in evolutionary terms. Nettle (2006) reviews possible positive and negative aspects of all the "Big 5" personality traits.

Other individual characteristics also appear to have both positive and negative aspects. Thus individuals high on "satisficing" will make better decisions (because they spend more time considering alternatives) but will be less happy with their decisions (due to regret about investigated alternatives). Similarly, high degrees of materialism maybe associated with high motivation to gain social status (i.e., good for fitness) as well as with negative outcomes such as anxiety, depression, and ill health (Kasser 2002). We give less attention to these latter outcomes, as little is known about within-individual variation in them.

Across-Individual and Within-Individual Variation: A Sampling Approach

In this final section we consider how sampling-based approaches to judgment and decision making have offered a perspective on both across- and within-individual variation on decision making. Orthodox approaches derived from economics approach traits, such as risk aversion and temporal discounting in terms of the curvature of stable utility functions. By extension, accounts of within-individual variation need to adopt a similar approach.

However, alternative accounts of the form of utility functions, and of concepts such as risk aversion and loss aversion, have recently been provided by rank-based sampling accounts such as the *decision-by-sampling* (DbS) model (Stewart 2009; Stewart et al. 2006). The DbS model does not assume the existence of utility functions even as a descriptive "as if" convenience, but instead assumes that behavior results from rank-based judgment based on a mental sample. Such an account offers a very different potential account of both between- and within-individual variation in decision making.

DbS was independently motivated by psychophysical research showing that people find it much easier to (and perhaps can only) make binary ordinal comparisons of stimuli (e.g., whether one sound is louder than another rather than how much louder it is; whether a light is brighter than another, rather than how much lighter it is). The model specifies the psychological processes involved in judgment formation. It assumes that, when making a judgment, participants sample from their long-term memories (and sometimes also the experimental context) to make a relative judgment based on "cognitively easy" binary ordinal comparisons. For example, when deciding whether a price of £1.20 for a cup of coffee is high, participants might think of half a dozen occasions when they have paid less for a cup of coffee, but only two occasions when they have paid more. The relative rank value of the coffee price of £1.20 would therefore be calculated according to that retrieved sample: (number ranked lower)/(sample size) = .75. It is hypothesized that both social quantities (e.g., health-related behaviors, energy consumption) and economic quantities (e.g., prices, wages, win probabilities) behave like basic psychological quantities (such as weight, brightness, and loudness) in that their subjective magnitudes are given at least partly by their relative ranked position within a comparison set. Considerable evidence supports this suggestion, including studies originally conducted to test the predictions of the earlier *range frequency theory* of judgment (Parducci et al. 1976; Parducci and Perrett 1971). Range frequency theory can be viewed as a descriptive account, in contrast to the process-level model offered by DbS.

Relative rank effects were observed initially in psychophysics (Parducci and Perrett 1971) and subsequently in domains as diverse as sweetness perception (Riskey et al. 1979) and perception of body image (Wedell et al. 2005). More recently, the same principles have been shown to be applicable to economic and social quantities. Thus, rank effects are seen in price perception (Niedrich et al. 2001, 2009). Judgments of "fair" allocations of wage and tax increases follow rank-based principles (Mellers 1982), as do judgments of other economic quantities (Smith et al. 1989) and event-rated death tolls (Olivola and Sagara 2009). Brown et al. (2008) found that wage satisfaction and well-being depended on the ordinal rank of an individual's wage within a comparison group; effects of ranked position of income on satisfaction with income have been found in other large datasets (Clark et al. 2008; Hagerty 2000), while ranked position of income impacts have been found on more general life satisfaction (Boyce et al. 2010a).

The DbS model has been used to show how a number of classic descriptive theories in economic psychology, such as prospect theory (Kahneman and Tversky 1979), can be understood at a deeper level as deriving from sample-based judgments of the type assumed by DbS (Stewart 2009; Stewart et al. 2006). How might it account for both across- and within-individual differences?

Specifically, according to DbS, an individual's attitude to the riskiness of a given event, or the expensiveness of a given price, will be given by the statistics of the retrieved or observed sample of events that provide the context for judgment. Thus an individual's beliefs about event distributions will determine their decision making and judgments regarding risks, losses, etc.

The skewness of the retrieved distribution will be particularly important. In an application to individual differences, Olivola and Sagara (2009) demonstrated that the level of risk seeking in mortality-related decisions is lower in countries in which high mortality events are observed relatively more often. Thus an event that carries an associated death toll of 500 will seem subjectively less disastrous in a country where higher event-associated death tolls are relatively more common. According to this account, the attitude toward risk (e.g., how much risk it is worth taking to save 1000 lives rather than 500) is derived not from stable underlying representations of value but from the retrieved distributions of relevant events and the rank-based nature of the evaluation process. Such an approach may, we suggest, hold promise as a more general account of both across- and within-individual variation in decision making and choice.

Further work is also consistent with the suggestion that individual differences in attitudes may be accounted for, at least partly, in terms of beliefs about distributions, with the implication that changes in individuals' beliefs about distributions may change their attitudes; this is consistent with much research on the effects on behavior of providing "social norm" information (Thaler and Sunstein 2008). We illustrate this point with research we recently completed (Wood et al. 2012) to study attitudes toward alcohol consumption. Regression analyses found that students' level of concern about their drinking, and their self-perceived risk of suffering from alcohol-related disease over the next twenty years, is predicted partly by their absolute level of drinking, not at all by the relation of their level of drinking to what they think the average level of drinking is, but strongly by the percentile ranked position that they perceive themselves as occupying within the community. This study supports the suggestion that, consistent with the rank principle and DbS, people's judgments about the acceptability of their own behavior (e.g., alcohol consumption) are influenced by perceptions of their ranked position within their community rather than their true position. Thus an individual who is in fact in the most heavily drinking 1% of the population, but who incorrectly believes that he or she is only in the top 20%, will be much less concerned than if they correctly perceived the distribution. This suggests a foundation for successful interventions through the provision of more psychologically salient social norms (Perkins 2003).

A sampling approach may therefore offer a process-level account of across-individual variation in judgment and decision making, although research is at an early stage. Next, we suggest that the same basic model can offer a useful perspective on within-individual variation.

A key source of variation in individual decision making is given by the context of choice options available at the time when a decision is made. For example, the amount of money that people will pay for a given product, or the size of food portion they choose, is highly dependent on the context of available options. Effects of choice context on judgments and real-world behavior are both ubiquitous and large. Thus food consumption is heavily influenced by both the context of available options and the amount consumed by social companions (for a brief review, see Wansink et al. 2009), leading to up to 30% differences in consumption. Removing a 12 fl oz option from a menu of soft drinks leads around 25% of consumers, who previously chose a 16 fl oz drink, to switch to a larger one; that is, the 16 fl oz option is chosen less often when it becomes the smallest available option (Sharpe et al. 2008). Consumption—not just purchase—is affected. Similar and substantial, rank-based context effects are seen in financial choices and decision making. Thus substantial within-individual variation in choice can emerge from changes in choice context. Such results argue strongly against the assumption of stable, context-independent within-individual preferences and are therefore difficult to account for within the framework of conventional accounts of decision making derived from economic theorizing. Sampling models such as DbS, in contrast, are starting to be applied to the within-individual variation in decision making and choice that results from shifts in the decision-making environment.

The predictions of DbS, as a rank-based alternative to conventional models derived from economics, can be tested experimentally. Recently, using an incentive-compatible procedure designed to elicit willingness-to-pay (the BDM procedure), we found that the amount of their own money that participants were willing to pay to purchase a bag of mixed sweets of a given size was strongly influenced by the relative ranked position of the bag's size in an array of nine choices (Becker et al. 1964). Participants paid up to 50% more for a given bag when it was the fourth smallest than when the same size of bag was the second smallest in the choice array (range and mean held constant). Moreover, given doubts about the true incentive-compatibility of the BDM procedure (Mazar et al. 2009), we replicated strong effects of rank when passersby were merely given the option of purchasing (again with their own money) a bag of chocolates from a display.

In summary: alternative models of decision making can offer a perspective on within-individual variation that is difficult for conventional models to grasp.

Conclusion

There is considerable evidence of within-individual variation in decision making even when the decision environment remains, in normative terms, unchanged. Such variation can result from a multitude of sources, such as fatigue, mood variation, or changes in choice context. Such variation is difficult to account for in a motivated fashion by models of decision making derived from economic approaches, despite the success of economic approaches in accounting for within-individual changes in strategy choice in game scenarios.

Evolutionary accounts of across-individual variation in decision-making personality are beginning to emerge and have been developed much further in nonhuman animals. There remains a need for further integration of descriptive approaches to human personality with ecological approaches to animal personality; more research is needed to provide evolutionarily plausible accounts of the adaptive function of mood variation.

Finally, alternative sampling models of decision making exist that have been derived from psychophysics rather than economics. Although at an early stage of application, these models may shed light on within- and between-individual variation in decision making, thus holding out the prospect of a more unified account of both within- and across-individual variation than has previously been possible.

Acknowledgments

This research was supported by the Economic and Social Research Council (UK) grant RES-062-23-2462. For extensive and helpful comments on the manuscript we thank Niels Dingemanse, Jeffrey Stevens, Rosemarie Nagel, and Yaakov Kareev.

First column (top to bottom): Sasha Dall, Peter Todd, Laura Schulz, Gordon Brown, Martin Kocher, Sasha Dall
Second column: Max Wolf, Franjo Weissing, Sam Gosling, Laura Schulz, Niels Dingemanse
Third column: Sam Gosling, Gordon Brown, Niels Dingemanse, Ido Erev, Max Wolf, Sasha Dall, Ido Erev

15

Variation in Decision Making

Sasha R. X. Dall, Samuel D. Gosling, Gordon D. A. Brown,
Niels J. Dingemanse, Ido Erev, Martin Kocher,
Laura Schulz, Peter M. Todd, Franjo J. Weissing, and Max Wolf

Abstract

Variation in how organisms allocate their behavior over their lifetimes is key to determining Darwinian fitness, and thus the evolution of human and nonhuman decision making. This chapter explores how decision making varies across biologically and societally significant scales and what role such variation plays when trying to understand decision making from an evolutionary perspective. The importance of explicitly considering variation is highlighted, both when attempting to predict economically and socially important patterns of behavior and to obtain a deeper understanding of the fundamental biological processes involved. Key elements of a framework are identified for incorporating variation into a general theory of Darwinian decision making.

Why Is Variation in Decision Making Important?

Decision Making from a Darwinian Perspective

How individuals behave is of profound importance in both the human-centered and the biological sciences. In the human-centered sciences, as well as in attempting to understand the human condition and what its key influences are, it is often important to be able to predict individual behavior and what will happen to it as a result of specific interventions. From an evolutionary perspective, for any evolved entity, how this entity behaves over its lifetime is central to determining how it accumulates resources and allocates them to the crucial demands of surviving to influence how its genes persist into future generations. From such perspectives, decision making—how individuals choose to allocate their behavior—is likely to be a key focus. Here we explore how decision making varies, what role such variation is likely to play when trying to understand decision making from a Darwinian perspective, and how we can use such an understanding to promote more effective models of human behavior.

Variation in Decision Making Matters!

Why focus on variation in decision making? The simple answer is that there is abundant evidence from a range of perspectives that the way in which the allocation of behavior varies really matters for understanding and predicting the human condition and in broader biological contexts. To briefly illustrate:

Variation Is Needed to Predict What Actual People Do

Comparing the most popular process-based models of human behavior with the most successful applications of models of human behavior in the "real world" reveals an interesting distinction. The popular models (e.g., prospect theory; Kahneman and Tversky 1979) focus on capturing typical human behavior and pay little attention to individual differences. In contrast, many of the successful applications of behavioral models are based on the assumption that observed behavior reflects robust individual differences in decision making. The most important examples of these applications are recommender systems that are used in e-commerce sites, like amazon.com (see Resnick and Varian 1997). These systems assume that future behavior of a target individual (e.g., a person's tendency to buy a specific product) can be predicted from the behavior of individuals that are similar to the target individual. In other words, these systems use past behavior to classify individuals to one of several classes, and use this classification to predict behavior. Another important set of examples involve psychometric and psychological tests. These tests measure individual differences and are effectively used to predict future behavior. For instance, performance in the SAT test correlates significantly (with coefficients between 0.36 and 0.65) with performance in college (Ramist et al. 1993).

Personality Variation Influences Life Outcomes

In the human domain, personality measures have been shown to predict consequential life outcomes, such as mortality, divorce, occupational choice, occupational attainment, health, community involvement, and criminal activity (Ozer and Benet-Martinez 2006; Roberts et al. 2007). For example, one 14-year prospective study found that the personality trait of trust predicted mortality with an effect size (r) of = –.22 (Barefoot et al. 1998). Another study found that girls' scores on a test of aggression predicted (r = .30) their likelihood of being divorced (vs. intact marriage) 28 years later (Kinnunen and Pulkkinen 2003). A recent meta-analysis of prospective longitudinal studies that controlled for background factors (e.g., existing health conditions, age, gender) found that the magnitude of effect sizes for personality on such important life outcomes were similar to those found for socioeconomic status (SES) variables and cognitive ability (Roberts et al. 2007). For example, in the domain of marital outcome, scores on conscientiousness, neuroticism, and agreeableness predicted divorce with average effects (r) of –.14, .18, and –.16, respectively; in the same set of studies, SES predicted divorce with average effects of .05 (Roberts et al. 2007).

Variation Influences Society

Differences among human societies can result from a range of factors. These include the fact that small groups often consist of heterogeneous decision makers whose composition can differ substantially due to stochastic sampling effects, specific decision makers self-select into certain types of groups (e.g., assortative matching), specific voting or decision-making rules vary within groups, and preferences often shift (group polarization; for an early overview, see Pruitt 1971) as a direct consequence of group membership. An example of the latter, the "risky shift" (Stoner 1961), implies that groups consisting of, on average, more risk-averse group members come up with even more risk-averse unanimous decisions, while groups whose members are, on average, more risk-prone make more risky unanimous decisions. Thus, existing individual predispositions can be reinforced in the group decision-making process.

In general, it is unclear whether unitary groups—groups which do not have any internal conflict in terms of objectives and that have to come up with a joint group decision—are "better" decision makers than individuals. The relative advantage of groups over individual decision makers depends on the nature of the task, on the organization of the group, and the inclination to fall prey to group decision-making biases such as "groupthink" (Janis 1972). One important advantage of group decision making is the possibility of aggregating knowledge if information is distributed heterogeneously among group members. An apparent disadvantage of group decision making is the decision-making process, which is supposed to be more complicated and slower in groups than in individuals. Both effects are likely to be significant for individuals that self-select into groups. Across different simple economic tasks, a surprisingly uniform proportion of about two-thirds of human decision makers self-select into a group decision-making mode, whereas one-third prefers to decide alone (Kocher et al. 2006).

Cultural Variation: Practices, Freedom, and Wasteful Consumption

The study of dress codes reveals two important variation-related problems. First, conservative religious groups use strict dress codes to control their members and reduce their ability to leave the group (Arthur 1999). The strict codes reduce the members' opportunities to interact with members of other groups and facilitate enforcement of the groups' rules. Thus, the among-group variation, implied by strict dress codes, helps reduce within-group variability and can impair the freedom of their members. At the same time, however, strict dress codes have many positive effects. Without dress codes, rich group members can benefit from signaling their high status with expensive clothes and/or with high within-person variability (e.g., changing and washing trousers every day by modern Israelis). It is easy to see that such outcomes can lead to wasteful cultural practices since "wastefulness" is often a reliable signal of wealth (e.g., modern Israelis use much more water than they otherwise would).

Variation Has Ecological Impacts

Behavioral variation is often correlated with resource-use and resource-exploitation strategies. Since competition is most intense among the most similar strategies, intraspecific competition may be reduced in the presence of variation. Variation may therefore be associated with a higher carrying capacity of the population. The same holds when different variants have a synergistic effect upon each other (such as in division of labor). For example, bold-shy pairs of great tits seem to have a higher reproductive output than either bold-bold or shy-shy pairs (Both et al. 2005). Indeed, such variation seems to be important for reproductive outputs even where such effects are more complex, such as in zebra finches where partner matching is only important for the most exploratory and aggressive individuals (Schuett et al. 2011).

Some recent theoretical and empirical work in the field of animal personality may be used to illustrate that variation may affect ecological processes (Cote et al. 2011; Fogarty et al. 2011). For example, Fogarty et al. (2011) modeled the spread of invasive species and concluded that the populations consisting of a mix of social and asocial individuals spread faster than populations consisting of either one. Furthermore, experimental work on invasive fish species showed that the average level of boldness in the population affected dispersal distance of individuals, implying that the social environment imposes selection on individual actions.

Variation Influences Evolution

It is well known to game theorists that the existence of even small degrees of random variation can have major implications for game theoretical predictions. The trembling hand approach (where optimal decisions are implemented with error) of Reinhard Selten (1975) is used in economic and evolutionary game theory to distinguish "unreasonable" equilibria (which only make sense in the absence of variation) from more reasonable ones. Indeed, such decision errors are crucial for solving the more computationally intensive game theoretic models such as state-dependent dynamic games, which are growing in popularity in evolutionary biology (Houston and McNamara 1999).

In addition to selecting among equilibria, new equilibria may become available if there is sufficient (stable) variation in behavior. Several modeling studies on the evolution of cooperation demonstrate, for example, that stable cooperation is achievable and stable once there is sufficient behavioral variation in the population (McNamara and Leimar 2010). McNamara et al. (2004) consider a finitely repeated prisoner's dilemma game where the interaction between two players is terminated once any of the players' defects. In the absence of variation, "always defect" is the only equilibrium. If, however, nonadaptive alternative strategies arise repeatedly in the population (e.g., by mutation), then it may actually be adaptive to "exploit" this variation by adopting a slightly more cooperative strategy than always defect. This may initialize a positive feedback process of "variation begets variation" that eventually leads to the establishment of cooperation. There is a second reason why variation can stabilize cooperation. Several models have stressed that partner choice and partner inspection are generally favorable for the evolution of cooperation. However, choosiness and inspection will typically be costly. To persist stably in the population, these costs have to be off-balanced by some benefits. Such benefits can only accrue if there is variation (choosiness does not make sense, if there is no variation to choose among). Indeed, McNamara et al. (2008) showed that in a snowdrift game with partner choice, the degree of cooperation achieved was positively related to the degree of behavioral variation. In a different setup, Wolf et al. (2011) showed that small degrees of variation can select for social sensitivity which, in turn, stabilizes and enhances this variation. Again, the outcome is qualitatively different than in the absence of variation. Finally, McNamara et al. (2009) have shown how small amounts of variation, in how trustworthy individuals are, select for conditionally trusting strategies which, in turn, select for further individual differentiation in trustworthiness.

In addition to changing the outcome of evolution in a qualitative way, the existence of variation can have a quantitative effect, since it can speed up evolution enormously. Evolutionary biologists have always been amazed at how rapidly populations can adapt to novel circumstances. In particular, the rapid evolution of a highly integrated phenotype (requiring the evolution of correlations among a multitude of traits) is, at least to some evolutionary biologists, an unresolved mystery; this is sometimes referred to as Haldane's dilemma (Haldane 1957). The existence of a "standing stock" of variation may provide an explanation. First, evolution is much less mutation limited in such cases. If selection can "feed" on existing variation, it does not have to wait for the appearance of the right type of mutations. In individual-based simulations with realistic parameter settings, the mutation process typically imposes severe limitations on the speed of adaptation. This is particularly true in cases where selection has to "solve" a design problem in a multidimensional trait space. West-Eberhard (2003) suggested that a correlational structure as found in personalities might enhance the adaptive potential of a population. These correlations might gradually be shaped by selection in reaction to small-scale fluctuations of the environment. When the environment suddenly changes more dramatically (in the direction of the earlier small-scale fluctuations), the required correlational structures are already in place and do not have to evolve from scratch.

In addition to these direct effects on the course of evolution, the phenomenon of personality differences (stable behavioral differences among individuals) may force scientists to reconsider evolutionary processes in a fundamentally new way. Until now, the trend has been to atomize behavior into different functional contexts (e.g., finding food vs. finding mates vs. caring for young etc.), which are subsequently tackled in separate evolutionary models. This may be quite misleading. If, for example, boldness in response to novel and potentially dangerous environments is fundamentally linked to aggressiveness in intraspecific contests, it does not make sense to investigate the "boldness game" as being separate from the "aggressiveness game." In fact, a behavior can appear maladaptive when viewed in isolation but make perfect sense (e.g., as a costly signal) when viewed from a more integrative perspective. There are several examples where this does indeed seem to be the case (e.g., Johnson and Sih 2005).

What Are the Scales at Which Decision Making Varies?

There is an interesting dichotomy between the human-centered sciences and biology in that there is a tendency in the former to assume stable individuality as the default. In contrast, students of nonhuman animal behavior who think about its evolution tend to presume that behavior is highly flexible and optimally tactical (from an evolutionary perspective) in every possible context. For the purposes of this chapter, it is important to provide a brief overview of the patterns of variation in behavior and decision making that are of interest to researchers of human and nonhuman behavior (i.e., that have significant economic, social, and ecological consequences).

Temporal patterns of variation can be distinguished by whether they operate on developmental versus intermediate versus moment-by-moment timescales. Evolutionarily significant units of variation range from within an individual, among individuals, to groups and populations. The latter are distinguished biologically by the fact that populations tend to be divided by gene flow, whereas groups are divided by behavior (interactions). In the human species, populations can be thought of as being synonymous with human cultures as they are both defined by limits to replicator spread (population: genes; culture: cultural replicators like memes). The key difference is that cultural barriers are very often permeable to highly adaptive (useful) variants (e.g., metal axes replacing stone axes), whereas this will not necessarily be the case for gene flow among populations.

Are all patterns of variation interesting? The term "animal personalities," for example, refers to variation in highly structured behavioral types: individuals differ in their behavior and these differences are correlated over time and across different contexts (e.g., aggression, foraging, mating). To judge whether a given correlation is surprising, it seems important to have a "null model" at hand; that is, a model of trait correlations (over time, across context) that serves as a standard of comparison. At present such a null model does not exist. This may be for good reasons: developing a nontrivial "general-purpose" null model in the context of behavioral types may be very difficult or even impossible. The trivial null model is that there are no correlations. This null model is not really interesting: rejecting it is like tearing down a straw man in which nobody really believes. The reason one does not believe in the trivial null model is that any "realistic" mechanism will create some correlations. However, the details of such correlations will depend very much on the mechanisms in question. As a consequence, a nontrivial null model will have to refer to a specific underlying mechanism. This is problematic, since we typically do not have a good idea of what the "real" mechanisms are and what kind of correlation structure they will generate. So, even if we ignore the fact that there is a multitude of such nontrivial null models, we are currently unable to characterize their statistical properties (specifically or in general). Therefore, for the time being we will have to continue using "biological intuition" when judging whether observed correlation structures are interesting or not.

Are there any general patterns of variation nature? Repeatability, the proportion of the observed variance in the population that can be explained by stable differences among individuals, is of key importance when studying labile phenotypic traits like behavior and physiology, and it has therefore been documented for a wide variety of behavioral traits across a wide range of species. Recent meta-analyses show that values of repeatability are on average 0.37, implying that a major proportion of the observed variation in behavior (e.g., up to 0.63) is within rather than between individuals of the same population (Bell et al. 2009). Some of the variation in behavior is evident at the population level, shaped by local adaptation to population-specific ecological conditions like predation risk. For example, work on three-spined sticklebacks, *Gasterosteus aculeatus* (Figure 15.1), revealed that 9.4% of phenotypic variation in predator-inspection behavior could be attributed to variation among 12 Welsh populations, 41.2% to variation among individuals within these populations, and 49.4% to within-individual variation (Dingemanse et al. 2007). To further illustrate how behavior and decision making varies at different scales, we now proceed by highlighting some interesting examples from the human-centered and biological literature.

Variation within Individuals

Standard economic theory relies on the assumptions that individual preferences are relatively stable over time (Stigler and Becker 1977) and that behavior is well-approximated by assuming a representative individual, especially if the decision-making situation has a unique optimum or equilibrium. The former assumption implies that decision makers choose the same option in the same decision-making situation (under the same conditions/states) over time. The

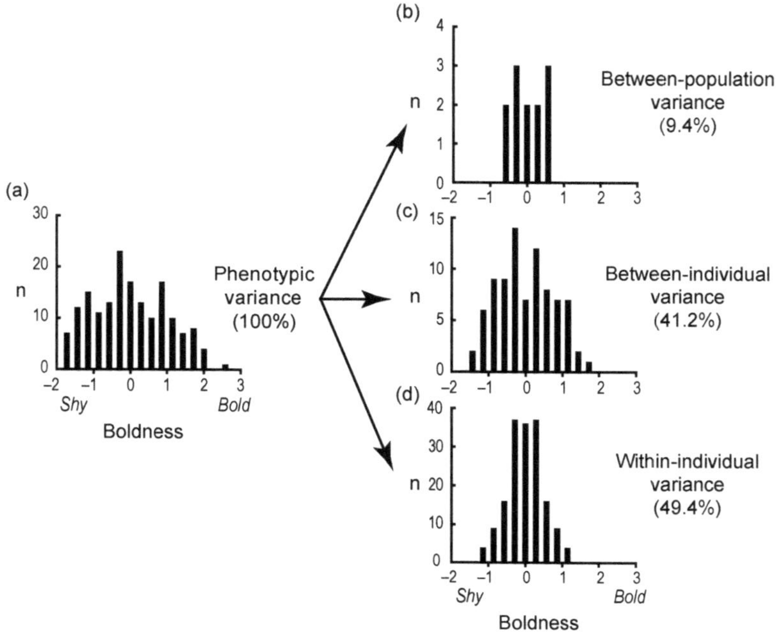

Figure 15.1 Variation in predator-inspection behavior (ranging from shy to bold) across hierarchical levels. Histograms show the variation in "boldness" at each level. (a) Distribution of the raw phenotypic data (n = 168 assays: 2 repeat assays for each of 7 individuals within each of 12 populations). (b) Distribution of population-average values (n = 12 populations). (c) Distribution of individual-mean values (n = 84 individuals) expressed in deviations from the population mean (i.e., we show here the distribution of individual mean values within populations). (d) Distribution of the deviations from individual-mean values (n = 168 deviations) (i.e., the within-individual variance). The data used in this example are from exploration of altered environments by three-spined sticklebacks (see Dingemanse et al. 2007).

latter assumption is not critical for the application of (game) theory, and more recent models often simply assume the existence of different types of decision makers with different optimal points (e.g., selfish types and reciprocal types) and specify their interactions formally. There is little dispute about the fact that the assumption of a representative individual is too restrictive in many situations. Different levels of other-regarding preference (e.g., Fehr and Schmidt 1999; Bolton and Ockenfels 2000; Charness and Rabin 2002; Fehr et al. 2007), different degrees of rationality or bounded rationality (e.g., Nagel 1995; Costa-Gomes and Crawford 2006; Crawford and Iriberri 2007) and different attitudes toward uncertainty (e.g., Abdellaoui et al. 2011) across decision makers lead to heterogeneity in final choices and behavior.

The intertemporal stability/plasticity of individual preferences in human decision makers is a much more controversial issue, and it affects the basic modeling methodology in standard microeconomics (e.g., crucial axioms associated with preference revelation; Samuelson 1938). No one denies that preferences could and would change over long time horizons (e.g., in the context of trust/trustworthiness and age; Bellemare and Kröger 2007; Sutter and Kocher 2007). However, studies that look at changes in behavior over short time horizons by the same decision makers have recently been conducted, and the results are not fully conclusive. Some find quite high levels of robustness in behavior over time (Volk et al. 2012), others report very low levels of robustness for the same behavioral tendencies in slightly different contexts (Blanco et al. 2011). Imprecision is also noted as one important reason for intertemporal instability (Butler and Loomes 2007). An empirical problem is that providing decision makers with exactly the same tasks over time is confounded with learning, and providing decision makers with similar tasks over time is confounded with potential context or framing effects. Hence, intertemporal stability of individual preferences is hard to measure precisely in the controlled environment of an experiment.

Recent studies of learning reveal an interesting general pattern: large within-subject variability, which can sometimes swamp species differences. One indication of this pattern is presented by Erev and Haruvy (2012). Their review of the classical learning literature shows that many of the phenomena, originally documented in studies with different animals, can be reliably replicated in the study of human behavior in simple computerized experiments. One example is the partial reinforcement extinction effect (Humphreys 1939). Human and nonhuman subjects learn less in a noisy environment (under a partial reinforcement condition), but learning in this noisy setting is more robust to extinction. At the same time, the results of these studies reveal large within-subject variability (see Shafir et al. 2008). For example, when faced with a repeated choice between "3 with certainty" or "80% chance to win 4; 0 otherwise" the typical subject does not learn to prefer one of the two options. Rather, the typical subject selects the risky prospect in about 60% of the trials. High within-subject variability was documented in human subjects even after 400 trials with immediate feedback. Erev and Haruvy (2012) show that the main experimental results can be captured with simple models which share the assumption that subjects tend to rely on small sets (about four) of past experiences in similar situations. Different samples are used in different trials; as a result, these models imply a payoff variability effect (Myers 1960; Busemeyer and Townsend 1993): large payoff variability increases within-subject variability.

Variation among Individuals

Variation in decision making among individuals can occur at different levels. Individuals can differ in (a) the mechanisms that underlie behavior (e.g., physiological or cognitive systems underlying behavior), (b) their evaluation systems, and/or (c) states that affect decision making (e.g., information about features of the decision problem). A "decision-making mechanism" may be viewed as a device that "chooses" an action (possibly in a nondeterministic way) for a given set of external stimuli and (external or internal) state variables. It makes sense to speak about "variation in decision making" if the underlying decision-making mechanisms vary (i.e., if in a systematic way different actions are chosen despite similar states and stimuli). Such variation can, for example, occur if individuals differ in the way they receive information (e.g., variation in receptors, as described for female sticklebacks), process that information, evaluate the processed information, translate this evaluation into action, and perform the corresponding action. One complication in the above description is the concept of "state." When do we consider a "state" to be part of the decision-making mechanism, and when is it something external to it? For example, "memory" may be viewed as a state that is external to the decision-making process (e.g., the memory of previous outcomes is an input of conventions like "winner-loser" effects), but the use of memory may be a distinguishing feature of a decision-making mechanism. Even if state variables are clearly external to the mechanism, they may still create considerable confusion. Let us assume that males and females differ substantially in the way they make decisions. Technically speaking, we could still consider this as the outcome of one and the same decision-making mechanism that has "gender" as one of its input variables. On the other hand, we can always assume that there are subtle, unobserved differences in state, when individuals are seemingly taking different decisions. So we have to be pragmatic. If we are interested in systematic variation in behavior that cannot in any "obvious" way be explained by differences in environment or differences in state, we should not forget that we are also interested in variation that is patterned, stable in time, and consistent across contexts.

Ecologists and evolutionary biologists have provided convincing evidence for the existence of individual variation in decision making in a wide range of species. Empirical evidence for individual differences comes from studies in which the same set of individuals are assayed for the same behaviors repeatedly, such that the variation in the sample can be decomposed into between- and within-individual variation. Values of repeatability, the proportion of total variance that comes from individual differences, suggest that the majority of the variation in behavior is within rather than among individuals in animal populations (Bell et al. 2009). Nevertheless, repeatable variation implies that the average level of behavior differs among individuals; however, it does not imply that individuals are completely stable in their decision making (see Figure 15.2a). Fleeson (2001, 2004) discusses similar issues in a human personality context. While the existence of repeatable variation is well documented, there is growing awareness that individuals may differ in how they change their behavior in response to variations in environmental conditions or age—phenotypic plasticity (Dingemanse et al. 2010b; Nussey et al. 2007). Plastic individuals alter their behavior as a function of context, whereas nonplastic individuals do not (Figure 15.2b). Individual variation in plasticity,

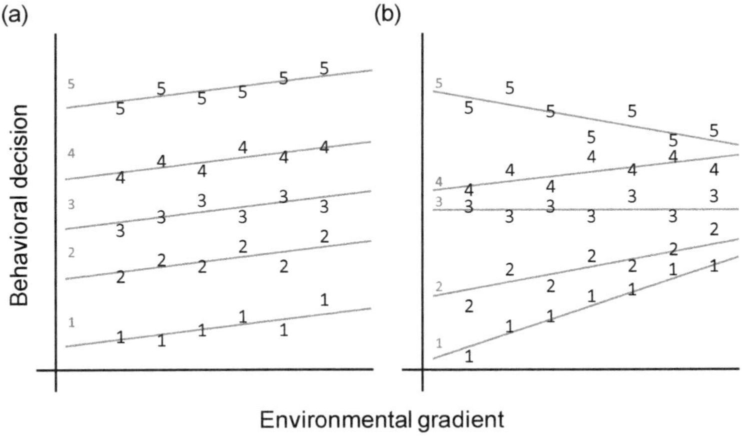

Figure 15.2 Graphic depiction of how between- and within-individual variance components are separated by plotting seven measurements of a behavioral decision (Y-axis) for five individuals (numbered) whose behavior was assayed over an environmental gradient (X-axis). For example, the Y-axis might represent aggressiveness and the X-axis conspecific density. (a) Gray lines represent the average phenotypic value of each individual; the variance among lines represents the between-individual variance. The variance in within-individual deviation from individual means represents the within-individual variance. (b) Individuals differ in average phenotype as well as in phenotypic plasticity.

termed "individual by environment interactions" in the evolutionary literature, usually explains about 5% of the variation in decision making within animal populations (Nussey et al. 2007).

As an example of long-term calibration of behavioral phenotype by some other characteristic, in humans, men who are physically large tend to be more aggressive. However, it turns out that the key parameter is not adult size, but size relative to others at a certain age in adolescence (e.g., Pellegrini and Long 2002). Men who are larger than their peers at this age adopt a more aggressive phenotype, and this persists even though their peers may subsequently grow to be as large as they are. This could be an example of what is referred to in the human personality literature as reactive heritability. That is, size variation is heritable, and aggression is calibrated to this in development; thus aggression ends up showing effective heritability in behavioral genetic studies. A similar effect occurs with extraversion, where it has recently been argued that much of the heritable variation is in fact variation in physical attractiveness. Attractive people are reinforced for initiating social interaction and become more extroverted as a consequence, causing an effective heritability of what is in fact a developmentally calibrated trait (Lukaszewski and Roney 2011; Meier et al. 2010).

More generally, dominance hierarchies provide good examples for systematic individual differences in behavior: individuals at the top of the hierarchy behave very differently in many respects (depending on the species in question) than individuals at the bottom. One might think that the position in the dominance hierarchy and, hence, behavior related to that position, just reflects individual differences in fighting ability ("resource holding potential"). In a modeling study, van Doorn et al. (2003) showed that the "winner-loser effects" on which dominance hierarchies seem to be based can evolve from scratch even if all individuals have the same fighting ability, irrespective of the outcome of previous fights. In other words, winner-loser effects (and, hence, dominance hierarchies) may not be more than a "social convention"; that is, a conditional strategy which allows individuals to coordinate their actions, thereby avoiding escalated contests. Of course, differences in fighting ability will often be reflected in an individual's position in a dominance hierarchy. However, there are various indications that this is not the whole story. First, experiments with a variety of organisms have revealed that the position on a dominance hierarchy can actually be quite arbitrary (Hsu et al. 2006). In these experiments, groups with an established dominance hierarchy are taken apart and the individuals are reshuffled over new groups. After reshuffling, the position of individuals in the newly establishing dominance hierarchies is often only loosely related to their position before reshuffling. Second, dominance is often site specific. Territorial animals are good examples: within its territory, an animal is dominant, although it is subdominant in the territory of another individual. The relative dominance status of two individuals *A* and *B* on neighboring territories can thus be shifted from high to low by observing these individuals on the transect from the center of territory *A* to the center of territory *B*. Similarly, site-specific dominance is also observed in nonterritorial species, like oystercatchers in the winter (Ens and Cayford 1996) or white-throated sparrows (Piper and Wiley 1989). Here an individual *A* can be dominant over *B* at site *X*, while *B* is dominant over *A* at site *Y*; in contrast to territory ownership, the individuals are only loosely attached to the sites where they are dominant. In these cases, dominance is apparently a social convention that is not explained by "obvious" differences, such as differences in fighting ability.

Beyond differences in competitive behavior, individuals can differ systematically in other ecologically important ways. For instance, lions manifest two types of social organization. Some are residents, living in groups, called prides. The pride usually consists of five or six related females, their cubs of both sexes, and a coalition of males who mate with the adult females. The number of adult males in a coalition is usually two. Other individuals are called nomads, who range widely, often in pairs of related males. Males have to go through the nomad lifestyle before they can become residents in a pride, which is always different from the pride into which they were born. The resident males associated with a pride tend to stay on the fringes, patrolling their territory. Both males and females defend the pride against intruders. The males associated with the pride must defend their relationship to the pride from nomad males who attempt to take over their residency position. When a group of male nomads oust the previous males associated with a pride, the conquerors often kill any existing young cubs, presumably because females do not become receptive until their cubs mature or die. A lioness will often attempt to defend her cubs fiercely from a usurping male, but such actions are rarely successful. In addition to these sex and social role differences, there are strong individual differences in the behavior of both male and female lions. The classical example concerning female behavior comes from Heinsohn and Packer (1995:1260):

> Female lions (*Panthera leo*) showed persistent individual differences in the extent to which they participated in group-territorial conflict. When intergroup encounters were simulated by playback of aggressive co-vocalization, some individuals consistently led the approach to the recorded intruder, whereas others lagged behind and avoided the risks of fighting. The lead females recognized that certain companions were laggards but failed to punish them, which suggests that cooperation is not maintained by reciprocity. Modification of the "odds" in these encounters revealed that some females joined the group response when they were most needed, whereas other lagged even farther behind. The complexity of these responses emphasizes the great diversity of individual behavior in this species and the inadequacy of current theory to explain cooperation in large groups.

Several follow-up studies have been conducted and similar differences were found in experiments with males (Grinnell et al. 1995). Again, the males within a coalition strongly differ in their tendency to launch an attack on rivaling males.

Variation among Groups

There have been many reports of systematic between-group differences in lion behavior. For example, there are pronounced differences in how prides of similar size hunt down prey in cooperative hunting efforts. There are also pronounced differences in the ways groups of male nomads try to conquer a reproductive position in a pride. These differences, however, are difficult to quantify, and it is not clear whether such consistent differences among groups corresponds to something like a "group culture" or whether it is just a reflection of the different "personalities" of their constituent members. The differences alluded to above have mainly been reported in books (e.g., Schaller 1972), more general reviews (e.g., Packer 1986), or—not unimportantly—in nature documentaries (e.g., National Geographic: Super Pride). Nevertheless, work by Sapolsky (1990) indicates that there may also be stable differences in behavior among baboon groups that persist over generations. In one group, Forest troop, an outbreak of bovine encephalitis led to the death of several of the dominant males in the group. Compared to neighboring control groups, social relations in the Forest troop were peaceful. There was less aggression and more reciprocal grooming. Interestingly, this difference persisted over a ten-year period in which all of the original males died and were replaced by immigrant males.

Group differences that emerge from simple interactions among individuals are well documented. Group composition is often influenced by locomotion performance (Krause and Ruxton 2002) and can be seen in flocks of birds, schools of fish, and herds of mammals (Couzin and Krause 2003). For example, herds of African ungulates are often structured according to walking speed, with faster individuals at the front and slower individuals at the back of the group. In the extreme, this can result in social segregation with lactating and nonlactating females in zebras forming distinct herds (Gueron et al. 1996). Another mechanism that can structure ungulate herds is active preference. For example, Thomson's gazelles, *Gazella thomsoni*, actively associate with Grant's gazelles, *Gazella granti*, because the latter are taller and have a greater ability to detect an approaching predator, such as a cheetah, *Acinonyx jubatus* (FitzGibbon 1990). Grant's gazelles appear to benefit, in turn, from the presence of the Thomson's gazelles because the latter are the preferred prey of cheetah. Given an attack on the group, the per capita risk of a Grant's gazelle is therefore lowered by the greater group size, due to the presence of Thomson's gazelles.

Variation may also exist at higher levels in a hierarchy. In particular, populations might be comprised of social neighborhoods differing in decision making. Variation among social groups has, for example, been observed in shoaling fish, where shoals within the same population differ in average behavior (Magnhagen and Staffan 2005). Indeed, between-group variation can emerge at equilibrium when members of different groups are motivated to behave like members of their group, and are punished when they behave like members of other groups. This can be illustrated in a multiagent market game that is played by 20 players: 10 Red and 10 Blue players. Each player has to select a location: left or right. Agent *j* gains 1 point for every other member of *j*'s group who selects like her, and loses 1 point for each member of the outside group that does likewise. At equilibrium, all Red players will select one location and all Blue players will select the other.

Factors that motivate group members to behave similarly are well documented. Social contagion ("peer effects" in economics) is typically thought of as imitative behavior that is not merely a result of homophily or of the fact that neighboring social agents are likely to be subject to common influences. Specifically, there has been much recent interest in research surrounding suggestions that imitative behavior evolves in networks over time, such that smoking cessation and levels of alcohol consumption propagate readily through social networks (Christakis and Fowler 2007, 2008). It is, however, difficult to separate correlated influences (Ioannides and Topa 2010; Manski 1993), so interpretation remains controversial (e.g., Cohen-Cole and Fletcher 2008). Why does social contagion occur? Imitation can serve a social learning role (Chamley 2003), may reduce cognitive effort (Epstein 2001), or may reflect a concern with relative social position (e.g., Clark and Oswald 1998). The social contagion literature has remained largely silent (or has made claims inconsistent with other literature) as to how exactly the relevant social comparisons occur mechanistically. Effects related to social contagion have been extensively studied in agent-based network models. Such approaches (e.g., Schelling 1978) have provided useful insights into areas such as collective behavior in ants and traders (Kirman 1993), swarming behavior (Reynolds 1987), crowd behavior (Dyer et al. 2009), population group size (Axtell et al. 2002), cultural dissemination (Axelrod 1984), and imitative voting (Bernardes et al. 2002). Most of these models incorporate an imitation parameter or mechanism of some kind, such that agents in a network are likely to change in the direction of or influence other (typically local) agents with whom they are connected. It is then typically shown that interesting emergent behavior or segregation arises as such networks evolve.

Variation among Populations/Cultures

Variation among populations has been given substantial attention in the animal behavior literature, particular with regard to the question of whether populations with different ecologies (e.g., that vary in predation risk, population density, food availability) differ in the average level of behavior expressed (Magurran 1998). For example, populations of fish which live in environments with a high predation risk are on average less bold compared to populations that experience a low predation risk (Huntingford et al. 1994).

Human personality research has implied that the structure of behavior, as captured by the Big 5 personality traits, is similar in different cultures: one might not always get exactly the same structure, but it is remarkable how often similar dimensions arise (e.g., Gosling and John 1999; Weinstein et al. 2008; McCrae and Allik 2002). In contrast, comparisons of behavioral syndrome structure (the direction and strength of correlations between behaviors) across animal populations do not generally confirm such ubiquity. For example, comparative work on three-spined sticklebacks mentioned above reveal that the ecological conditions of populations, particularly predation risk, predict syndrome structure. Specifically, populations of sticklebacks may generally be characterized by positive associations between aggressiveness, activity, and exploratory behaviors, but these associations are tighter in populations living sympatrically with predatory fish, both in European (Dingemanse et al. 2010a; Dingemanse et al. 2007) and North American populations (Bell 2005; Bell and Sih 2007). Therefore, it appears that local selection regimes shape trait associations, which may explain population differences in genetic correlation structure across North American populations (Bell 2005). At the same time, there is also evidence that certain genes are only expressed in the presence, and others only in the absence, of predators (Dingemanse et al. 2009), implying that any population differentiation at the phenotypic (observable) level is not necessarily underpinned by population-genetic variation.

In studies on humans, there is an enormous body of literature which assesses the robustness of results to cultural variation first found in simple lab-based economic decision-making situations in Western countries. Attitudes to uncertainty, negotiation behavior, cooperative behavior, reciprocity, trusting behavior, sanctioning behavior, and many more have been compared in controlled and incentivized studies with human decision makers across different cultures. Overall, differences between student populations in different countries are much smaller than expected a priori. Exceptions to this trend seem to be for punishment behavior (Herrmann et al. 2008) and perhaps conditional cooperation (Kocher et al. 2008); other differences have also been found, but they are not always robust. Sometimes the variation in behavior within a given culture is greater than the variation in behavior across cultures (Kocher et al. 2008), but defining the boundaries can be difficult. Numerous studies for the ultimatum game, for instance, have reported small or nonexistent differences in the behavior of student participants in experiments across different countries around the world (for an overview, see Camerer 2003).

In contrast, larger and persistent differences emerge when comparing standardized behavioral tests in small-scale societies. Henrich et al. (2004) find that the behavioral variation across 15 selected small-scale societies when playing standard economic games (ultimatum game, public goods game) is extremely high, but it is consistently related to two factors: the higher the degree of market integration and the higher the payoffs to cooperation, the greater is the level of cooperation in experimental games. Nevertheless, recent studies show that potential cross-cultural differences in norms are not simply changed in different cultural environments. While humans, for instance, tend to adjust their tipping behavior when they travel (Azar 2004), they do not make (full) adjustments in other domains. Fisman and Miguel (2006) show that UN diplomats from high corruption countries accumulate significantly more parking tickets in New York. Note that diplomatic immunity implies that there is almost zero legal enforcement for these violations. Hence, a standard model of decision making would not predict any differences, and the results indicate that norms are transported to different situations.

What Proximate Mechanisms Underpin Variation in Decision Making?

Patterns of human and nonhuman behavior have been the subject of substantial research for many years. It is evident from our brief tour of behavioral variation at different scales that such patterns can vary substantially. Therefore, it is important to review the range of mechanistic processes that can give rise to variation in decision making over the psychologically, economically, and ecologically significant scales that we have focused on in this chapter. Such processes can be thought of as falling within the remit of the so-called "proximate" explanations (Tinbergen 1963) for the existence of variation in decision making.

Within Individuals

At the most fundamental level, the physiological mechanisms that control behavior (e.g., neural processes) are likely to be subject to stochasticity in the basic biochemical processes that underpin them (e.g., due to quantum fluctuations at the molecular level). The details of such "essential stochasticity" (e.g., synaptic noise) are likely to be important as they may generate substantial within-individual variation in behavioral outcomes.

Beyond neurophysiological noise, cognitive processes can also generate behavioral variability. The decision-by-sampling model (Boyce et al. 2010; Stewart et al. 2006)—a cognitive model of human judgment and decision making—assumes that attributes are judged in terms of their ranked position within a retrieved sample context (for further details, see Brown et al., this volume). Thus variability within individuals can result from different samples being retrieved from memory on different occasions, or from changes in choice context. Cognitive models have addressed the role of cognitive (and potentially variable) samples in drawing inferences and making judgments (see, e.g., Fiedler and Juslin 2005). For example, the MINERVA-DM model (Dougherty et al. 1999) has been used to explain judgments of likelihood within a memory model, and limitations of and individual differences in the capacity of general working memory have been referred to in accounting for bias in probability judgment (Dougherty and Sprenger 2006). Another tradition of research relevant to variability has suggested that participants represent distributions (e.g., of the location of an item in a sequence) but use samples from these distributions to make judgments, which can be repeated leading to improved average estimates such that successive estimates have uncorrelated errors (Vul et al. 2009; Vul and Pashler 2008).

Models of decision making via heuristics (e.g., Kahneman et al. 1982; Payne et al. 1993; Gigerenzer et al. 1999) posit that many choices and judgments are made using simple heuristics or rules of thumb, which often yield "good" decisions when used in appropriate contexts. These heuristics typically use little information and process it in a limited manner to allow rapid (or "fast and frugal") decision making, making them an evolutionarily plausible alternative to traditional axiomatic approaches to rationality. Many heuristics have been explored that are algorithmically specified in terms of their exact information-processing steps. Less well-explored and understood are the means by which different heuristics are selected to apply to different tasks, which could be a considerable source of both within- and between-person variability in decision making. Studies have found evidence for appreciable variation in the use of heuristics across individuals (Bröder 2012; Gigerenzer et al. 1999), often because different heuristics can produce approximately the same quality of decisions in particular tasks. Indeed, it is considered important to compare the ability of multiple heuristics to account for participant data, so that this between-person variability can be discovered (Brighton and Gigerenzer 2011); less is known about variability in each individual's use of particular heuristics over time or in different situations. Variability in the decisions made by an individual (as opposed to variation in which heuristics they use) can be accounted for by specific heuristics. Many heuristics are deterministic in their operation and thus will not produce this variation—one example is take-the-best, which chooses between two options (e.g., which of two desserts has more calories) by comparing one cue or feature at a time until the first cue is found that points to one or the other option (e.g., first considering how much butter each dessert contains, and then if butter content is equal, considering how much sugar each has, and so on). In this case, the cues are searched in a fixed order, which can vary between people based on their learning history (Gigerenzer and Goldstein 1996). Other related heuristics incorporate external environmental influences or stochasticity to create decision-to-decision variability. For instance, the take-the-last heuristic operates similarly to take-the-best but uses cues that were previously successful in producing choices (introducing variability through contingencies of learning), whereas the minimalist heuristic uses cues in a random order. Other heuristics base their decisions on aspirations determined by samples experienced over time, which will vary from one set of experiences to another (as in sequential mate search by Miller and Todd 1998; see also the sample-based heuristics in Pachur et al. 2011).

From the standard economic perspective, it is constructive to distinguish between two sources of variability in behavior. The first class involves variability that can be observed even if individuals are fully informed of the incentive structure and are fully rational (i.e., behavior is at a Nash equilibrium). Within-subject variability emerges at equilibrium in situations in which the agents do not want to be predictable. One set of examples of this comes from constant sum games with unique equilibria, like the matching penny game presented in Table 15.1.

A second source of variability includes variation that is likely to emerge when the agents do not know the incentive structure and have to rely on their past experiences (Erev and Haruvy 2012). One source of within-subject variation in this setting is exploration. Some level of exploration is necessary (and assumed by all learning models) to collect information. Another source of within-subject variability is assumed by sampling models of learning. These models imply that different choices are expected to be based on different

Table 15.1 A two-person matching penny game.

	A2	B2
A1	1, –1	–1, 1
B1	–1, 1	1, –1

samples of past experiences. A third source of individual variability involves behavioral changes that can be described as adjustments to changes in incentive structure. One set of examples comes from the effect of aging, which can lead to within- and between-subject variation in behavior. Indeed, there is an extensive literature on age changes in personality, but to link this to changes in "incentive structure" it is probably worth distinguishing the "internal" (e.g., puberty) and "external" (e.g., having children, getting a job) influences that can impact age changes (e.g., Roberts et al. 2006; Roberts and Wood 2006; Soto et al. 2008, 2011; Srivastava et al. 2003), and translate these into incentives.

Among Individuals

Continuing to think economically, between-subject variability can emerge at equilibrium in situations in which the payoff for each rational agent decreases with the number of other agents that behave like her. A simple subset of this set of situations is represented in the market game, presented in Table 15.2, in which each of two sellers has to select between two locations, and payoffs are maximized (for both players) when each seller select a different location. At the pure strategy equilibrium of the game, the sellers select different locations (i.e., there is variation in the choice of position).

More generally, there are a range of processes that can underpin variation among individuals in how behavior is expressed. On the one hand, it is possible for individuals to be employing exactly the same decision-making process (e.g., the same rule for allocating behavior as conditions vary) while experiencing different local conditions from one another (e.g., being more or less hungry due to idiosyncratic—chance—recent history of access to food). This can cause individuals to differ from each other in how they behave at any given moment in time. Indeed, one source of learning-induced between-subject variability is implied by the fact that different agents are likely to experience different outcomes. Alternatively, it could be that individuals follow distinct decision-making processes (e.g., due to genetic variation in condition-dependent behavioral rules), in which case differences in behavior among individuals will be evident even if individuals have had exactly the same experiences. For instance, some models of learning assume between-subject variability in learning parameters.

There can be substantial impacts of early developmental conditions on how individuals behave. Such ontogenetic programming (e.g., early environment and maternal effects) can amplify slight (e.g., stochastic) initial differences

Table 15.2 A two-person market game.

	A2	B2
A1	−1, −1	1, 1
B1	1, 1	−1, −1

in experience or nutrition and lead to profound differences in the way that individuals behave. For example, Bateson et al. (2004) have argued substantial variation in adult human eating behavior can be explained by early life physiological "programming" by maternal nutritional status during pregnancy. Mothers that are nutritionally stressed while pregnant induce fetuses to "expect" food to be limited and so switch on a "thrifty phenotype" in their progeny, who crave energetically dense foods and store fat whenever possible. This effect is argued to underpin a significant amount of the variation in patterns of obesity and Type II diabetes being documented in modern, food-rich developed societies.

Among Groups

As in between-individual variation, it is important to appreciate that variation among groups can emerge in two distinct ways. First, different groups may be formed by distinct types of individuals. For instance, in many species, males and females often have different basic (e.g., energetic) demands, and therefore they behave differently and form separate groups (Ruckstuhl 2007). Second, individuals may not necessarily vary in any fundamental sense among groups (i.e., they all follow the same decision rules) but the group-conditions may vary idiosyncratically (e.g., different groups just happen to be different sizes, which cause different levels of antipredatory vigilance). A good illustration of this: mechanisms for individual social learning—essentially, copying the behaviors of others—can result in behavioral differences between groups as a result of within-group social convergence (e.g., conformity) after different starting conditions. That is, if asocial learners in one group hit upon a particular behavior, and those in another group discover a different behavior, social learners in both groups may copy those asocial learners, as well as each other, to the point where the two groups diverge in their behavioral profiles. A number of social learning rules have been explored, which can be roughly broken down into those determining when copying should happen and others directing whom to copy (Laland 2004). "When" rules trigger the application of a social learning strategy and include copying when a current behavior is unproductive, copying when learning asocially (e.g., through trial-and-error) is too costly, and copying when the environment, and hence the appropriate behavior, is uncertain. "Who" rules specify the other model individuals whose behavior should be copied, and different rules can determine whether or not and how quickly a population will converge on a shared behavior. Such conformity is promoted by the widely studied copy-the-majority rule (Boyd and Richerson 1985), as well as by copying successful individuals, and possibly copying others if they are just doing better than oneself. Within-group convergence is unlikely if individuals use a copy-the-rare rule, adopting uncommon behaviors of others, though this can be individually advantageous if having a novel behavior (e.g., mating display) enhances local (or even group) competitiveness. Rules for copying others in one's social network can lead to within-group homogeneity or heterogeneity depending on the network structure (Lieberman et al. 2005).

Language acquisition can be a pervasive social force for group differentiation in humans. Infants come into the world prepared to be part of any culture and language group. At birth, humans (like other primates) can distinguish any phoneme produced in any language around the world (Kuhl 1991; Kuhl et al. 1992). However, during the first year of life, perceptual tuning results in a loss of function: sounds that a newborn can distinguish (e.g., phonemic distinctions present only in Mandarin) are no longer distinguishable by, for example, an English-speaking one-year-old. Even brief exposure to a nonnative language can keep this perceptual window open. However, preservation of this ability depends on live social interaction. A 10-month-old from an English-speaking household who interacts with a Mandarin speaker for an hour a week will continue to distinguish the Mandarin phonemes at 12 months; a 10-month-old exposed to identical input from a video display will not (Kuhl et al. 2003). Case studies of hearing children of deaf adults also suggest that live human interaction is critical for language learning. A hearing child (1 years 8 months) who interacted only with his deaf mother but had abundant exposure to spoken English through television had no productive spoken language before intervention; his brother (3 years, 9 months), who briefly attended an English-speaking preschool, also exhibited severe delays (Sachs et al. 1981). These findings suggest that although widespread cultural transmission through media may reduce cross-cultural variability in human populations, specific aspects of human cognitive development may tend to maintain cultural differentiation. In particular, as long as there are distinct language groups among adults, those distinctions will be conserved insofar as children only learn language by interacting with the human beings in their immediate vicinity.

How Can Variation in Decision Making Evolve?

Given our focus on decision-making agents that are subject to Darwinian selection, it is appropriate to consider the evolution of variation in decision making, the so-called "ultimate" explanations (Tinbergen 1963) for its existence. Indeed, for many of the scales of variation we have discussed so far, there has been significant research effort devoted to understanding how variation in decision making can be favored by Darwinian selection.

Within Individuals

It is possible that individual behavior will vary in different contexts because there are distinct modules in the central nervous system that operate independently to control behavior for functionally (biologically) distinct types of problems (e.g., when foraging vs. choosing mates vs. caring for young, etc.).

Such modularity may be adaptive because the fundamental (e.g., statistical) properties of the problems posed in the different contexts may differ quite substantially, such that completely different control processes are required to maximize performance in each biological "domain." It is also possible that modularity could act as a within-individual "hedging of bets" when there is substantial, irreducible uncertainty about the environment on a moment-by-moment timescale (Dall 2010), such that knowing what the appropriate set of responses is likely to be at any moment in time is too difficult (and requires too broad a range of responses) to allow a general-purpose decision rule to evolve. Nevertheless, the existence of module-general common resources (e.g., information, energy) or control traits (e.g., sensory systems) will erode modularity and could structure individual behavior across contexts (functional domains). Wolf and Weissing (2010) discuss how the latter can select for cross-context correlations in behavior.

Individual behavior often varies over time as a result of learning. The conditions under which we would expect learning to evolve are relatively well understood (Stephens 2007). For prior experiences to improve decision making (i.e., allow for a better fit between behavior and current conditions) there has to be some degree of patterning of ecologically important events over time. However, because learning can be costly in terms of effort, time, and the maintenance of complex decision-making "machinery," if environments are too patterned in their states (i.e., do not change enough over generational time) then "hard-wired" behavior (e.g., that does best in the average environment) is likely to evolve. From such a perspective, Stephens and colleagues (reviewed in Stephens 2007) have developed a framework for predicting variation in the adaptive value of learning/behavioral plasticity by specifying uncertainty and reliability of experience on orthogonal axes. Their predictions were corroborated using experimental evolution in Drosophila and setting blue jays operant tasks in the lab. In a similar way, ontogenetic changes in reproductive value and ability (more generally: changes in fitness trade-offs with state) can drive longer-term variation in behavior as a function of the judicious use of information over a lifetime, when such phenotypic plasticity is adaptive.

One influential approach to analyzing the behavioral consequences of information use (including learning) from an evolutionarily adaptive perspective involves the application of statistical decision theory (Dall et al. 2005), which makes extensive use of inductive (Bayesian) inference. Most organisms, regardless of cognitive complexity, have to make guesses about the world: Is that a bright red berry in dim light or a dull red berry in bright light? Is that prickly looking animal going to eat me or can I eat him? Should I go to this restaurant or that one? The problem of inductive inference is the problem of trying to make a decision under uncertainty; many conclusions are possible given the data, and the organism must choose just one. Bayesian inference provides a computational-level, ideal (adapted) observer and analysis of how background knowledge should be integrated with statistical data to narrow the hypothesis space. These models can be applied to problems of induction across species and cognitive domains. They have been used to describe foraging decisions in a wide range of species (Krebs et al. 1978): problems of visual perception (Yuille and Kersten 2006), decision making in sensorimotor control (Körding and Wolpert 2006), language acquisition (Chater and Manning 2006) and many aspects of abstract, higher-order cognition (Tenenbaum et al. 2011). Specifically, Bayes's law provides a rule for how an organism might evaluate a hypothesis, h, about the process that produced some data, D. Bayes's law states that the probability of the hypothesis given the data, $P(h|D)$, depends on both the prior probability of the hypothesis, $P(h)$, and the likelihood of the hypothesis, $P(D|h)$: the probability that the data would have been observed if the hypothesis were true. That is, $P(h|D) = P(h)P(D|h)$. We can borrow a simple example to illustrate (see Tenenbaum et al. 2011): Imagine you observe some data D—your child is coughing. You can consider many hypotheses, including, ($h1$) your child has a cold, ($h2$) your child has lung cancer, ($h3$) your child has the stomach flu. Colds and stomach flu in children are common; lung cancer is not. Thus prior probabilities favor $h1$ and $h3$. However, if your child has either a cold or lung cancer, it is very likely that she will cough; it is less likely that she will cough if she has the stomach flu. Thus the likelihood, $P(D|h)$, is higher for $h1$ and $h2$ than $h3$. Integrating both the prior and the likelihood suggests that you should decide that your child has a cold. From where do prior beliefs originate? In our example, they came from common cultural knowledge about the prevalence of childhood diseases. The transmission of such knowledge is critical to cultural learning in human cognition. However, constraints due to prior knowledge can come from many sources, including both the individual organism's own past experience and evolutionary adaptations (i.e., selective "experience" of one's reproductive lineage; Dall et al. 2005). In solving problems of visual perception, for instance, prior knowledge may include an evolved constraint to assume that illumination comes from a single, overhead source (i.e., the Sun). Bayesian inference, therefore, formalizes the claim that background knowledge is integrated with new data to affect judgment and decision making. This ability to integrate prior knowledge and statistical information supports rapid, accurate, inference across a range of problems and goals. Particularly relevant to decision making, formal analyses suggest that organisms whose decisions approximate the output of Bayesian inference models will out-compete organisms using other strategies when animals are contending for resources (Ramsey 1926/1980; de Finetti 1937/1980). Critically, however, Bayesian inference models provide computational-level accounts of cognition (Marr 1982): they describe optimal decision making given a particular set of constraints and data, but they do not prescribe the ways these computational outcomes are instantiated. Many different algorithms can approximate the output of Bayesian inference models, and these algorithms might, in turn, be implemented by many kinds of neural systems. Thus although evolution might favor organisms that approximate the output of normative models, there might be substantial variability in the mechanisms underlying decision making across species.

Another major source of behavioral variability in many species comes from variation in "self-control" in different contexts. Nonhuman animals commonly prefer immediate rewards. This is paradoxical in the sense that they often prefer smaller, sooner benefits even when they could achieve a higher overall intake rate by choosing a more delayed but larger alternative. There is, of course, a large literature on this topic that goes under several headings: delay discounting, intertemporal choice, self-control, and failure to delay gratification. Superficially, at least, this so-called self-control problem resembles naturally occurring patch exploitation because when an animal chooses to stay longer in a food patch, it is "choosing" an alternative in which it takes longer and acquires more food. In the "self-control" literature, however, the time between trials typically does not affect preference, whereas in patch exploitation the so-called travel time is a consistently powerful variable: animals spend longer in patches when travel times are longer (Stephens and Krebs 1986). This problem has been studied by Stephens and colleagues by creating two types of choice situation (e.g., Stephens and Anderson 2001). One test situation involves a typical self-control situation: blue jays were offered a binary, mutually exclusive choice between a smaller-sooner and larger-later option. The second situation sought to create an economically similar situation that was more "patch like." To achieve this, a situation is created in which the jays had to choose between leaving and staying. The "stay" option led to a "larger-later" option, whereas the "leave" option resulted in a smaller-sooner option. The key result is that jays in the self-control situation behaved in a typically impulsive way, and they performed relatively poorly; in contrast, jays in the patch situation achieved high rates of intake (nearly optimal). In short, in patch-like situations, jays achieve a high level of performance, whereas in the self-control situation they perform relatively poorly. The hypothesis put forth by Stephens et al. is that the choice rules used by the jays evolved to make decisions that are more like patch exploitation than binary mutually exclusive choice; thus the rules perform better in the situation that more closely resembles the problems they have faced in their evolutionary past. They remark that it does not necessarily follow that jays are less impulsive in the patch situation; instead, the patch situation could be a case where an "impulsive rule" performs well (for the mathematical reasoning to support this claim, see Stephens and Anderson 2001). This has been termed the ecological rationality hypothesis of impulsivity. Here, the phrase ecological rationality means, crudely speaking, that the jay's decision rule "makes sense" in an ecological context. The phrase ecological rationality has come into use because it is associated with research on decision heuristics conducted by Gigerenzer and the ABC group (e.g., Todd et al. 2011b). However, some biologists (Kacelnik et al. 2006) have criticized Gigerenzer's use of this term, because when applied to human heuristics it is used to describe a fit between a decision mechanism and the current environment. From an evolutionary perspective, however, the environment that matters is the ancestral selective environment (sometimes called the adaptively relevant environment). To make this distinction, Kacelnik has suggested the term biological rationality to describe a rule that is rational in the adaptively relevant environment. Of course, if we are willing to assume stationarity (e.g., the statistical properties of patches have remained the same since the Pleistocene), then this distinction would be relatively unimportant. The distinction is clearly quite important for the evaluation of human decision heuristics because the environment of modern humans is often thought to be quite different from the adaptively relevant environment for humans (although recent molecular-genetic evidence suggests that humans have been subject to multiple bouts of substantial selection significantly more recently than the Pleistocene; e.g., Bustamante et al. 2005).

Patterning of behavior over time can often be selected for directly. Indeed, variability in behavior can be selected for when being predictable would facilitate exploitation by competitors and natural enemies. Such selection pressures can be particularly acute in antipredatory contexts, where exposure to certain types of predators (e.g., stalkers) can favor individuals that behave as randomly as possible while at risk (Bednekoff and Lima 1998).

Among Individuals

When thinking about individual differences in behavior from an evolutionarily adaptive perspective, we have to distinguish at least three questions (Wolf and Weissing 2010): First, what factors favor the evolution of variation in behavior, and/or the mechanisms underlying behavior, among individuals? Second, what factors favor the evolution of consistency in behavior within a functional context? And third, what factors favor the evolution of behavioral correlations across different functional contexts?

In terms of the first question, substantial effort in the evolutionary biology literature has been devoted to understanding the causes of adaptive variation. Evolutionary biologists typically recognize three major processes in this context:

1. Frequency-dependent selection: Competition for limiting resources, for example, can often select for mixtures of tactics to be expressed within populations, causing the increasingly common use of a tactic or strategy to render it less effective. Thus the fitness returns from expressing it depend negatively on its frequency within the population.
2. Spatiotemporal variation in the environment: Adaptive behavior can often vary within populations because different things are selected for at different times and in different places within the larger niche that a population occupies.

3. Nonequilibrium dynamics: The constant influx of variants associated with nonadaptive processes, such as random dispersal and mutation, can give rise to substantial standing variation in traits (including behaviors) within populations.

Concerning the second question (consistency within contexts), it is important to consider that, from an adaptive perspective, the optimal action that an animal should take typically depends on its current state (e.g., energy level, experience) and thus rarely remains constant from one moment to the next. When positive feedbacks between state and behavior occur, they can amplify stochastic variation among individuals and lock individuals into distinct regions of state space, causing individuals to differ consistently while such positive feedbacks persist (Dall et al. 2004). Indeed, learning can generate such positive feedback as initial differences in early experience can encourage individuals to develop different skills and behave differently (Tinker et al. 2009). Finally, recent work suggests that behaving consistently can be selected for directly in social situations when doing so can cause social partners to respond more favorably than they otherwise would. This can occur whenever there is coevolution between social responsiveness (social information use) and behavioral consistency, which may be the case during aggressive competition over resources (Dall et al. 2004; Wolf et al. 2011) and in some types of cooperative interactions (McNamara et al. 2009).

In terms of the third question (correlations across contexts), it is important to examine the factors that are likely to select for adaptive differences in underlying mechanisms that affect several traits at the same time: What should select for differences in metabolism or physiological commitments (e.g., neural tissue) to information processing? Wolf et al. (2007), for example, addressed the question of why, in many organisms (ranging from octopuses to chimpanzees) is boldness in novel and potentially dangerous situations often associated with aggressiveness in intraspecific interactions, whereas shyness is associated with less aggressive behavior. Based on the asset protection principle of life history theory (Clark 1994; Houston and McNamara 1988), they argue that individuals of high reproductive value (i.e., that have high future fitness expectations) should be risk-averse because they have much to lose, whereas individuals with low fitness expectations should be more risk-prone because they could hardly end up worse off than they already are. Since this basic principle applies to behavior in all kind of risky situations (Dall 2010), differences in future fitness expectations should give rise to correlated differences in all kinds of risk-related behaviors despite limited evidence for such domain-general attitudes to risk in humans (e.g., Weber et al. 2002). In the meantime, various empirical researchers have concluded that this principle can indeed explain a diversity of phenomena, ranging from the risk-prone "personalities" of wild guinea pigs that are born relatively late in the season and, hence, have lower fitness expectations (Groothuis and Trillmich 2011), to the risk-averse behavior of oystercatchers breeding on high-quality territories (Goss-Custard 1996). More convincing than such indirect evidence is the outcome of a recent experiment on house mice, which was designed to test the a priori prediction that female house mice of genotype t/+ should be more risk-averse than their wild-type +/+ counterparts, while the opposite should be the case in males. Here, t refers to a certain gene locus (the t-complex) that does not directly affect the behavior of house mice (e.g., it is associated with sperm motility). The prediction was based on the observation that heterozygous t/+ females have higher expected fitness than homozygous +/+ females, while heterozygous males have lower fitness expectations than homozygous wild-type males. Auclair et al. (in preparation) performed several personality tests, with outcomes generally in line with the predictions: in comparison to wild-type individuals, heterozygote females are more risk-averse, while heterozygote males seem to be more risk-prone. Finally, Wolf et al. (2008) provided a theoretical explanation for the observation that in many populations some individuals readily react to changes in their environment, while other individuals exhibit more rigid, routine-like behavior that is much less affected by environmental cues. Moreover, Wolf et al. predicted that responsiveness and rigidity should be relatively stable over lifetime and consistent across contexts. Both features are consistent with various experiments, for example in ducks (Shettleworth 1998) and spice finches (Mottley and Giraldeau 2000), where environmental conditions were repeatedly changed by the experimenters.

Among Groups

As discussed above, group differences can be an important source of variation in decision making when there is selection for social information use and/or conformity. Social information use can be adaptive when evolutionary conflicts of interest are minimal within groups (e.g., where group members are highly related or when competition for resources is reduced), and an individual's uncertainty about how to allocate behavior can be significantly reduced by observing the behavior of group mates (Dall et al. 2005). Copying the behavior of social partners, or conformity, is a particular form of social information use; it is often the cheapest form of information use but carries substantial risks, stemming from the erroneous copying of inappropriate behaviors or "informational cascades" (Dall et al. 2005). Cultural learning (both within and across generations) facilitates the tracking of intermediate rates of environmental change ("red noise" environments) and will be often be selected for in species that are dependent on the populations of other species (e.g., predators), especially if they are marine (Whitehead 2007). This form of learning will maintain very substantial differences among groups, even without conformity, but conformity will exacerbate this type of variation.

Variation and a Theory of Darwinian Decision Making

We have discussed variation in decision making by individuals that have evolved by Darwinian selection from a range of perspectives: Does this really matter? What are the interesting scales of variation? What kind of mechanisms can generate it? How could it evolve? In the process, we hope to have highlighted the importance of explicitly considering variation both when attempting to predict economically and socially important patterns of behavior as well as to obtain a deeper understanding of the fundamental biological processes involved. Our approach so far has been rather piecemeal, with little attempt at any general understanding of the role that variation should play in an evolutionary account of decision making. Thus we finish by sketching the key elements that we feel should be included when incorporating variation into a general theory of Darwinian decision making:

- Game theoretic approaches. The underlying structure of the problems that evolved entities is essentially coevolutionary. This is because major components of the selective environments, to which individuals are subject, are biological in origin and will thus evolve. Therefore, specifying the outcomes of evolution naturally lends itself to applying evolutionary game theory. Moreover, variation is emerging as a key factor in determining outcomes predicted by evolutionary game theory. This has been illustrated a number of times in this chapter (e.g., how variation and socially acquired information use can coevolve to maintain individual variation in trust and trustworthiness; McNamara et al. 2009).
- Does one general, all-purpose mechanism exist? Or does it reflect diversity of forms and function? The specific set of problems faced by individuals is likely to be different for different species (set by their specific ecological circumstances or niche, which will include their social environments). Nevertheless, selection acts on existing variation so evolved systems are likely to share the basic components of a decision apparatus in proportion to their phylogenetic distance. Observable behavioral variation must therefore reflect the constraints on the system. Thus the evolutionary history of a lineage must be considered explicitly when incorporating variation in Darwinian accounts of decision making.
- Statistical decision theory (Bayesian inference). All evolved systems are contingent, yet such contingencies will represent the prior experience of ancestral and developmental environments (contexts) as a result of prior adaptation. For adaptive decision making, such genetically or developmentally induced priors will often need to be updated by more current experience. Specifying when or how such updates occur should be a major focus of an evolutionarily sensible theory of decision making, and statistical decision theory offers a formal framework that incorporates

this intuition very comfortably (Dall et al. 2005). Variation in behavioral allocation emerges naturally in Bayesian decision makers.

- The performance of alternative models of decision making should be assessed according to their fitness consequences, which may or may not include explicit analysis of potential persistence over evolutionary time. Because of this, such assessments must include explicit consideration of the fitness costs and benefits of the rules considered. Moreover, performance must be assessed using invasibility analysis, which means that the competitive environment must be specified, including the set of possible rules (as well as by assessing the evolutionary stability of rules). Thus standing variation, or the potential for variation, in decision making will be crucial when determining which rules are likely to evolve in particular systems.
- The statistical properties of the ecological (including social) context in which decision making evolved must be explicitly considered. What are the limits to potential specific models of decision making? (What is the ecological problem set?) While this is often difficult to specify in full detail for any given system, even identifying key statistical properties of typical ecological problems would help to identify the kinds of decision-making process that we should expect to see, and how they can vary in different contexts.
- For organisms subject to epigenetic development (i.e., develop from a totipotent cell), ontogeny will likely have a strong impact on individual variation (earlier environments will matter disproportionately). This suggests that attention should be focused on detailing such early influences and how persistent they are likely to be.
- The adaptive tinkering and contingency to which evolved systems are subject will tend to limit the dimensionality of control mechanisms (key state variables), which can account for imperfect optimization across contexts of higher dimensionality. The challenge will be to identify the key control mechanisms (e.g., sensory, cognitive processes) and how likely they are to vary both individually and phylogenetically.
- For humans (and some other animals), cultural evolution is likely to play an influential role in driving variation in decision making. This suggests that social contingency and/or historical influences are likely to be pervasive when attempting to predict specific outcomes. Furthermore, cultural variation will be bounded by genetic adaptation (e.g., many cultures have cleanliness-based practices that are likely to have evolved to mitigate the risk of infectious disease) and so studying the interplay between these factors will prove fruitful. Finally, basic psychological processes (e.g., salience) are likely to play a key role in determining how behavior varies over time and across cultural units (e.g., oddity effects, appeal to basic biological “drives,” etc.).

Conclusions

It is clear that variation is central to understanding how individuals should and do behave, from both biological and human-centered perspectives. Perhaps this should not be too surprising since evolution is at the heart of any scientific (i.e., natural) account of biological (including human) systems, and there is no evolution by Darwinian selection without variation. Moreover, how individuals allocate their behavior is key to determining the evolutionary success, and hence existence, of their specific (heritable) traits. This interplay between behavior, variation, and evolutionary outcomes was a major theme of this chapter. Indeed, we hope to have illustrated how this interplay is central to promoting models of human behavior that can successfully predict individual behavior and identify interventions that are going to be effective in determining specific outcomes. To this end, our recommendations for incorporating variation into a theory of Darwinian decision making will hopefully draw attention. To paraphrase Theodosius Dobzhansky: We feel that no behavior, human or otherwise, makes sense except in the light of variation and, therefore, evolution!

Evolutionary Perspectives on Social Cognition

16

The Cognitive Underpinnings of Social Behavior

Selectivity in Social Cognition

Thomas Mussweiler, Andrew R. Todd, and Jan Crusius

Abstract

To navigate their social worlds successfully, humans must coordinate their own behavior with the behaviors of others. Because human behavior takes place in a complex social world, it imposes high demand on cognitive capacity. Yet the cognitive resources available to humans to meet this demand are relatively limited. Selectivity is a crucial element in social cognition. Only through informational selection are humans able to make decisions that are simultaneously adaptive and efficient. This chapter reviews evidence from social cognition research which demonstrates that humans are selective in the social information they attend to, the manner in which they process this information, and the behaviors they ultimately enact. This selectivity in social attention, social thinking, and social behavior is an adaptive tool that helps humans successfully maneuver through their complex social worlds.

Introduction

Humans are fundamentally social beings. From the cradle to the grave, their survival depends on how they interact with others. Without receiving direct support from others, humans are unable to survive infancy and reach reproductive age. Without interaction with others, humans cannot reproduce, and without interpersonal coordination, they cannot master most major life challenges. Humans are thus essentially interdependent: to survive and reproduce, they necessarily have to interact and coordinate with others. Social interaction and coordination, however, are dauntingly complex endeavors. To navigate successfully through their social worlds, people first have to make sense of their social counterparts; they have to interpret the behaviors of those around them, anticipate what those others might do next, and decide on the most promising course for their own action. Is the young man who is staring at me friend or foe? Is he going to approach me or leave me alone? Should I stay or should I go? These are issues that people entertain on a daily (if not an hourly) basis. Despite the complexity of these social judgments and the ensuing social decision, humans are remarkably well versed at maneuvering their social worlds. They appear to be well adapted to the social complexity that surrounds them.

How do people make sense of and react to the behaviors of others? This is a simple question but it begs a rather complicated answer. For more than forty years, social cognition research has tried to answer this very question about the cognitive underpinnings of social behavior. Summarizing this research in a short chapter is more than a challenge; it is impossible. We can thus only highlight what we see as core principles of social information processing and refer readers to more comprehensive treatments of social cognition, along with its affective, motivational, and neuronal underpinnings, as well as its cross-cultural variations (Fiske and Taylor 2008).

Our chapter is guided by three basic premises: First, social behavior builds on social thinking. How people behave toward others depends largely on how they interpret others' behaviors. Clearly, whether you run away from or eagerly confront an approaching stranger depends on how you judge this stranger's behavior and what you infer about underlying motives. Even a rudimentary understanding of social behavior necessarily entails a basic comprehension of how people think about others and how they process social information more generally. Accordingly, in this chapter, we attempt to shed light on the cognitive underpinnings of social behavior. The cognitive mechanisms humans engage to process social information are partly similar and partly different from those used to process nonsocial information (for a detailed discussion, see Fiske and Taylor 2008).

Second, social information processing can be fruitfully differentiated into three conceptually distinct processing stages (though we readily admit that these processes might operate in parallel): Attention determines which novel social information enters our cognitive apparatus. This incoming information is then cognitively processed by relating it to stored social information. Finally, this processed information is transformed into a behavioral response. This chapter is structured around these three basic steps of social information processing.

Third, each of these three steps is characterized by selectivity in information processing. The amount and complexity of social information that people routinely encounter dramatically outnumbers their limited processing resources (Taylor 1981). Although many human cognitive capacities appear almost unrestricted in comparison to the capacities of other primates, these capacities—especially those that require some kind of parallel conscious processing—are far from limitless. For most people, for example, navigating a car through traffic while text-messaging with one's mother and attempting to resolve a conflict with one's partner in the passenger seat will result in suboptimal performance on at least one of these tasks. In light of these processing constraints, people must necessarily be selective in the social information they attend to, the stored social information they activate, and the behavioral options they examine. This selectivity in social information processing represents a fundamental mechanism through which people make sense of their social worlds.

Social Attention: What Social Information Enters the Cognitive System?

As social perceivers, we are routinely inundated with information-rich stimuli in our immediate social environments. Indeed, the human visual system must contend with about one megabyte of raw data every second. Given the inherent capacity constraints of the human cognitive system, however, people are limited in how much information they can attend to at any given time. Thus, it is critical for people to be able to parse their surroundings quickly for the information that is most useful for confronting the challenges of day-to-day life, while simultaneously disregarding information that is not particularly relevant. That is, to navigate their social worlds effectively, perceivers must select from countless possibilities the information that will undergo additional processing. Here we examine what types of social information initially capture attention (for a more detailed treatment, see Bodenhausen and Hugenberg 2009), as this information profoundly shapes everything else that subsequently unfolds in the course of social interaction. We contend that social attention is guided by at least two basic principles:

- Which information people notice depends on a combination of bottom-up, stimulus-based and top-down, perceiver-based influences.
- People selectively attend to information that is perceptually salient (i.e., contextually distinct) and relevant to their immediate and long-term goals, values, attitudes, and expectancies.

Features of Environmental Stimuli: Capturing Attention from the Bottom Up

Which features of environmental stimuli make them particularly noticeable? At the most basic level, stimuli that are novel, unique, or otherwise contextually distinct are among the most likely candidates selected to receive preferential attention. The bright blinking light in a dark room, the lone guy wearing casual attire at a formal dinner party, and the scorching hot day in the dead of winter all tend to be particularly attention grabbing. Also receiving attentional priority are stimuli which require an immediate behavioral response; for instance, those that appear suddenly, without warning, or that are moving toward oneself (Franconeri and Simons 2003). Along these same lines, the human perceptual system is adaptively tuned to notice selectively stimuli that have clear functional implications with respect to reproduction and survival. Although people selectively orient to extremely positive stimuli (e.g., attractive potential romantic partners) quite readily, being able to recognize quickly and avoid potentially dangerous or untrustworthy stimuli confers enormous survival advantages. Thus, it makes sound evolutionary sense for extremely negative stimuli—especially those signaling potential physical danger (e.g., snakes)—to be particularly attention grabbing (Öhman and Mineka 2001).

On the more social end of the spectrum, people selectively attend to functionally relevant signals conveyed by the human face. For instance, faces expressing anger automatically draw attention. In one study (Hansen and Hansen 1988), participants were presented with a 3 × 3 array of faces, one of which had an emotional expression that was discrepant from the surrounding faces (e.g., one angry face surrounded by eight neutral-expression faces); their task was to identify the discrepant face as quickly as possible. Results indicated that participants were faster in detecting the lone angry face in a happy or neutral-expression crowd than they were in detecting the lone happy or neutral-expression face in an angry crowd. Subsequent research has revealed that this anger superiority effect is especially pronounced when faces are displaying direct (rather than averted) eye gaze (Adams and Kleck 2003), thereby signaling that the expresser's anger has direct implications for oneself. Similarly, because out-group members are often thought to be more threatening and less trustworthy than in-group members, faces of the former tend to be more attention grabbing than faces of the latter, a tendency that, again, is most pronounced when targets display direct eye gaze (Trawalter et al. 2008). Interestingly, at later points in the attentional stream, when vigilance concerns have been relaxed, people often selectively attend to in-group targets more than out-group targets (Richeson and Trawalter 2008), perhaps because the former are more frequently encountered in daily life and thus are more likely to be directly relevant for people's ongoing activities and future cooperative endeavors.

Perceiver-Based Characteristics: Capturing Attention from the Top Down

Although it is evident that particular environmental stimuli can capture attention based on their features alone, the accessible mental contents that perceivers bring to bear in a given environment profoundly influence what they notice in that environment. We now turn to a brief discussion of several top-down influences on selective attention.

First, it appears that highly personally relevant stimuli are especially likely to attract attention. A classic example of this is the "cocktail party phenomenon," whereby the sound of one's own name uttered from the opposite side of a crowded room seems to rise above the commotion and capture one's attention.

Similarly, insofar as people possess or attach personal meaning to a particular personality trait, their attention is more likely to be drawn to behavior episodes that exemplify that trait (Bargh 1982). For instance, the behaviors of a dishonest politician will undoubtedly be attention grabbing for a person who values trustworthiness. Another source of personal relevance stems from people's attitudes. Just as maintaining a low threshold for detecting biologically threatening stimuli in the environment is clearly functional, it also makes sense that perceivers would automatically orient to other hedonically relevant stimuli: stimuli that they personally (but not necessarily other people) find appealing or unappealing (Roskos-Ewoldsen and Fazio 1992).

A related top-down influence on selective attention involves whether stimuli are relevant to perceivers' ongoing goals. Insofar as a particular stimulus is goal relevant, it is significantly more likely to be selected for additional processing by the cognitive system. Indeed, perceivers attend more closely to behaviors performed by people on whom the perceivers' outcomes depend (e.g., significant others, professional colleagues) than to behaviors performed by people who are irrelevant to perceivers' current goals. Even temporary motivational states can substantially alter which stimuli attract attention. For example, after experiencing an instance of social exclusion, people are highly motivated to seek reconnection with others; to facilitate this affiliative goal, people selectively attend to signs of social acceptance (e.g., smiling faces; DeWall et al. 2009).

A final class of top-down influences on selective attention derives from perceivers' expectancies. Numerous studies have shown that information and behaviors that are inconsistent with one's expectancies are granted preferential attention (Roese and Sherman 2007). For instance, if everything you know and have heard about a colleague leads you to expect him to be an honest and noble person, any evidence suggesting that he may actually have questionable integrity will likely capture your attention and prompt attempts to reconcile this behavior with your initial expectancy. The functionality of noticing extremely negative stimuli notwithstanding, another reason why such stimuli capture attention is that they are normatively rare, unexpected, and thus attention grabbing.

In combination, these mechanisms of top-down and bottom-up selectivity allow social perceivers to master the abundance and complexity of social information they routinely encounter. It is only with the help of this selectivity that perceivers can focus their limited processing resources on the information that is most relevant to them.

Social Thinking: How Is Incoming Social Information Processed?

Once social information has entered people's cognitive system through these attentional mechanisms, it is subjected to additional processing vis-à-vis the stored information representing people's knowledge about the social world. What are the central mechanisms that operate at this stage of information processing?

1. Comparative thinking allows people to relate knowledge about different stimuli or options to one another.
2. Categorical thinking allows people to apply their stored social knowledge to novel social stimuli.
3. Accessibility influences which cognitive content primarily influences social judgment and behavior.

Notably, the selectivity that influences which information people attend to and which information thus enters their cognitive apparatus is also apparent in these mechanisms of social information processing. Selectivity shapes how incoming information is processed.

Comparative Thinking: Relating Stimuli to One Another

When processing social information, when trying to make sense of others' behaviors, or when trying to decide on a sensible course for one's own actions, people invariably engage in comparative thinking. Social judgment typically involves a comparison of the target person with one or more other people. With whom do people compare? How do such comparisons influence social judgment and behavior? And, what are the downstream advantages of the ubiquitous tendency to engage in comparative thinking?

When deliberately trying to find the most diagnostic standard for comparison possible, people typically seek those standards that are similar to the target person on dimensions related to the critical characteristic (Festinger 1954). When trying to evaluate the trustworthiness of a colleague, for example, you would likely seek comparison standards that are similar to the target in terms of profession, age, gender, and other characteristics. Assuming your colleague is a generally trustworthy person, a comparison with a convicted swindler is not likely to yield an optimal amount of diagnostic information. Clearly, locating a standard that is perfectly matched on multiple related attributes is a rather arduous task. Consequently, people often rely on a less deliberate standard selection process. Oftentimes, people utilize those standards that they routinely use for comparison (Mussweiler and Rüter 2003). Alternatively, they may simply rely on standards that just happen to come to mind. In social judgment, the self constitutes a particularly prominent routine standard; consequently, judgments about others are often egocentric in nature. When trying to evaluate the trustworthiness of another person, for instance, people routinely use their perceptions of their own trustworthiness as a comparison standard. Notably, such egocentric comparisons also contribute to perspective-taking (Todd et al. 2011a) and may thus play a critical role whenever we attempt to coordinate our actions with the experienced or anticipated actions of others.

How do comparisons with these various standards influence judgment and behavior? Recent research has established that the consequences of comparison critically depend on whether judges focus on similarities or differences between a target and a standard (Mussweiler 2003). When comparing your own trustworthiness to that of your colleague, for example, you could either focus on ways in which the two of you are similar (e.g., you work in the same office, you have the same friends) or on ways in which you are different (e.g., you are at different stages of your career). Whether you focus on similarities or differences critically shapes the consequences of the comparison. Similarity-focused comparisons yield assimilative effects, whereas difference-focused comparisons yield contrast. Whether judgments about the trustworthiness of another person are assimilated toward or contrasted away from the self as a result of egocentric comparison depends, therefore, on whether similarities or differences between the self and this person stand in the foreground. Furthermore, whether comparisons are primarily similarity-focused or difference-focused depends on a variety of contextual, intrapersonal, socio-structural, and even physical factors.

Comparison is a versatile mechanism that is ubiquitous in both social and nonsocial information processing. What are the advantages of using this all-purpose tool? Recent research demonstrates that, based on mechanisms of information focus and information transfer, comparison yields two invaluable benefits for the social thinker: It increases the efficiency of information processing and simultaneously reduces judgmental uncertainty. First, comparisons allow judges to focus selectively on a subset of potentially judgment-relevant information and to ignore information that does not bear on the comparison itself. This selectivity in what information is considered allows social thinkers to make judgments more quickly and efficiently (Mussweiler and Epstude 2009). Judges who are induced via an initial priming task to process information in a more comparative manner, for example, are subsequently able to solve complex decision problems more quickly without becoming less accurate (Mussweiler and Epstude 2009). Second, comparisons allow judges to use easily accessible information about well-known standards to compensate for missing information about unknown targets. This information transfer reduces judgmental uncertainty (Mussweiler and Posten 2012).

Categorical Thinking: Using Group-Based Information to Judge Individuals

As is the case with comparison, categorical thinking is a ubiquitous element of social cognition (Macrae and Bodenhausen 2000). When interpreting the behavior of others and forming impressions of them, people are guided by their stored knowledge about the targets' social categories. Gender, race, and age are among the most frequently used social categories (Macrae and Bodenhausen 2000). The act of flagging down a car on a dark street, for example, can be interpreted quite differently depending on whether it is carried out by an elderly woman or a young man. The use of categorical knowledge allows social thinkers to render judgments of individuals they have not previously encountered and about whom they have very little stored information. Categorical thinking allows judges to form expectations about others' behaviors and thus makes the world appear more meaningful and orderly (Macrae and Bodenhausen 2000).

Categorical thinking simplifies social judgment in multiple ways. First, it allows judges to forgo the arduous search for individuating information about a target person by simply using categorical knowledge (i.e., stereotypes) instead. Stereotype use has efficiency-enhancing properties that make it a particularly helpful heuristic tool for social thinkers, especially when cognitive resources are in short supply. People whose circadian rhythms limit their ability to process complex information early in the day (i.e., "night owls"), for example, rely more heavily on stereotypes in the morning than at night, whereas "morning birds" exhibit a reverse pattern (Bodenhausen 1990). Second, categorical thinking typically encourages judges to focus on the most relevant of the multiple category memberships applicable to a given person and thus involves selectivity in information activation. The one category that is particularly salient, accessible, and relevant to the perceiver's current goals becomes dominant and is thus called upon to guide social information processing, whereas alternative category memberships are inhibited (Bodenhausen and Macrae 1998).

Accessibility: Using What Comes to Mind First

How activated stereotypes guide information gathering and processing is one example of the more general tendency for accessible knowledge to dominate social thinking (Higgins 1996). For many of the social judgments people make, they actually have too much information available that is potentially useful. For example, when trying to determine whether your colleague is a trustworthy person, you could, in principle, consult an endless amount of potentially relevant information. Integrating all of this information into a coherent impression is far too complex and time consuming if one intends on reaching a judgment in a timely manner. To maximize the speed with which this judgment is reached, people are typically selective in their information search and thus rely primarily on what is most accessible in their minds. If the first things that come to mind about your colleague are several episodes in which he behaved dishonestly, you are likely to judge him accordingly. Accessibility fosters selectivity in social thinking in that it focuses attention on attributes that are associated with the accessible concept, enhances memory for such aspects, and inhibits the activation of competing concepts (Förster and Liberman 2007). Social judgments and behaviors tend to be consistent with accessible concepts, so that activating the concept of aggressiveness, for example, leads people to judge another's behavior as more aggressive and to behave more aggressively themselves.

Social Behavior: Which Action to Choose?

After environmental information has been attended to and processed, the ultimate task of the cognitive system is to prepare and to enact behavior. What are the central guidelines that determine which actions people take? We contend that the way in which environmental input is transformed into behavioral output is characterized by the following three principles.

1. Reflective behavior is guided by intentions that are shaped through accessible information about the desirability and feasibility of behavioral outcomes.
2. Impulsive, nonintentional behavior selection is shaped by associative links.
3. Impulsive and reflective factors interact to produce behavior.

Just as the cognitive system is selective in what information it focuses on and how this information is processed, selectivity is also a central attribute at the behavioral output stage. In principle, every action is preceded by a selection of one of several behavioral alternatives, even if this selection involves only the decision to act or not to act.

Reflective Behavior Selection: Thinking About What to Do

One way to determine which action to take is to reflect consciously about the advantages and disadvantages of each behavioral alternative and form a behavioral intention (e.g., Ajzen 1991). Importantly, because of the selectivity of cognitive processing, this process is not exhaustive. It will not entail all potentially relevant information. Rather, it will be shaped by the limited subset of information that has been attended to and the way this information has been construed and made readily accessible.

What kind of information is used for behavior selection? A key issue concerns whether the behavior will yield desirable consequences. The attitude toward a particular behavior is determined by the sum of the expectancy that an action will lead to certain outcomes weighted by subjective values of those outcomes. Naturally, this calculation will comprise the intended consequences as well as the unintended repercussions of an action. Furthermore, actions that are carried out collectively or that may affect other people involve the mentally complex task of predicting how these coactors will act and react. In deciding how to treat a colleague, one should be predisposed toward the behavior that is associated with the most positive attitude—the one with the maximum utility. However, action is also influenced by the subjective norm—beliefs about the social pressure to act in a certain way and the motivation to comply with this pressure. In other words, another part of the equation includes a person's beliefs about whether socially relevant others would approve or condemn an action, weighted by the subjective motivation to comply with the referents in question, even if these referents are not physically present or if the action does not have direct consequences for them. One might abstain from cooperating with a convicted swindler, for instance, because his behavior is met with disapproval by important referent persons or groups, such as friends or colleagues. Finally, enacting a behavior also rests on beliefs about whether one has the requisite resources and ample opportunity—such as relevant skills, time, or the cooperation of others—to complete it. The perceived behavioral control over an action is determined by factors that could likely facilitate or impede it, weighted by their subjective importance. In sum, if the attitude toward an action is positive, if it is perceived to be normative, and if one assumes that one will be able to carry it out, an intention to perform the action should be formed. Then, if an opportunity presents itself, the action will likely be taken. These causal factors can account for much variance in behavior (Armitage and Conner 2001).

Impulsive Behavior Selection: Just Doing It

Not all behavior is intentional and shaped by conscious reasoning (for a review, see Strack and Deutsch 2004). Often, it is neither possible nor functional to engage in a relatively slow and effortful reasoning process. In many situations, it is sufficient to act on the first association that comes to mind about a particular target. In other situations, quick, impulsive action may even be necessary for survival. Imagine a sudden encounter with a seemingly dangerous stranger in a dark alley. Should you carefully weigh your options, or should you act on your gut feeling?

Evidence suggests that in addition to a reflective pathway, there is also an impulsive pathway that directly connects environmental input and behavioral output. This impulsive pathway relies on spreading activation in the memory network, which contains evaluative and semantic associations and preformed response patterns (habits), and elicits basic motivational orientations of approach versus avoidance. It can operate quickly and effortlessly, and its influence on behavior may occur outside of conscious awareness (Strack and Deutsch 2004). If, as a consequence of categorical thinking, one associates a particular social category with distrust, an impulsive avoidance reaction may result upon encountering an individual member of the social category.

Abundant research investigating the workings of impulsive precursors of human behavior has relied on sophisticated methods that assess behavioral inclinations indirectly, without the need for introspective self-report (Petty et al. 2008). For example, Correll et al. (2002) had U.S. participants play a videogame in which they were instructed to "shoot" only at targets carrying a gun. Participants' responses revealed a strong racial bias: They were quicker to shoot an armed black target than an armed white target, and they were more likely to shoot erroneously an unarmed black target than an unarmed white target. Because race was not diagnostic in the task, and because participants had to react extremely quickly, it is reasonable to assume that this bias is largely caused by impulsive influences on behavior.

Interaction of Impulsive and Reflective Precursors of Behavior

In most situations, reflective and impulsive precursors of behavior do not operate independently; instead, they interact in complex ways (Strack and Deutsch 2004). For example, propositions entertained by the reflective system can increase the importance attached to certain associations by making them selectively accessible. Similarly, associations activated in the impulsive system can influence propositional reasoning in the reflective system. Thus, impulsive and reflective precursors may hold overlapping behavioral implications; however, these behavioral inclinations may also be incompatible. For example, an encounter with a member of a particular ethnic group may prompt a distrustful impulsive response because of culturally shaped stereotypical associations. Many people are motivated not to act on this impulse because they consciously endorse the goal of responding without prejudice. The reflective system can enforce this conscious goal by implementing behavioral decisions that override the unwanted impulse. Because the operation of the reflective system depends on the availability of cognitive resources, a self-control dilemma ensues. Impulsive precursors will govern behavior when these resources are constrained (e.g., Crusius and Mussweiler 2011; for a review, see Hofmann et al. 2009).

Thus, selectivity characterizes the behavioral output stage both in terms of the informational content and in terms of the mental processes that shape behavioral decisions. The cognitive system is selective in deciding which information it uses and in determining how this information will form the basis of reflective behavioral choice and impulsive responding. Furthermore, provided that the cognitive system is endowed with ample cognitive resources, perceivers can be selective in deciding whether impulsive or reflective processes dominate behavior.

Conclusion

Humans may well be the most complex stimuli that humans encounter. Because social behavior is—by its very nature—directed toward and coordinated with complex others, it is necessarily complex itself. To cope with this complexity and to maneuver through their social worlds successfully, humans have to be selective in what social information they attend to, in how they process this information, and in which behaviors they engage. The foremost advantage of this selectivity in social cognition is that it reduces social complexity, thus allowing social thinkers to make efficient use of their limited cognitive processing capacity. At the same time, selectivity produces reasonably adaptive outputs.

In fact, each of the facets of selectivity that we have described typically allows social thinkers to focus on the information that is likely to be most relevant. Selectivity thus ensures that less is more in social thinking. By focusing on the most relevant aspects of their social contexts, humans are able to make reasonably good decisions and to produce reasonably adaptive behaviors in an efficient manner.

Here, we have highlighted a number of central cognitive mechanisms that allow humans to be selective in how they process information. These mechanisms are ubiquitous elements of human information processing that contribute to every decision people make. Social cognition research has identified the basic cognitive building blocks of decision making. To date, however, decision theories in other disciplines have not sufficiently taken these building blocks into account. To fully understand how humans make decisions, their basic cognitive architecture—the fundamental cognitive mechanisms upon which they rely when making sense of the world—must be considered. After all, it is this subjective understanding of the world that forms the basis for every decision: from the most trivial to the most consequential.

17

Early Social Cognition

How Psychological Mechanisms Can Inform Models of Decision Making

Felix Warneken and Alexandra G. Rosati

Abstract

Many approaches to understanding social decision making use formalized models that account for costs and benefits to predict how individuals should choose. While these types of models are appropriate for describing social behavior at the ultimate level—accounting for the fitness consequences of different patterns of behavior—they do not necessarily reflect the proximate mechanisms used by decision makers. It is argued that a focus on psychological mechanisms is essential for understanding the causes of decision making in a social context. Focus is on the behavior of human children to elucidate the psychological capacities that are foundational for the developmental emergence of social decision making in humans. In particular, evidence is presented across a wide range of contexts to indicate that young children appear to focus on the underlying psychological states of potential social partners in cooperative contexts. This suggests that many types of social decisions may be driven by intention attribution, not explicit utility calculations. It is proposed that a comprehensive theory of social decision making must address both questions about ultimate function as well as integrate empirical studies of the psychological instantiation of these processes. Developmental approaches are particularly informative, as they elucidate the origins of decision making as well as the factors that shape them into their mature form seen in adults.

Introduction

How do humans (and other animals) behave in social contexts? This fundamental question is relevant both in the biological and social sciences, and many different disciplines have attempted to address this question through theoretical models of behavior and experimental inquiries. Research traditions in economics and biology, in particular, have utilized formalized models from game theory to understand social behavior (Hammerstein and Hagen 2005). These models describe social behavior as a series of decisions between multiple options as social actors attempt to sort out an optimal course of action. In evolutionary biology, these models are often used to describe behavior at an ultimate level of analysis; from this perspective, social decisions are evaluated in terms of fitness benefits, so value can be differentially assigned to different types of social interactions or behavioral strategies in terms of how these behaviors impact reproductive success in the long term. In economics and other social sciences, such models are used both to explain behavioral outcomes and to elucidate the psychological processes in which social agents engage. The underlying assumption of many of these approaches is that people use a function of expected utility, or some internal measure of "goodness" or desirability of various options (Fehr and Camerer 2007). For example, recent neuroeconomic approaches have integrated economic and neurobiological models to form a physiological utility theory—the hypothesis that some desirability function is computed in the brain (e.g., Glimcher et al. 2005). However, even if the behavioral outcomes of social decisions could be described in these terms (i.e., if people computed the expected utility of different social outcomes), whether such utility calculations actually reflect the processes going on in the minds of individuals remains an open question. That is, even if evolution acts on a value-based currency (fitness), this does not necessarily mean that decision makers actually employ these sorts of utility computations when making social decisions. Similar examples of divergence between ultimate and proximate levels abound in biology. For example, while kin selection at the ultimate level hinges on the degree of relatedness between social partners, individuals appear to use cues such as physical similarity (phenotypic matching) or early familiarity to decide how to interact with others (Widdig 2007). They do not actually calculate genetic relatedness, but rather use a set of proxies that are easily detectable to recognize kin. Consequently, in the domain of social decision making it is critical to investigate the actual psychological mechanisms that individuals use. These will then provide important insights at a proximate level of analysis into the behavioral goals that humans and other animals are attempting to reach.

In this chapter we examine to what extent utility-based models can capture the essence of how humans make social decisions at the proximate level. We propose that while such models of decision making may be appropriate to describe social behavior at an ultimate level, these types of analysis may not be as useful when examining the diversity of actual motivators that drive behavior. We claim that at its core, social decision making depends on representations of the psychological states of other social agents. In particular, we claim that humans do not appear to engage in value calculations in many social contexts but are instead primarily attuned to the intentions, dispositions, and motivations of social partners. This can already be witnessed in young children, who represent others' actions in terms of psychological states (especially intentions) and not just concrete behavioral outcomes for themselves and others. Thus, representing social interactions in terms of the psychological states of partners may be an adaptive way to make social decisions. This type of social cognition can occur in the absence of utility calculations or even concrete payoffs in a given situation, yet it still may have important long-term fitness consequences. For example, even if no significant resources are at stake during the decision-making process itself, displaying cooperative intentions and detecting others who possess cooperative dispositions might lead to long-term beneficial outcomes through positive assortment (McNamara et al. 2008).

We present evidence to support our claim that many types of social decisions hinge on intention reading, understanding of social relationships between individuals, and attribution of dispositional traits to potential social partners. We focus on the basic capacities that are already present in young children, as these provide the foundation for further social interaction and ontogenetic development more generally. In particular, we examine three types of social problems which we claim pose unique demands on decision makers relative to nonsocial choices. First, we examine how children detect opportunities for cooperation during social interactions. How do children decide whether or not to have a social interaction at all? In contrast to many economic tasks, in more naturalistic contexts actors do not face particular cooperative situations with predefined roles, but rather must infer that the opportunity for payoffs even exists and engage potential partners. Second, we examine how children identify appropriate social partners for various cooperative activities. We emphasize the importance of partner choice models of social decision making (McNamara et al. 2008; Schino and Aureli 2010), here based on attributions of psychological traits to others. Finally, we examine the roots of how young children form enduring social bonds with good partners. More broadly, how do children decide who should be their friend or long-term cooperative partner? The formation and maintenance of long-term bonds is a serious social problem for gregarious animals like humans, and the psychological processes that support these relationships may be very different from those recruited in more short-term interactions.

Models of Social Decision Making

Traditional economic theory assumed that humans make social decisions much like they make nonsocial decisions: by focusing on one's own potential payoffs. These types of analyses derive from game theoretic models and operate under the assumption that rational actors will decide based on their own self-interest. However, experimental studies have revealed that humans often do not choose in accordance with these models. These studies typically involve economic tasks in which subjects engage in a social interaction with a predefined structure and a concrete set of behavioral alternatives (described in terms of financial payoffs for the self and others). Tasks including the ultimatum game, trust game, and the prisoner's dilemma create structured interactions with quantifiable social payoffs in a similar fashion. However, although classical predictions suggest that proposers should act to maximize their own payoffs in these contexts, experimental studies have revealed that humans often exhibit "social preferences." That is, their choices do not depend only on their own payoffs, but also integrate a concern for the welfare of others. For example, people may have a motivation to behave altruistically (Andreoni 1990), be concerned about the equity of the final payoff distribution (Fehr and Schmidt 1999), or may be motivated by fairness and reciprocity: wanting to be kind to nice people and unkind to mean people (Rabin 1993).

As such, newer models also assume that social decision making involves utility comparisons, but they differ in that they assign additional value to certain types of social outcomes (Levine 1998). Different models also vary in how this "social utility" is calculated. Some consequentialist models focus on the final distribution of payoffs alone; whether or not someone is "kind" could be evaluated simply by comparing two partner's outcomes to assess whether they are equitable (Fehr and Schmidt 1999). Other economic models, in contrast, emphasize the importance of intention reading in social decision making. For example, people appear to account for the intentions of their partner when choosing whether or not to accept an offer in the ultimatum game (Falk et al. 2003). Accumulating evidence from neuroeconomics further supports the importance of intention reading in social decision making (Delgado et al. 2005; Sanfey et al. 2003). Importantly, however, even in models that emphasize intention reading, the underlying social cognitive skills are often viewed as an "input" to the broader utility calculation. That is, intention reading is used to assess how valuable different courses of actions would be, and it is this value assessment which actually drives the decision maker's choices. For example, in one model of reciprocity (Falk and Fischbacher 2006), another player's "kindness" is represented as a function of the outcome distribution and an "intention factor" (whether the other player's actions were intentional or not in causing this distribution). As such, the desirability of a given outcome is modified by the intentions of the other player—a bad outcome is not so bad if he didn't do it on purpose!—but a utility calculation still lies at the heart of the behavioral decision. However, most social interactions do not involve concrete payoffs explicitly assigned to a particular choice. Thus, a critical question is: How do humans act when they are faced with social problems that are not easily described by such models? What happens when the costs and benefits are not quantifiable (either in an absolute sense, or in terms of the relative value for the different participants) or, more importantly, when the costs and benefits are potentially quantifiable, but are not readily apparent to the individuals involved?

Studies of social-cognitive development can fill this void, as such studies often examine cooperative situations that are ubiquitous and pervasive in everyday life, but may not be easily captured by game theoretic approaches. Thus, examining the cognitive skills that humans use in such situations can help identify the way people make social decisions more broadly. We focus in particular on social cognition because it is essential for our lives as social animals, most prominently captured in the *social intelligence hypothesis* (Dunbar and Schultz 2007; Humphrey 1976; Jolly 1966). The uses of social cognitive capacities to attribute intentional states to other agents are manifold, including the domain of cultural learning and the attempt to outsmart others in competitive encounters. Here we focus on the social-cognitive capacities which underlie our ability to cooperate with others in various ways, structured by the three main tasks that social agents face: the decisions to cooperate when, with whom, and with whom over the long term. We show that many of the social-cognitive capacities have roots early in ontogeny, highlighting their foundational role for the development of social interaction.

Social Cognition in Young Children

Humans are sensitive to cues of intentional action from early on in development. Infants already begin to look behind the surface of people's behavior, quickly developing the ability to represent actions in terms of goals and intentions rather than mere behavioral outcomes alone. The rudiments of this developing capacity can be seen in 6-month-old infants: when they observe the simple action of a hand grasping different objects, they selectively encode the goal object of the agent's reach rather than the spatiotemporal movement of the other's hand (Woodward 1998). Infants also seem to represent the most efficient means to achieve a goal, taking into account the situational constraints of an agent's action (Gergely et al. 1995). Last but not least, young children are able to differentiate between intentions from behavioral outcome, distinguishing between situations in which a person is either unwilling to hand them a toy (such as offering but then withdrawing a toy in a teasing manner) or unable to provide the toy (trying but failing because it slips out the agent's hand). Thus, even if the outcome is the very same—the child does not receive a toy—infants are responsive to the different intentions leading to this outcome (Behne et al. 2005). Similarly, young children selectively imitate those aspects of an action that a person did on purpose over those which happened by accident (Carpenter et al. 1998; Nielsen 2009) or as a result of some behavioral constraint (Gergely et al. 2002). Finally, young children can infer what a person was attempting, even if they never saw the outcome that the person was trying to achieve (Meltzoff 1995; Nielsen 2009). Taken together, these studies show that from an early age, children seem to represent not only observable outcomes of behavior, but rather they try to understand the psychological processes driving the behavior. This helps children not only in discovering new ways for interacting with the physical world, but also how to interact with others socially. We now examine how these developing abilities to represent other's behavior, in terms of underlying psychological states, impact social decision making.

How Do Children Detect Opportunities for Social Interactions?

In contrast to existing experimental tasks derived from game theory, in more naturalistic contexts children do not face well-defined social choices with predefined outcomes. Rather, they must infer that the opportunity for an interaction even exists and engage potential partners as appropriate. Children's emerging social-cognitive capacities are therefore critical for enabling them to recognize when the opportunity to cooperate has arisen. This becomes apparent in two classes of cooperative behaviors: helping and collaboration. In helping, one individual is acting on behalf of another person's need or problem (potentially driven by altruistic motivations), whereas in collaboration, social partners act collectively to produce a mutually beneficial outcome. Importantly, the behavior that children exhibit in many of these contexts may not be that mysterious from a game-theoretic perspective; for example, the benefits provided may be mutualistic and thus the motivation to engage in such behaviors may be clear at an ultimate level. However, children would not be able to engage in these various behaviors without sophisticated social-cognitive skills in the first place. That is, the way in which children detect (and create) opportunities for cooperation is based on their ability to infer goals and intentions from observing other's actions. More specifically, these types of cooperation may be based on two different kinds of social cognitive representations: the ability to represent individual intentions as a prerequisite for helping behaviors and the ability to form joint intentions with a social partner in collaborative activities.

Concerning helping behaviors, from early on in their lives children are able to help in various contexts with different kinds of problems that require different types of intervention. Specifically, toddlers spontaneously help unfamiliar individuals in a variety of contexts, including handing over objects that a person dropped accidentally, removing obstacles that block another person's path, producing the desired outcome even if they never witness a successful completion of the task, opening a novel box if the other person fails, or pointing to the location of a misplaced object for which another person is searching (for an overview, see Warneken and Tomasello 2009). In these situations, children use their intention-reading capacities to detect whether the person has succeeded with her goal and help is needed or not. Second, the variety of situations in which children are able to help highlights that children utilize social-cognitive capacities to help flexibly. Importantly, an altruistic motivation to help other people may ultimately benefit kin, be part of a reciprocal exchange, or contribute to the success of the group. However, these actions would not be possible without the social cognitive capacity to read intentions which vastly increases the opportunities for cooperative behavior.

Young children not only act for others to achieve individual goals, they also act with others collaboratively to achieve joint goals. This enables individuals to produce outcomes that lie beyond the means of any one individual. However, collaboration can require fairly sophisticated behavioral and psychological capacities, as individuals must recognize the potential for a mutualistic situation and be able to act jointly to some degree to benefit mutually. For example, humans may have the unique ability to form joint intentions, a specific cognitive representation that includes a representation of the partner's actions and their interrelationship (Tomasello et al. 2005). Briefly put, participants of a joint collaborative activity represent their own and the partner's action as part of a joint plan of action, including the commitment to coordinate their mutual actions in pursuit of the joint goal. Developmental research shows that during the second year of life, children begin to collaborate successfully with others: first with adults, and somewhat later with peers (Brownell et al. 2006; Eckerman and Peterman 2001; Warneken et al. 2006; Warneken and Tomasello 2007). This includes problem-solving tasks that involve complementary roles to retrieve a reward, as well as social games in which two individuals coordinate with no external goal. Several studies indicate that children not only adjust their behaviors to that of the partner on a superficial level, but seem to grasp the joint intentional structure of these activities.

There are three pieces of evidence in favor of this claim: First, young children engage in role-reversal, flexibly switching between two complementary actions (Carpenter et al. 2005). Second, 18-month-old children do not just follow the partner's lead, but actively participate; when an adult partner interrupts the joint activity, children often try to reengage the partner (Warneken et al. 2006). Importantly, children respond differently to a partner who interrupted because she was either unwilling or unable to continue; this demonstrates that they do not simply respond to the behavioral outcome (partner does not act), but rather the reasons behind this inaction (Warneken et al. 2012). Finally, children appear to form joint commitments to collaborate. From around three years of age, children are more likely to try to reengage a person after their joint commitment to collaborate together ("Let's play!") than when no such commitment existed and both individuals just play in parallel (Gräfenhain et al. 2009). Joint intentions impact payoff distributions as well, as peers are more likely to help and share obtained rewards as part of a collaborative activity rather than individual problem solving (Hamann et al. 2011, 2012). Taken together, these studies demonstrate that when children engage in joint tasks, they do not simply focus on mutualistic outcomes but conceive of self and other as genuine collaborative partners, who have the intention to act collectively toward a joint goal in a committed fashion. Thus, while many collaborative tasks involve mutualistic benefits, children would not necessarily be able to realize the mutualistic opportunity was present without their emerging social-cognitive skills.

How Do Children Identify Appropriate Social Partners?

Social agents must not only be able to decide when to cooperate, but also with whom. This raises the question of what kind of information guides the decisions that children make as developing cooperators. We distinguish two classes of information that can serve as the basis for this decision: children's previous direct interaction with potential social agents and their observation of third parties interacting with each other. Once again, evidence is accumulating that when deciding with whom to cooperate, young children are attentive to and perhaps even prioritize information about another person's intentions over the mere behavioral outcome of a cooperative interaction.

Concerning direct social interaction, research shows that two-year-olds tend to help "nice" people over "mean" people. Specifically, children interacted with a person who was unwilling to give them a toy as well as another person who failed to do so because she was unable. After the interaction, children tend to help the previously unable individual over the unwilling individual, even if they had not received a toy from either person (Dunfield and Kuhlmeier 2010). Similarly, children also prefer to help someone who handed them a toy intentionally compared to a person who transferred a toy to them as a kind of side effect of their action, even though the outcome was again the same. Finally, when children are confronted with a person who successfully handed over a toy and another person who tried but failed, they are indifferent in their subsequent helping; this indicates that beneficial outcome does not necessarily trump good intentions (Dunfield and Kuhlmeier 2010).

Children can also learn about potential social partners from the observation of third-party interactions. Before their first birthdays, infants differentiate between "helpers" and "hinderers" in short animations (where agents either facilitate or obstruct another agent's path up a hill) and expect the helpee to approach the helper over the hinderer (Kuhlmeier et al. 2003). The fact that these infants form expectations about behavior that carry across different contexts indicates that they might attribute dispositions to these agents. Moreover, not only do infants differentiate these events and draw inferences, they display a preference for the helper over the hinderer, as measured by their approach (Hamlin et al. 2007). Toward middle childhood, these social evaluations then begin to guide their own cooperative behaviors. For example, three-year-olds selectively direct their helping toward "nice" people, based on their assessment of how others had previously interacted. In one study, children preferentially helped a neutral person over a person who had previously harmed another individual; they even show this preference when harm was intended but not actually realized (Vaish et al. 2010a). That is, the intentions underlying the observed behavior are particularly important, and children use this information to cooperate selectively with specific individuals. These results are corroborated by more explicit measures of children's judgment, with 3.5-year-olds advising that a character should share more with those who previously shared reciprocally, and should preferentially share with others who had shared with third parties (Olson and Spelke 2008).

Why might children focus on the underlying intentions of different potential social partners, even sometimes to the exclusion of the actual outcomes they received with those partners? One possibility is that intentions are a better predictor of future behavior than are payoffs, assuming that intentions underlying a behavior are more stable than the behavioral outcome which is influenced by situational factors (Fishbein and Ajzen 1975). For example, someone who struggles (but fails) to cooperate may be a better bet for future interactions than someone who just happened to create a positive outcome without any sign that this resulted from genuine cooperative intentions. Thus, although at the proximate level children may not be calculating a utility associated with interacting with different partners, at the ultimate level focusing on intentions may serve as good proxy for identifying the most advantageous partners. This suggests that children should perceive cooperative intentions in others as a stable trait that persists over time. Next we examine the evidence concerning whether children actually use their social-cognitive skills to attribute personal characteristics to others in this way.

How Do Children Maintain Enduring Bonds with Good Partners?

Social agents living in stable social groups do not have to decide anew each time whether another individual is an appropriate social partner or not. Individuals interact with many other social agents over repeated encounters and thus form social bonds and animosities which shape social decision making over longer time frames. Humans expect other agents to act consistently over different social encounters and different situations, attributing dispositions as enduring traits that are expressed across multiple behaviors. Young infants already seem to attribute dispositions rapidly to agents, extrapolating from one context how agents will act in a different context. This is evidenced by studies that examine helpers and hinderers (Kuhlmeier et al. 2003): when 12-month-old infants observe one agent acting as a hinderer and another agent acting as a helper when an agent is trying to travel up a hill, they can extrapolate from one type of situation to another both to predict how others will act and use this information to decide with whom they should interact. This phenomenon is also reflected in tasks used with three-year-olds who, after witnessing how one person was mean to another person, are then less likely to help the mean person than the unfamiliar person with an allegedly clean record (Vaish et al. 2010a). Thus, children seem to attribute enduring characteristics to other agents that transcend the immediate context or situation.

These abilities are potentially important for the decisions that children need to make in real life to determine with whom they should interact repeatedly. Indeed, children develop enduring relationships with people outside their family early in life. From at least three years onward, children differentiate between friends and non-friends, preferentially interacting with certain peers over others (Rubin et al. 2005). When interacting with friends, they show more positive and more elaborate social behaviors, including cooperation and social games (Hartup 1996; Rubin et al. 2005). Interestingly, episodes of conflicts and even aggression are more likely to occur among friends than non-friends, with the crucial difference that friends are also more likely to resolve these conflicts than non-friends, settling things—as the word already implies—amicably. Finally, friendship relations mediate the specific decisions that children make in cooperative contexts as well. Three-year-old children suggest that friends should give more to each other than agents who are strangers to each other (Olson and Spelke 2008), and they are more likely to share with friends than with non-friends when it involves an actual cost to themselves (Moore 2009). Moreover, when children enter school, they begin to show parochial tendencies, acting more generously toward in-group members than out-group members (Fehr et al. 2008a). Thus, children are prepared not only to decide on the fly with whom to cooperate or not, they encode information about more enduring traits of individuals which transcend the interpretation of behaviors as single episodes and intrepret them instead as expressions of dispositions underlying other people's acts.

The Evolution of Social Decision Making

Can this view of the development of human cooperation be extended to encompass a more general explanation of social decision making across species? The social world has long been thought to be a major force shaping cognition, especially in primates (Byrne and Whiten 1988; de Waal 1982; Jolly 1966). More recently, empirical investigations suggest that at least some species of primates possess social-cognitive skills mirroring those seen in young children (Rosati et al. 2010), including intention-reading skills (Call et al. 2004; Phillips et al. 2009). While many of these empirical investigations of primate social cognition have focused on competitive contexts, the basic thesis that primates' sophisticated cognitive abilities evolved for a social function also encompasses cooperative contexts. For example, wild chimpanzees engage in several complex cooperative behaviors including meat sharing, group hunting, mate guarding, and boundary patrols (Muller and Mitani 2005). Thus, a major question for generalizing our view of human social decision making is whether social-cognitive factors, including intention reading, are also a major determinant of cooperative interactions in nonhumans.

Several lines of empirical evidence address this question. First, apes know the quality of the relationships they share with group-members: chimpanzees will spontaneously cooperate to acquire food in a mutualistic task with conspecifics with whom they share a tolerant relationship, but not with intolerant partners (Hare et al. 2007; Melis et al. 2006b). Similarly, some types of cooperative behavior (such as helping) appear to be driven by intention cues from the recipient, as in human children, not the expectation of explicit rewards (Melis et al. 2010; Warneken et al. 2007; Warneken and Tomasello 2006). Second, chimpanzees appear to engage in selective cooperation, like human children, using their social-cognitive skills to identify appropriate social partners. In particular, they prefer skillful partners over unskillful partners (Melis et al. 2006a) as well as those who cooperated with them in the past over those who did not (de Waal and Luttrell 1988; Koyama et al. 2006; Melis et al. 2008). Chimpanzees also use information about whether a conspecific was the cause of them loosing food access when deciding whether to punish (Jensen et al. 2007b). Finally, many species of primates appear to engage in reconciliation to maintain long-term bonds with good social partners (de Waal 2000), in some cases using their social knowledge about the relationships of others (Wittig and Boesch 2010). While it is not clear for all cases how "deep" the attributions of psychological states reach, animals are likely using attribution of agency or even intention reading when making social decisions.

This type of social-cognitive analysis also suggests that some forms of social-decision making may be limited to humans—not because other primates would not potentially benefit from the interactions, but because they lack the opportunity to detect potential cooperative opportunities. For example, while chimpanzees do engage in helping behaviors, they appear to depend on overt goal cues—such as reaching or attempting to enter a room—much like very young children at 14 months (Warneken and Tomasello 2007). By 18–24 months, however, children can already engage in flexible helping across a diverse set of contexts, involving goals that are likely more complex to understand (e.g., placing items in particular locations or failing to open a container with the correct method). In addition, there is currently no evidence that chimpanzees form joint intentions when engaging in mutualistic interactions in the way that young children do (Greenberg et al. 2010; Hamann et al. 2011; Melis et al. 2010). This suggests that chimpanzees may miss out on potential opportunities for beneficial social interactions because they lack the cognitive ability to detect that the opportunity is present.

In general, children's developing social-cognitive abilities suggest three main ways in which intention reading might be a good psychological strategy from an ultimate perspective, relative to utility-based models of psychological processes. First, when payoffs are not explicitly defined, intention reading might be the most salient cue for detecting opportunities to cooperate. Individuals who are skillful at reading other's intentions might therefore have the best behavioral outcomes, because they have successful cooperative interactions (including mutualistic interactions) whereas others who do not engage do not. Second, intention reading may be a good proxy for predicting good social partners. Utility-based models often require previous behavioral experiences with a partner to assess the desirability of difference courses of action. In contrast to repeatedly engaging in costly interactions with another individual to learn whether they are a good partner, assessing cues to another's intentions can allow decision makers to find good partners more cheaply. Finally, individuals can use social cognitive skills to assess other individuals' enduring behavioral patterns, differentiating between those who are likely to be a friend or foe.

Conclusion

Different research traditions have approached similar phenomena from different ends, either beginning with the potential fitness costs and benefits of social behaviors, or starting with the psychological processes involved in social decision making. Looking at concrete payoffs and ultimately fitness consequences is important to understand behavior in terms of its evolutionary significance; however, the psychological relevance of payoff is still unclear. This highlights the difficulty of connecting the analysis of ultimate function and proximate mechanisms. What we are facing could be called "Tinbergen's fifth question." That is, how do we connect the four questions of biology that Tinbergen (1963) identified, especially ultimate function ("survival value") and proximate mechanisms ("causation")? We have provided examples of the psychological capacities that enable agents to navigate the social world, focusing on the basic capacities seen in young children. We suggest that social decision making is social in a deep sense, not only as a decision about other social agents but about their agency itself, because humans pay particular attention to intentions as causes of other people's behavior. Many of these capacities can be traced back into early ontogeny, which highlights the foundational role that they must play in the development of social decision making.

Acknowledgments

We wish to thank Peter Hammerstein and Jeff Stevens for the opportunity to contribute to this Ernst Strüngmann Forum. We also extend our thanks to the members of the working group "Evolutionary Perspectives on Social Cognition" for their inspiring discussions, especially Jeff Stevens, Rob Boyd, Rosemarie Nagel, and two anonymous reviewers for their feedback on an earlier version of the manuscript.

18

Who Cares?

Other-Regarding Concerns—Decisions with Feeling

Keith Jensen

Abstract

The abilities to feel distress at the suffering of others and to share in their joy are indicative of positive other-regarding concerns. The value to social decision making is intuitive: helping others feels good, harming them feels bad. Less intuitive are negative other-regarding concerns, taking satisfaction in the misfortunes of others and feeling sad at their successes and joys. Yet these sentiments also play a role in the choices humans make when interacting with others. This chapter explores other-regarding concerns and how they influence social decisions. The nature of other-regarding concerns is discussed, with an emphasis on the role of emotions in guiding human other-regarding preferences. Possible origins of the emotional cornerstone of human sociality are suggested based on animal research, particularly nonhuman primates, and studies on children.

Introduction

> Your cat knows exactly how you feel. He doesn't care, but he knows.—A get-well card greeting

Cat owners will recognize their companions in this greeting. If cats do know how we feel, they appear not to care. (Dog owners, on the other hand, will say that their best friends care about how we feel, but do not know.) What does it mean to care about others? Various philosophical, psychological, and economic views suggest that we do not truly care about others, but only care about ourselves and use other people as means to our ends. This is a rather bleak view, and it may not be true. Even so, other-regarding concerns are not all for the good; there is a dark side to knowing and caring about how others feel.

Other-regarding concerns motivate us to seek outcomes that are for the good of others but they can also lead us to bring about their suffering. In this chapter, I raise the point that emotions which are tuned to the welfare of others are an important—perhaps essential—part of social decision making. I contrast the human case with our closest living relatives to highlight the importance of other-regarding concerns and raise the question of how they might have evolved in our species. By reviewing some recent work on children, I ask how other-regarding concerns might develop.

Other-Regarding Concerns: The Heart of Social Decision Making

Preferences guide choices between outcomes. As a simple example, most of us prefer chocolate over broccoli because of the flavor (as opposed to the health benefits), and given a choice, we will choose the sweet instead of the vegetable. We will eat broccoli if we are hungry and there is nothing else available, but we will generally ignore it if more preferred alternatives are available. I say generally, because we might not show this preference consistently; for example, when thinking about our diets, if satiated with chocolate, when trying to impress our mothers, and so on (for a discussion of consistency in preferences, see Bugnyar et al., this volume). What we are unlikely to do is to take into consideration the effects of our preference on the food. Broccoli will not feel grateful if eaten and shunned if unconsumed.

However, if the choice involves other individuals, we take the effects of our actions on them into account. So if we have one gift to give—say a box of chocolates—we will consider the happiness of the recipient as well as the hurt feelings of someone else who had hoped for it. *Other-regarding* (social) *preferences* guide choices between outcomes that are sensitive to the consequences for other individuals. At the simplest level, there is a choice between outcomes affecting the actor (like eating the chocolate yourself) and another person. Choosing outcomes that benefit others can indicate positive other-regarding, or prosocial, preferences, such as giving away our last box of chocolate because we know it will make our friend happy to receive it. People can also have negative other-regarding (antisocial) preferences and choose outcomes that harm others; for instance, eating all of the broccoli, even though we do not like it, because we know it will upset the hungry vegan who is watching.

Preferences are inferred from behavioral outcomes, but it is difficult to ascertain their *motives*, namely the desires, intentions, or judgments (Clavien 2012). Other-regarding preferences are tested by giving individuals a choice between selfish outcomes and outcomes affecting others; other-regarding preferences are supposedly what remain after selfish preferences have been ruled out. Measuring other-regarding preferences is not always straightforward. Just because another individual benefits from or is harmed by an actor's decision does not mean the actor had as a goal that the other individual should experience that effect. Outcomes may be unintended by-products of self-regarding motives. Giving chocolate to someone else because we do not like chocolate and would not eat it anyway, or giving up in the face of their insistent begging are examples of unintended prosocial outcomes; the effects on others are means to selfish ends. It is difficult, particularly in naturally occurring behavior, to tease apart prosocial motives from the multitude of ulterior motives. Evolutionary biology encounters the same sort of problem of behaviors that were not selected for, producing the illusion of altruism, such as a squirrel running along a branch dislodging a nut that falls to the forest floor to be eaten by a hedgehog. Ruling out pure self-regard is surprisingly difficult. Psychological hedonism (egoism) assumes that every action is ultimately for the benefit of the actor, just as evolution by natural selection requires that, on average, individuals—or at least their genes—thrive as a result of their actions.

Knowing the *motivation*—the emotion—behind an act, such as altruism, can bypass the problem of multiple (and egoistic) motives. Having the appropriate emotions gives credence to the motive (Clavien 2012; de Sousa 2004). A large amount of recent research in developmental and comparative psychology has focused on theory of mind, the ability to infer the beliefs, thoughts, and desires of others (Call and Tomasello 2008; Doherty 2008; Jensen et al. 2011). Whether theory of mind is necessary for other-regarding preferences is an open question. Perhaps just as important, if not more so, is affective perspective taking, taking the emotional perspective of others (e.g., Vaish et al. 2009).

Other-regarding concerns are emotions that are sensitive to the emotions and welfare of others (Ortony et al. 1988 refer to this as "fortunes-of-others" emotions; Smith 1759/2000 speaks of "moral sentiments"). These can be aligned or misaligned (Table 18.1). *Positive other-regarding concerns* are emotions that are aligned with the emotions and well-being of other individuals, namely empathy[1] (sadness at the sadness and misfortunes of others) and symhedonia (happiness with their happiness and good fortunes). These emotional states would motivate the actor to reduce the suffering of others and to seek their happiness as a goal. *Negative other-regarding concerns* are emotions that are misaligned with the emotions and welfare of others. They occur when the emotional state of the subject is opposite to those of another individual, such as anger or disappointment at the happiness and good fortunes of others (envy) and joy at their sadness and misfortunes (schadenfreude). As a result, individuals should be motivated to reduce the happiness of others. If an individual's emotions are unaffected (tangential) to the well-being of others—as in the cat lampooned in the get-well card greeting—then the emotions are unaligned.

[1] Empathy is often defined along the lines of having the emotions appropriate to the feelings and circumstances appropriate to another individual (e.g., Hoffman 1982). The German word *Einfühlung—feeling into others*—better captures this meaning. Empathy is usually taken to refer to feeling sadness at the distress of others (e.g., Ortony et al. 1988), and this is the sense in which I use it. Symhedonia is the other example of an other-regarding concern that is appropriate to the feelings and circumstances of another individual.

Table 18.1 Other-regarding concerns can be positive or negative. The emotions of the subject can be positive (+) or negative (–), and these can be aligned with the emotions or welfare of others (+ , + and – , –) or misaligned (– , + and + , –). Note the similarity to Hamilton's (1964) payoff matrix. There is, however, no change in fitness (Gadagkar 1993). Emotions and welfare are proximate (emotional) payoffs, as in Ortony et al. (1988), rather than fitness costs and benefits, as in Hamilton (1964).

		Other Individual's Emotions or Welfare	
		Positive	Negative
Subject's Emotions	Positive	(+ , +) Symhedonia	(+ , –) Schadenfreude
	Negative	(– , +) Envy	(– , –) Empathy

Unaligned emotions can come in different forms, but they are all forms of indifference. Indifference to the happiness and success of others is apathy; indifference to their suffering or misfortunes is callousness. The subject can experience emotions regardless of how other individuals feel or fare, such as states of happiness (blithe indifference) and sadness (malaise); any absence of emotion is impassiveness. It is important to point out that indifference is not an other-regarding concern, regardless of any consequences for other individuals. The main features of the underpinnings of other-regarding preferences are that they have an emotional component, rather than being solely self-serving, and they are motivated by their effect on others rather than only on the self.

Philosophers have long viewed emotions as the enemy of reason, though this view has been changing (de Sousa 2008). Early cognitive research tended to avoid emotions as being messy. However, people with emotional impairments such as frontal lobe damage perform well on standard cognitive tests, but have impaired sense of social judgment and take risks that nonimpaired individuals would not make (Damasio 1994). Frank (1988) has argued that emotions are an essential part of human decision making and cooperation by honestly signaling intent. The role of emotions in decision making is garnering an increasing amount of attention (e.g., McElreath et al. 2003). Second-order, or social, emotions which are tuned to the emotions of others (e.g., shame and pride) play an important role in social decision making (Fessler 1999; Fessler and Haley 2003). While the "primary" emotions such as fear and anger are likely homologous in many animals, including humans (mammals, at least, and certainly primates; Darwin 1872), many of the social emotions such as gratitude and guilt may be uniquely human. For example, while dogs exhibit "guilty looks" when they do something they should not have done, this is due to responses to owner cues (such as anticipating punishment) rather than guilt (Horowitz 2009). At present, little is known about the evolutionary origins of social emotions. What is clear is that they are central to human social decision making.

Positive Other-Regarding Concerns

Positive other-regarding concerns would logically seem to be important for adaptive social decision making, adaptive in both the functional (payoff maximizing) and evolutionary (fitness maximizing) senses. If helping others, even if it is costly in the present, brings about delayed direct benefits, benefits to kin, or benefits to others in one's group, then a mechanism that provides immediate, though intangible rewards, will serve to motivate acts altruism. The difficulty is that there is the temptation to cheat, to seek as much for oneself, and hope (if one cares) that any benefits that go to others are guided by an invisible hand (Smith 1776/2005). I will first consider the conflict between self-regard and positive other-regarding preferences before delving into the minds of altruists. Thereafter I discuss the evidence for the ontogeny and phylogeny of positive other-regarding concerns.

Positive Other-Regarding Preferences: More than Maximizing

Economic models make clear predictions of how people ought to behave. The core assumption is that individuals should seek outcomes that maximize their utility (this utility can be anything, but is usually money). Purely self-interested people (*Homo economicus*) will try to get as much as possible for themselves (Frank 1987). However, people do not behave in a straightforward, self-regarding manner. Economically, in experiments such as the dictator game, a player given a monetary endowment willingly gives a portion of it away to another person whom they do not know, often around 20% (Camerer and Thaler 1995; Camerer 2003). This behavior is puzzling because a self-maximizer (a greedy miser, in layman's terms) should keep all of the money for himself since there is nothing to be gained by giving any away. On an evolutionary scale, altruistic acts, such as giving one's life so that others can live or forfeiting reproduction so that others can breed, have clearly been documented in a wide range of organisms. Even humbler acts of altruism that do not have immediate fitness consequences, such as grooming another individual, sharing food, and even helping others from different species, are pervasive and puzzling.

There are various theories for both the economic and evolutionary anomalies that attempt to explain away these phenomena. To maintain the conventional assumption of behavior guided by self-interest, generous behaviors are apparently attributed as being due to errors of judgment, such as mistakenly believing that the interaction is not truly anonymous, or as being a by-product of our ancestral past in which the artificial conditions of one-shot, anonymous interactions did not apply (Boyd and Richerson 2005; Hagen and Hammerstein 2006). Evolutionary biologists look for personal fitness benefits either at the level of the gene (kin selection: Hamilton 1964) or delayed benefits (reciprocal altruism:

Trivers 1971). In other words, standard economic and evolutionary theories scratch away at the altruist to watch the selfishness bleed out (Ghiselin 1974).

There is a lot to be said for these approaches, but they appear to be insufficient in explaining the degree to which humans help each other. People do come to the aid of others in nonexperimental settings, sometimes at great personal cost, when there is no clear expectation of direct or indirect benefits. A tribal, kin-based, psychology would seem to be a weak substructure on which to build large-scale societies assembled around large groups of strangers which interact, often indirectly, to produce benefits that are distributed among the group. Such a kin-centered psychology would seem to be the basis for the social structure of chimpanzees and may certainly have been sufficient for very early hominid societies, but does not go far enough to explain human sociality (see, e.g., Langergraber et al. 2011). The degree to which people help nonkin and the limits of direct reciprocity in sustaining cooperation in large groups suggest that there is something more to cooperation in humans than can be explained by standard models (Fehr and Fischbacher 2003; Richerson and Boyd 2005). Some economists and evolutionary theorists suggest that the *something more* is group-beneficial behavior. At the level of individual decisions in experiments, these are other-regarding preferences. At the evolutionary level, cultural group selection has been proposed to explain large-scale cooperation (Richerson and Boyd 2005). There is currently considerable debate around both of these explanations. The main point is that people do genuinely act in ways that benefit others and that there is some sort of concern for their well-being.

Inner Workings of Positive Other-Regarding Concerns

Deciding to forfeit lucre for others may be due to a sense of fairness or, more precisely, an aversion to advantageous inequity (e.g., Fehr and Schmidt 1999). It is a mechanism to restore equity, or some degree of it. This sense of inequity aversion is tuned to outcomes, namely measuring one's own gains relative to another individual's. In the dictator game, people compare how much money they receive compared to some other individual who has nothing, judge this as disproportionate, and give up a portion of their wealth to the other player. The amounts given are highly variable and very sensitive to framing, namely the context surrounding the game (Hagen and Hammerstein 2006). Even at the best of times, offers are fairly low: 20% is a far cry from a fifty–fifty split. So, while people might have some positive inclination toward improving the lot of someone else, they are also self-regarding.

It is hard to imagine the decision-making process in cool, cognitive terms. Does a player in the dictator game reason the following?

> I have been given €10. There is (presumably, if I trust the experimenter) someone else in the next room who has nothing. No one (presumably) is watching what I

> do. I can keep everything for myself, but in all fairness (so I've been taught), the other person deserves something, so I will leave €2 for him.

In another experimental tradition, social psychologists, notably Batson et al. (1981), place people in more life-like, though deceptive, settings in which a participant observes someone else—an experimental confederate it turns out—in need of help. For instance, the participant is told that the other person will receive painful electric shocks. Given the opportunity to take the confederate's place and receive shocks or to leave and thus avoid having to witness a potentially distressing scene, many—but not all—people make the self-sacrificing choice and volunteer to receive the shocks. The conclusion is that altruistically motivated acts should have the benefit to another individual as an end in itself and not as a means to a selfish end (psychological egoism). Thus, if the participant volunteers to receive the shock to gain standing as a nice person and so receive future benefits (perhaps the phone number of the confederate), then the other-regarding motivation is questionable. Certainly leaving the situation is motivated out of a selfish concern, yet even this may have an emotional basis that is sensitive to the suffering of others—an individual who is indifferent to, or takes pleasure in, the suffering of others might elect to stay to watch someone else suffer.

People are likely motivated out of empathy, namely having the emotions appropriate to the circumstances of another individual (Hoffman 1982). This is distinct from emotional contagion in which someone "catches" another individual's emotions. Seeing someone visibly in distress is personally distressing, and fleeing is one way to counter that unpleasant feeling. Positive other-regarding preferences involve some degree of taking the needs of others into account, even in the absence of emotional cues. This affective perspective taking need not be the only motivation for helping another person, or giving them money, since one can imagine how sad it would be to be the other person. One can have selfish motivations as well, even if they are intangible, such as feeling good about oneself for helping an elderly gentleman cross the street, or imagining the reputation one is garnering from friends, or avoiding guilt. The important point is that people are not *only* self-regarding. They appear to exhibit genuine concern for the suffering of others. In fact, it would seem very strange to encounter altruism (in the behavioral sense) in the absence of empathy or symhedonia. If it felt bad to help someone (e.g., if someone gave money to charity and cringed at every penny given), it would be hard to call the person charitable. If you were being aided by someone who did not feel distress at your suffering, but administered aid in a cool and dispassionate manner, you would likely feel as if you were being treated by a robot. From a motivational perspective, an altruist is someone who exhibits concern, who *feels*, for the recipient. Contrary to those worried about psychological hedonism, the primary motive should not be the "warm glow" (e.g., Andreoni 1990). The primary (or ultimate, in the psychological sense) objective is the well-being of others. The positive feelings (warm glow) and reduction in empathic concern would be secondary by-products that come via the effect on the recipient.

The Development of Positive Other-Regarding Concerns

Concern for others and a tendency to engage in prosocial acts emerge early in childhood. Children just past their first birthdays will help others when needed, such as handing adults clothespins that have been "accidentally" dropped (Warneken and Tomasello 2006). Children will also share rewards, such as candies and stickers, with peers and experimenters, and they strive to achieve fair divisions (Thompson et al. 1997; Rochat et al. 2009; Fehr et al. 2008a), though they may not always do this spontaneously (Brownell et al. 2009). They will also communicate information to others, such as by pointing (Liszkowski 2006). Infants as young as six months of age may even be able to distinguish between characters (geometric shapes) that help others achieve their goals from those which hinder (Hamlin et al. 2007), suggesting that an understanding of prosociality is innate or at least present from shortly after birth (Wynn 2008).

Certainly at some point in their development, children are capable of recognizing the goals of others, not only in helping contexts (e.g., Gergely et al. 2002). They also appear to be motivated to help others achieve these goals, demonstrating positive other-regarding concerns. For instance, helping is intrinsically rewarding—children do not need praise or some other reward (Warneken and Tomasello 2008). Whether children are motivated to help or are completing actions, something they are proficient at in other contexts, is an open question (Prinz 2007). More direct evidence for prosocial behavior in children comes from studies of empathy. Young infants will cry in response to the cries of others (emotional contagion). In their first year of life, they show more differentiation in their responses, but it is not until they are in their second year that their response to the distress of others is appropriately other-directed (Eisenberg et al. 2006; Hoffman 1982). To tease apart emotional contagion and actual concern for the welfare of others (empathy), children were shown one adult "harming" another (such as tearing up a victim's drawing) with the victim showing no emotional cues (Vaish et al. 2009). Children as young as 18-months-old not only showed looks of concern after witnessing a harmful act, they also responded to the victim, for example, by giving up one of their own balloons more often if the victim (rather than a neutral person who was not harmed) lost hers. The suggestion is that children are capable of affective perspective taking and that this motivates helpful acts. Additionally, children have fairness concerns; older children (around eight years of age) show a preference for fair divisions of candies, even if this means forfeiting a larger amount for themselves (Fehr et al. 2008a; see also Hook and Cook 1979). How these positive other-regarding concerns emerge in development and are sensitive to their cultural environments (i.e., social learning) are interesting questions for future research.

Positive Other-Regarding Concerns in Other Species

Given the possible importance of positive other-regarding preferences in motivating prosocial behavior, it is plausible that we would see their antecedents in our ancestors. Since I am focusing on the evolution of other-regarding concerns and other-regarding preferences in humans, the discussion will emphasize nonhuman primates, due to their phylogenetic proximity to us and to the fact that these are the species on which most experimental work is done. It is important, of course, to study other taxa since they may have evolved other-regarding concerns to solve their adaptive challenges (e.g., through cooperative breeding; Burkart et al. 2007). Darwin (1871), for one, believed that a sense of morality would have its phylogenetic origins in other species. Some researchers have suggested that other animals have a sense of empathy (Preston and de Waal 2002) as well as some components of morality (Flack and de Waal 2000). For example, mice (*Mus musculus*) will show a more pronounced response to a painful stimulus if another mouse is seen to be in pain (Langford et al. 2006), lab rats (*Rattus norvegicus*) will release conspecifics trapped in a tube (Bartal et al. 2011), greylag geese (*Anser anser*) show increased heart rates if they see their partner or kin (but not other geese) in fights (Wascher et al. 2008), and chickens (*Gallus gallus*) have also been suggested to have something akin to empathy (Edgar et al. 2011). Chimpanzees (*Pan troglodytes*) will express distress when another is distressed and will comfort (console) the individual (de Waal and van Roosmalen 1979). It has been suggested that chimpanzees may experience grief at the loss of a companion (Brown 1879), sympathy and pity for a sick conspecific (Yerkes and Yerkes 1929), and empathy for injured group members (Boesch 1992).

However, as difficult as it is to ascertain the role of empathy in altruism in humans, it is even more difficult to speculate on the motivations and subjective experiences of nonlinguistic species. Even something as rudimentary and vital as food sharing in chimpanzees appears to be due to succumbing to harassment (Gilby 2006; Stevens and Hauser 2004) rather than to empathy. Instances of *failing* to help, as in Goodall's observation of chimpanzees failing to accommodate individuals afflicted with polio (Goodall 1986), are likely to be underreported (de Waal 1996). The debate over whether nonhuman animals, chimpanzees, and other apes, in particular, have prosocial motivations has been active for some time and continues to be. However, as was pointed out in the context of language (Pinker 1995), our closest relatives (the Neanderthals, *Homo erectus*, and the Austalopithocenes) are extinct. If apes had also gone extinct long ago, leaving monkeys as our closest living relatives, would it be necessary to find other-regarding preferences there, or in insectivores? It is sensible to peer back along the branch on which we evolved to see how various traits emerged and changed through time, but in reality we can only look at parallel branches and assume that the paths connecting them are similar. This approach to comparative psychology has an air of nostalgia about it, as if by looking at our neighbors we can discover something of our childhood. It is, however, the best we can do.

What has been found so far in recent experiments is not overly encouraging for those hoping to see the roots of positive other-regarding concerns. One experimental paradigm in the spirit of the dictator game gives one subject a choice between two outcomes: one that is mutually beneficial (a 1/1 payoff) or selfish (a 1/0 payoff). (The number before the slash is the actor's payoff; the other player's is after the slash.) This is a very weak version of the dictator game in that there is no cost to helping. If the subject is empathic or averse to being better off than another member of one's social group, the mutualistic outcome should be preferred as opposed to when there is no partner. This was not the case for chimpanzees (Jensen et al. 2006; Silk et al. 2005; Vonk et al. 2008); they were neither mutualistic nor selfish, but chose indifferently. This was also true when there was no benefit for the subject; there was no preference for 0/1 over 0/0. It may be that our closest living relatives are too competitive to be nice, but if that were true, they should choose *against* the mutualistic or altruistic outcomes, which they failed to do, even when this was the only way of preventing the partner from getting food (for recent reviews, see Jensen 2012; Silk and House 2011).

It may be that some features of these experiments, or the species studied, can account for the utter indifference to outcomes affecting others. Horner et al. (2011) suggest that simplifying experimental paradigms allows the prosocial choices in chimpanzees to emerge. Others have suggested that a more socially tolerant species, such as cooperatively breeding New World monkeys, show positive other-regarding preferences (Burkart et al. 2007; Cronin et al. 2010), but this finding has been contradiced (Cronin et al. 2009; Stevens 2010). Competitive drive over food is not the issue, at least for chimpanzees, because whether food or a nonfood item can be delivered to a conspecific did not negatively affect choices. In fact, chimpanzees were more likely to release food for a partner, because of the attractive value of the food and the beseeching of the partner (Melis et al. 2010). It may be that chimpanzees need to be able to recognize clearly the intentions—the goals of the recipient—before they will help (Warneken and Tomasello 2009; Yamamoto and Tanaka 2009). This may be, but given that chimpanzees perform poorly at role reversals, which require some sort of attending to the actions of others (Warneken et al. 2006), it seems that they get caught up in their own goals and only incidentally attend to the goals of others. It is also not clear why recipients failed to indicate their desires (not a very demanding requirement) or, when they did signal, desires were ignored in some of the experiments.

Given the lack of consensus among the various studies, it is not possible to confidently assert that our closest living relatives and other animals have positive other-regarding preferences. It is also difficult to proclaim that they do not. There is no irrefragable evidence hinting at shame, guilt, symhedonia, or other other-regarding concerns. Still, chimpanzees and other animals do respond to distress in others and may have emotional contagion, or something other than empathic concern (Koski and Sterck 2010). Many animals are highly social and engage in a variety of activities that can benefit others in their group. However, they do not necessarily employ the same system of other-regarding concerns as used by humans in motivating prosocial behaviors. Given the difficulties in ascertaining whether humans even have other-regarding concerns (there is no consensus that they do), it is not surprising that it is more difficult to determine whether other animals do. Future testing will help to determine this, largely by using tools adapted from game theory and experimental economics where the costs and benefits of actions are clear. Here, large sample sizes and replications from different laboratories will be especially important. Furthermore, using additional measures to complement behavior responses (e.g., facial expression, physiology, heart rate, and brain imaging if possible) may help pinpoint the role of emotions. Getting to the heart of other-regarding concerns will require inferring something of the motivations of animals.

Negative Other-Regarding Concerns

Negative other-regarding concerns, taking pleasure in the misfortunes of others, would seem to be counterproductive to any sort of adaptive social decision making, at least in cooperative contexts. However, free riders and cheats degrade any benefits from cooperation, and one way of dealing with them is punishment (Boyd and Richerson 1992; Clutton-Brock and Parker 1995). Since, like altruism, punishment is costly at the time it is performed, an immediate reward, such as satisfaction at seeing perpetrators suffer, can serve as a bridge to delayed material and fitness benefits. Instead of an invisible hand, an iron fist guides social decision making. Perhaps even more counterintuitively than punishment, harm for harm's sake can motivate a degree of competitiveness not seen in any other animals. I will first look at the evidence for negative other-regarding preferences in adults and children, after which I will probe their underlying other-regarding concerns and describe the search for these in other species (primates in particular).

Negative Other-Regarding Preferences: Harm for Harm's Sake

As with positive other-regarding preferences, negative other-regarding preferences fly in the face of self-regarding maximization. As little sense as it makes to suffer a cost to benefit someone else, it makes even less sense to do so to cause someone else to experience a loss. The same reasoning applies to evolutionary biology: reducing one's own fitness to cause others to suffer greater fitness costs—biological spite—would appear to be maladaptive. However, biological spite can be explained if a costly self-sacrifice indirectly benefits others carrying the actor's genes by destroying rivals who are less related (negatively related) to the actor than the population average (Gardner and West 2004). The cases for biological spite are few. One example is "bacteriocidal warfare," where some bacteria sacrifice themselves by releasing a toxin that is lethal to bacteria from another colony (Gardner and West 2004). Other cases that would appear to be biological spite, such as reproductive interference (Brereton 1994) and food destruction (Horrocks 1981) in monkeys, are dismissed as being selfish since the actor may receive benefits in the future (Foster et al. 2001). However, as with cases of spiteful acts in humans, there is reason to take interest in these behaviors, even if they are ultimately selfish; the end results may be the same as indifference and selfishness, but the path to getting there is very different (Jensen 2010).

In economic experiments, people will sacrifice their monetary endowment to cause other players to lose theirs. One of the best examples of this phenomenon comes from the ultimatum game (Güth et al. 1982). Like the dictator game, which was developed later, the first player—the proposer—is given a cash windfall to divide with the second player. Here, however, the second player—the responder—has a role to play by accepting or rejecting the offer. It the responder accepts, both players get the proposed division. When the responder rejects an offer, both players get nothing. The rational choice is to accept, because any offer is better than none at all. As a result, proposers should make the smallest offer possible. However, this is not what people do. Responders routinely reject nonzero offers (around 20%), and proposers tend to make offers of around 50%. There is considerable variation across cultures: people in some societies accept any offer, whereas in others, generous offers are rejected (e.g., Henrich et al. 2006). Despite the variation in the specifics, people tend to be willing to pay some cost to inflict a loss on others. In a repeated game setting, this would make sense: rejecting sends a signal to the proposer that the responder is not to be trifled with, resulting in offers approaching equity. However, economic games are played anonymously and only once between any pair of players, so this signal has no value. (Again, concern has been raised as to whether people act as if they are truly in anonymous, one-shot interactions; Hagen and Hammerstein 2006.)

As another example, the public goods game has groups of players contributing their endowments to a public pool, which is then divided equally among them after being topped up by some proportion by the experimenter. The self-interested thing to do is to hold onto one's endowment and free ride off the division of everyone else's donations. Naturally, if everyone takes this selfish stance, no one contributes anything and leaves with less than they could have had all contributed. This is what happens: people start off contributing something, though less than what they think others will contribute (Fehr and Gächter 2002), and after several rounds, always played with different players, contributions plummet. If people are allowed to punish, however, in the form of giving up some of their money to cause a larger loss in the target (the free riders), cooperation is maintained at a high level. This is called *altruistic punishment* because the punisher does not benefit directly due to the shuffling of the group members; the reformed target of the punishment moves to a different group, after having learned his lesson (Fehr and Gächter 2000). Punishment, however, does not always lead to an increase in cooperation; cycles of revenge can cause any group benefits to disappear (Dreber et al. 2008). In some cases (e.g., in a caste system in India), spiteful preferences can interfere with mutually beneficial cooperation (Fehr et al. 2008b).

Inner Workings of Negative Other-Regarding Concerns

The term "altruistic punishment" does not imply that the punisher has an altruistic motive. Any benefits to other group members may be unintended; that is, any altruism that arises from punishment may be a by-product of punishment, not a reason for it.[2] "Antisocial punishment" (Herrmann et al. 2008) or "spiteful punishment" (Jensen and Tomasello 2010) might better capture the spirit of the punishment in which the motivation is to cause suffering rather than to produce benefits. Certainly, except perhaps in the case of a parent disciplining a child, the driving force behind the sentiment, "this hurts me more than it hurts you," is unlikely to be the well-being of the "punishee." The motive behind punishment, altruistic or otherwise, should be the deterrence of a behavior that is contrary to the goals of the punisher. What motivates people to punish non-cooperators?

Spitefulness and other negative other-regarding concerns such as schadenfreude might provide the immediate benefits in terms of motivational rewards. The suffering and misfortunes are the goals. Spitefulness is not instrumental, as is punishment, in having the enforcement of cooperation as an end with the suffering of another as a means to that end. Spitefulness has some form of other-regard; consequences inflicted upon another are not just by-products but are themselves desired end-states. Theory of mind or some form of perspective taking might come into this, as well as emotional sensitivity to the well-being of others. However, unlike empathy, which is aligned with the well-being of others and motivates acts of altruism, spitefulness, like schadenfreude, is misaligned in that it involves positive feelings at the distress or misfortunes of others and leads to harm for harm's sake. Just as feeling good about helping others can act as a bridge for future benefits, feeling good about the misfortunes of others can make punishment—which might not provide benefits for the punisher except at some point in the future—worthwhile. Empathy and

2 The term altruisim does not—or at least should not—imply that altruism has fitness costs for the actor and fitness benefits for the recipient (or third parties) in the sense that evolutionary biologists use the term.

schadenfreude may share a common root, and this ability to feel *into* the emotions and well-being of others can motivate both good and bad. Empathy and altruism have their evil twins in schadenfreude and spitefulness. Yet these malicious motivations can be just as important for cooperation as their virtuous counterparts.

In a social context such as an experimental game or an exchange of goods or other cooperative venture, unfairness—at least at the receiving end—should feel bad. Trivers (1971) and McGuire (1991) call the motivational component of punishment "moralistic aggression" whereas Price et al. (2002) refer to "punitive sentiments." People report feeling angry at unfairness in the ultimatum (Pillutla and Murnighan 1996) and public goods (Fehr and Gächter 2002) games. Offerman (2002) suggests that people are more sensitive to—and reciprocate more strongly against—negative intentions (harmful intent) than for positive intentions (altruistic intent). It may be that a self-serving bias (Hastorf and Polefka 1970)—where one attributes good events to internal (personal) causes and bad events to external causes—justifies why people react more strongly to perceived harm than perceived kindness. This is consistent with the "wounded pride" explanation for anger reported by Pillutla and Murnighan (1996). In addition to self-report measures, van't Wout et al. (2006) found that responders receiving unfair offers from another person, but not a computer, had elevated skin conductance indicating increased arousal. In addition, if a non-cooperator appeared to receive a painful electric shock, the duped subject showed neural signatures of pleasure, at least if they were male (empathy in females might override any satisfaction at seeing harm; Singer et al. 2006). Unfairness—whatever the cause—makes people feel bad and will motivate them, when possible, to alleviate this negative feeling. Sharing pain can feel good. Altruistic punishment, then, might not be motivated out of empathy or other altruistic concerns, but out of a need to see others suffer (especially if they deserve it).

Consistent with moral outrage and other-regard, people do not only punish violators who harmed them; they will impose costs and inflict harm on others for third-party violations. Witnessing someone play unfairly in an economic game (Henrich et al. 2010), or otherwise violating cooperative norms, will incite some people to punish the violators. The value of this non-self-serving behavior has important implications for the maintenance of cooperation. Recent work suggests that third-party punishment is a key part of enforcing cooperation in small-scale societies (Mathew and Boyd 2011), though it might be more important in large societies than in small ones (Marlowe and Berbesque 2008).

It is hard to say what motives people have when they punish others: Do they reject unfair offers or behave spitefully? There are likely numerous motives, even for single cases. For instance, if someone tries to walk across a street against a red light, even when there are no cars present, he may be chided by others at the crosswalk (an example that will be familiar in some cities and not others). One might scold the jaywalker out of concerns for his well-being, or because children are watching and might learn a bad behavior, or to impress upon friends that the scolder is a fine, rule-following citizen, or perhaps to ensure the scolder gets across the street to get a seat on the tram ahead of the rule-breaker. The underlying motivations (emotions) can be telling: If the person feels a warm glow, it may be that the motive was altruistic. If anger and then smug self-satisfaction are the primary emotions, it is more likely that it was upsetting to witness a rule being violated, and that the goal was to put the violator in his place. People do take pleasure in the suffering of others, and suffer at their pleasure, and these motivations might underscore much of our harm-causing behavior, including meting out just punishment as well acts of vengeance and malice.

The Development of Negative Other-Regarding Concerns

Past research on antisocial behavior in children and adolescents has focused on aggression as a problem that needs to be solved. For instance, classic studies by Bandura et al. (1961) had children punch and threaten a punching bag doll (called Bobo) after they witnessed adults treat it aggressively (though, more encouragingly, they would also treat Bobo nicely if this was the modeled behavior). However, the development of negative other-regarding concerns has received scant attention. The same is true for punishment. The focus of research has been on how punishment affects children, not on how they use this as a strategy to modify the behavior of others.

Recent studies are filling that void, primarily in the realm of inequity aversion. In these studies, children will pay a cost to prevent others from being better off. In one study mentioned earlier (Fehr et al. 2008a), children could choose between egalitarian outcomes of candies in a prosocial game (1/1 vs. 1/0), an envy game (1/1 vs. 1/2), and a sharing game (1/1 vs. 2/0). Between the ages of three and eight, children showed an increase in egalitarian choices. The preference for equity was greatest in the envy game across all ages, which would suggest that children were particularly averse to disadvantageous inequity. It may also be that children were spiteful (Hauser et al. 2009). However, they did not pay a cost to impose a loss. In an inequity game, Blake and McAuliffe (2011) had children choose between divisions created by the experimenter. This is akin to responder rejections in the ultimatum game, except that the outcomes were not the result of the other player; this makes it more like the money burning game in which people give up money so that others lose more (Zizzo and Oswald 2001). From four to eight years of age, all children rejected disadvantageous allocations of candy—an effect that increased with age—whereas rejection of advantageous inequity (4/1 vs. 1/1) appeared only at eight years of age. In ultimatum games, children as young as five years old will pay a cost to reject unfair divisions (see also Güroglu et al. 2009; Sally and Hill 2006; Sutter 2007; Takagishi et al. 2010). It is not clear to what degree children are intentionally punishing unfair behavior or reacting spitefully (i.e., with malicious intent). Theory of mind has been suggested to play a role in the fairness of offers but not in rejections (Takagishi et al. 2010). However, nothing is known about the role that affective perspective taking plays on ultimatum game rejections and spiteful behavior in general.

Recent work has shown that children will punish others for infractions. When three-year-old children witness a puppet perform the "wrong" action on an object, they will protest, correct the puppet, and show the correct approach (Rakoczy et al. 2008). Children of this age will protest moral violations; namely, a puppet that destroys another puppet's clay sculpture or picture. They will also tattle when the actor—who had been absent during this event—returns and will behave more prosocially to the wronged actor (Vaish et al. 2010b). Even at three months, infants appear to recognize harmful third-party acts (one animated character—a square—pushing another one—a circle—down a hill that the latter character is struggling to ascend), and they show an aversion (based on shorter looking times) to antisocial characters (Hamlin et al. 2010). From an early age, then, children show some concern for violations of norms and morals. Future work should address whether they punish these violations.

Negative Other-Regarding Concerns in Other Animals

Like prosocial motivations and positive other-regarding preferences, researchers have been looking for antecedents or parallels for antisocial motivations and negative other-regarding preferences in other species, especially other primates. Punishment has been documented in other animals (Clutton-Brock and Parker 1995; Raihani et al. 2012b), but it is not clear whether the punisher has as a goal a change in the behavior of the target (Jensen 2010; Jensen and Tomasello 2010). For example, client reef fish will punish cleaner fish (*Labroides dimidiatus*) who sneak a bite of mucous rather than glean off ectoparasites (Bshary and Grutter 2005). While it may be the case that the future result is more cooperative behavior, there is no evidence yet to suggest that this is the goal of the punisher, and it is difficult to disentangle punishment from dominance behavior and other acts of aggression (Jensen 2010).

Naturalistic observations have provided no suggestive evidence for disadvantageous inequity aversion. However, experiments have raised the possibility that other animals are sensitive to outcomes affecting others. Brosnan and de Waal (2003) suggest that capuchin monkeys (*Cebus apella*) reject cucumbers, a less-preferred food, and refuse to exchange tokens (rocks) with the experiment if another monkey gets better food (grapes) without having to work (insofar as passing back an object received by an experimenter counts as work). The suggestion is that monkeys are averse to disadvantageous inequity. One of the disadvantages with this study is that it is, in effect, an impunity game (Bolton and Zwick 1995; Henrich 2004) in which there is no sensible reason to reject an unfair "offer" because doing so has no consequence for the other individual. Furthermore, unlike the impunity game, the partners in these studies did not create the unfair outcomes, so it makes even less sense for the subject to express frustration at the inequity. Some replications of this paradigm suggest that rejections are better explained by self-regarding motivations, namely frustration at getting a poorer quality food than in the past (Roma et al. 2006) or an expectation at getting the better food based on its mere presence (Bräuer et al. 2006). At last tally, there were three studies using this paradigm that supported inequity aversion in other primates and seven which failed to replicate this effect (Jensen 2012). The mixed results and the theoretical shortcomings of the method weaken claims for disadvantageous inequity aversion in animals other than humans.

The standard tool for fairness sensitivity in humans is the ultimatum game. Getting nonhuman animals to divide a resource that can then be accepted or rejected is a challenging task. However, a variation of the ultimatum game has proven to be useful. In the mini-ultimatum game in humans (Falk et al. 2003), the proposer is faced with a pair of options; one is always unfair and is typically rejected (an 8/2 split). If the alternative is fair (5/5), rejection rates are high. However, if the alternative is generous (2/8), if there is no alternative (8/2), or if the option is even more unfair (10/0), responders are more accepting. One of the strengths of the mini-ultimatum game is that it distinguishes between sensitivity to unfair outcomes as well as the intentions that led to the outcomes (e.g., in the 8/2 game, the outcome is unfair but the intentions of the proposer are not). Chimpanzees faced with this paradigm neither chose fair options nor rejected any nonzero offer (Jensen et al. 2007a); they behaved in a purely self-interested manner. This result is consistent with the prosocial choice studies described earlier, in which the chimpanzees did not attempt to prevent a conspecific from getting food (Jensen et al. 2006). Also, in a study in the spirit of the money-burning game, chimpanzees did not react spitefully to "unfair" outcomes (i.e., having their food given to a conspecific) even though they would cause the partner to lose their food if the partner had stolen it (Jensen et al. 2007b).

An important piece of evidence on negative other-regarding concerns will come from studies on third-party punishment. Chimpanzees and other primates (as well as other animals, including fish and social insects) will inflict harm on others in response to harm received. There may even be some "policing" in the sense that certain individuals, through their dominance rank and preservation of their group interests, will redirect aggression to others (Flack et al. 2006; Frank 1995). However, there is no clear evidence yet that other animals are motivated to harm others for any other reason than asserting dominance or claiming a resource. To date, only one study has been conducted on fish, and here there was no third-party; a plastic plate stood in for a client that the cleaner fish (*Labroides dimidiatus*) were gleaning from (Raihani et al. 2010). If the female of the pair took a bite of the preferred food, the food plate was withdrawn and the larger male attacked her. While this bears some superficial resemblance to third-party punishment, it is likely that the male received direct benefits through dominance assertion by suppressing the growth of the smaller fish, which would switch sex if it became larger (Raihani et al. 2012a). There is nothing to suggest that the goal of the punishing fish was directed toward benefits to the third-party or even toward the modification of the behavior of the targeted fish. Future work on nonhuman primates that will directly test third-party punishment are necessary to answer questions about the role of negative other-regarding concerns and their evolution in humans. Jealousy, envy, schadenfreude, and spitefulness are not the most endearing human characteristics, and they may even be unique to our species. However, they may serve as the backbone for prosocial behaviors by acting as the enforcers for cooperation.

Conclusion

An important, perhaps essential, component of social decision making in humans is a concern for the well-being of others. Other-regarding concerns allow us to tap into the emotional perspective of others. When our emotions are aligned with theirs, we can be motivated to help them. When our emotions are misaligned, we can be motivated to harm them or seek to better ourselves relative to them. Social emotions such as empathy, symhedonia, jealousy, and schadenfreude as well as guilt, shame, pride, and so on are not bystanders to our social decisions; they may be the driving force behind them.

If it turns out that other-regarding concerns are unique to humans, this raises questions as to how they evolved in our species and what adaptive value they might have. I suggest that positive other-regarding concerns are the psychological mechanisms which allow humans to be ultrasocial (Richerson and Boyd 2005; Hill et al. 2009) and cooperate on a large scale with nonkin. By experiencing distress at the unhappiness of others and by sharing their joy—even in the absence of emotional cues—we can engage in acts that are costly now, but which may serve our personal interests in the future, or allow those benefits to go to others in our groups. Without positive other-regarding concerns, we would likely be restricted to reacting in an almost reflexive manner to eliciting cues, like infant's cries, in very narrow contexts. That is not to say that these cues are unimportant for sociality or that they do not constitute part of other-regarding concerns; the point is that these more primal responses, such as emotional contagion, would be insufficient for ultrasociality.

The same is true for negative other-regarding concerns. Not only might they be part and parcel of their positive cousin, they might also form another cornerstone of human sociality. Negative other-regarding concerns might be the cognitive substrate for hyper-competitiveness (Jensen 2010). Not only do people assess their gains and losses relative to past experience or expectation for the future, they measure these against the successes and failures of others. With social comparison (Festinger 1954), people measure their worth—and their happiness—relative to those around them. If we cannot succeed, we can at least ensure that our rivals do not. If the mighty fall, we can gloat. This hyper-competitiveness raises the bar so that we aim to achieve more.

Humans care about how others feel, for better and for worse. Researchers should care about other-regarding concerns because they are central to how we make social decisions. Because other-regarding concerns are built upon emotional, rather than rational, systems, they may have a strength and immediacy over rational decision making. Even though it is not logical or adaptive to do so, you will give your cat everything he needs to be happy, and his happiness will bring you joy, even if he cannot share your concerns.

19

Learning, Cognitive Limitations, and the Modeling of Social Behavior

Peter Hammerstein and Robert Boyd

Abstract

Learning experiments with rats and other animals have impressively documented that learning is "prepared" by evolution. Numerous specificities of learning mechanisms reflect the circumstances under which these mechanisms evolved. For social animals like humans, learning mechanisms are specifically adapted to social environments. In particular, humans are prepared to learn socially. This preparedness for social life is the key to understanding human sociality. It explains, for example, why humans often do smart things without having a clue of why they do them. It also explains why humans maintain behavioral habits over generations when these habits are no longer useful. To be prepared does not mean to be perfect. The bubbles that irritate our financial markets demonstrate this as well as the fact that winners in competitive economic interactions often experience what is called the winner's curse. In behavioral experiments, subjects do not learn to avoid the winner's curse. This reflects the downside of otherwise adaptive mental mechanisms described by learning direction theory. The bounded rationality captured through this theory has also positive aspects. It allows cooperation to occur in games where conventional game-theoretic wisdom would not give it a chance. Similarly, emotions play a crucial role in human decision making. They sometimes lead to irrational behavior but, at the same time, enable us to maintain cooperative institutions essential for human sociality.

Introduction

In the early days of modeling social behavior, cognition usually received little attention from behavioral ecologists and sociobiologists. Modelers typically addressed the question of what would be best to do under given circumstances and left open how a theoretical optimum could be achieved in practice by mental machinery (see Kacelnik, this volume). It seemed beneficial to treat this mental machinery as a black box and to separate almost completely the questions of "why" and "how." Even in this agnostic spirit, however, cognition could not be fully ignored because, for every mathematical model, the set of behavioral phenotypes needed to be specified upon which natural selection was assumed to operate. Early empirically based models in evolutionary game theory, for example, had to invoke limitations in perception (Brockmann et al. 1979) and cognition (Hammerstein and Riechert 1988) to explain the facts generated by quantitative fieldwork. Today, cognition itself, including learning, has become a major theme addressed by modelers in evolutionary biology, anthropology, and economics. Rather than viewing it as a source of constraints on optimality, cognition is now regarded as a sophisticated machinery shaped by natural selection—the deeper theme of this chapter.

We begin with a review of how the mechanisms of learning are fine-tuned by natural selection. This review sets the stage for a discussion of human social learning and its role in cultural evolution. We then demonstrate how certain ways of learning, even though they would make perfect sense when used in the appropriate contexts, constrain our ability to master the challenges faced by traders on modern markets. Despite their drawbacks, these ways of learning enhance the human potential for cooperation beyond the limits suggested by game theory. We go on to show that some forms of social learning may be highly beneficial at the level of individual agents but problematic at the level of populations, where they tend to induce crises like the "bubbles" that have repeatedly been observed in financial markets. Finally, we investigate some of the constraints that emotions impose on human decision making. These constraints need to be considered, for example, if one wants to understand the phenomenon of deterrence, which often escapes the wisdom of classical game theory. We also argue that, in general, emotions can act as both enhancers and inhibitors of sociality.

All Learning Is Prepared by Evolution

Learning is a salient feature of life in general and is of particular importance to the study of animal cognition. Genetic evolution can only track environmental variation when timescales are long enough. Since environments often change quickly, learning is an important and very efficient tool for generating adaptive decisions in unpredictably varying environments. As for all evolved tools, we expect learning to be tailored to the environmental problems that animals face in their natural habitats. Learning must, therefore, differ between species because no tool can be optimal for all purposes. In this sense, all learning should be prepared. The *preparedness of learning* was clearly recognized in the psychological work of Thorndike (1911) but was otherwise largely ignored in his day. Garcia and Koelling (1966), Breland and Breland (1966), and Seligman (1970) brought the idea back to the attention of the scientific community and, as shown below, there is now overwhelming evidence for the preparedness of learning in animals and humans.

How Rats Are Prepared to Cope with Novelty

Norway rats (*Rattus norvegicus*) are social animals that probably originated in Asia and which have spread almost all over the world. Understanding the prerequisites for their global expansion requires a close look at their learning abilities and social skills. How, for example, did they manage to avoid poisoning themselves with toxic foods and liquids in novel environments? Many other mammals, including humans, react to toxic effects sensed with some delay after ingestion by vomiting. Rats, however, cannot vomit for physiological reasons and thus have to rely on preventive measures. They only take tiny bites of novel food and rapidly develop taste aversion when toxic effects are sensed within a certain time window. Famous experiments by Garcia et al. (1974) demonstrated some of the specificity of the learning mechanism. Norway rats can easily learn the association between taste and nausea after a long delay, but not between a sound cue and nausea. Rats find it hard to associate an acoustic signal with nausea despite the fact that they are well able to develop adaptive responses to such signals. For example, they have no difficulty learning an association between sound cues and skin pain.

Generally speaking, rats have taught us that to be efficient, learning must rely on the cues that really matter; sound is obviously less likely to cause nausea than food. Social environments offer many such relevant cues, and rats are well prepared to make use of them. Galef and Wigmore (1983) showed, for example, that when a naïve rat interacts shortly with a conspecific, it develops a preference for the food that the conspecific ate before they met. In other words, rats have the ability to learn about dietary choices of social partners without observing any act of eating. The effect of this learning is very strong and can even lead to a reversal of poison-induced aversions (Galef 1989; Galef and Whiskin 2003). Experiments by Galef et al. (1988) identified a subtle cue by which the rats pick up the hidden information about each other's eating habits. It takes the joint occurrence of carbon disulfide, a normal constituent of rat breath, and the odor of a food to induce an enhanced preference for that food. Humans, who also have carbon disulfide in their breath, can enhance a rat's food preference in the same way.

Comparative Perspectives on the Specificity and the Use of Cues

The rat dietary choice example makes it easy to understand why all animals must have some difficulties in learning the novel association between events and certain cues offered to them in the laboratory. Although such difficulties can certainly be seen as cognitive limitations, they are often necessary consequences of the sophisticated and highly adaptive way in which cognitive machinery focuses its attention on the features relevant to success in an animal's natural habitat. This becomes particularly evident when comparing species that differ substantially in their foraging styles. While it seems advantageous for nocturnal foragers, such as rats, to learn the aversion of toxic substances primarily on the basis of how food smells and tastes, diurnal foragers might benefit strongly from relying on visual cues. To test this idea, Wilcoxon (1971) explored the relative salience of a visual and a gustatory cue in bobwhite quail (*Colinus virginianus*) and laboratory rats. For both quail and rats, drinking blue-colored or sour water was followed by a later injection of an illness-inducing drug. When acidity was the cue, rats learned to avoid drinking sour water but quail did not. Conversely, when the blue color was tested, quail avoided it but rats did not.

In the same way that different cues are used by nocturnal and diurnal foragers, learning should also differ between species that exhibit different degrees of food specialization. For rats and other dietary generalists, figuring out what to eat and what to avoid poses a great challenge. In contrast, extreme dietary specialists should almost by definition "know" how to recognize items of nutrition. One should thus expect dietary specialists to be rather poor at learning to associate a novel flavor with aversive gastrointestinal events. Ratcliffe et al. (2003) tested this idea by comparing three dietary generalists among bat species with the common vampire bat (*Desmodus rotundus*). As suggested by its name, the vampire bat is an extreme specialist that feeds exclusively on blood. While the generalists readily acquired taste aversions, the vampire bat did not learn to associate a novel flavor with gastrointestinal problems.

Apart from food choice, cue specificity has been demonstrated to play a role in many other behaviors. Dangerous animals like snakes are part of the challenges that humans, apes, and monkeys have had to face in their natural history. It is therefore often said that humans have less of a problem acquiring the fear of snakes than the fear of cars in modern urban environments. Similarly, since snakes can bite and flowers cannot, it should be easier for rhesus monkeys to learn that snakes are dangerous than to acquire a fear of flowers. In a social learning experiment, Cook and Mineka (1989) tested this hypothesis by allowing naïve monkeys to observe others who (due to manipulation) exhibited fear toward a snake or a flower. The observing monkeys quickly learned to fear the snakes but could not be taught to fear the flowers.

Taken together, these results confirm the idea that the evolved cognitive design of animals is adjusted to the problems they have to solve under natural circumstances. Cognition is not an all-purpose machine.

Social Animals Are Prepared to Learn Socially

Social learning is special both in the opportunities it offers and in the cognitive skills it requires. As David W. Stephens maintained at a recent workshop in Switzerland (pers. comm.):

> In order to understand the specialness of social learning, one needs to understand the preparedness of this learning.

We agree fully with this statement and think that the rat dietary example presented above, where information is conveyed through the rat's breath, nicely illustrates his point.

Aimee Dunlap (in preparation) has experimentally investigated how nectar-foraging bumblebees weigh the relative value of social and individual sources of information. These bees individually acquire information about floral attributes through trial and error learning. In addition, they use social cues in their floral choices. Dunlap varied the reliability of social and floral cues as predictors of reward. Simple models from economics would suggest that the bees should rely primarily on the more reliable source of information. However, because all learning is prepared, the bees react more sensitively to the unreliability of social cues than to unreliability of personal information. Thus it would appear that in their evolutionary past, bees had cause to question the value of social cues.

Similar evidence comes from an evolutionary experiment on individual (nonsocial) learning in *Drosophila melanogaster*. Here, Dunlap and Stephens (2009) varied the reliability of cues indicating the value of two alternative egg-laying media: one flavored with orange juice, the other with pineapple. In every generation, the flies first went through an "experience phase" where they had the opportunity to visit both media. During this phase, one of the media was paired with the aversive chemical quinine. Later the flies were offered the choice between the same media for egg laying, but this time the media did not contain quinine. The experimenter then used the eggs from one of the media to start a new fly generation. This permitted the experimenter to control the reliability of quinine as a cue for reproductive success. The experiment showed that evolution can favor or disfavor learning, depending on the reliability of the cue involved. It also demonstrated that the way in which learning is prepared can change within 30 generations under selection. Thus, evolution can modulate the cognitive prerequisites for learning quickly when needed.

Returning to sociality, Tomasello has performed many comparative studies with human infants and chimpanzees in an attempt to reveal the preparedness of social learning. He and his coworkers came to the conclusion that the crucial difference between human cognition and that in other species is the ability to participate with others in activities with shared goals and intentions (e.g., Tomasello et al. 2005). In their view, great apes understand the basics of intentional action but do not engage in activities involving joint intentions. Though the categorical way in which this view is expressed may be somewhat overstated, it seems very convincing that human children are well ahead of chimpanzees in their inclination to share attention and intentions with others—a cognitive prerequisite for human sociality.

Cultural Learning and Human Cognition

The human species is uniquely dependent on cultural learning. Over the last 50,000 years, *Homo sapiens* have expanded across the globe to occupy a larger ecological and geographic range than any other terrestrial vertebrate species. This unprecedented expansion required the rapid development of a vast range of new knowledge, tools, and social arrangements. While humans are probably smarter than other mammals, much evidence suggests that increased cognitive ability *alone* does not account for human adaptability (Boyd et al. 2011): it is our unique ability to learn from each other that has enabled us to accumulate information gradually across generations and to develop well-adapted tools, beliefs, and practices that no individual could invent on their own.

Social Learning Does Not Increase Adaptability if the Only Benefit Is Spreading the Costs of Innovation over More Individuals

Some authors (e.g., Barrett et al. 2007; Pinker 2010) have argued that such cultural accumulation can be understood as being due to communication of useful information. As emphasized in endogenous growth models in economics (Romer 1990), knowledge is a "nonrival" good: one individual's "consumption" of knowledge does not affect the amount left for others. This means that by communicating useful knowledge from one individual to another, the average cost of innovation can be greatly decreased. Barrett, Pinker, and others seem to believe that this reasoning explains the role of cultural learning in human adaptation.

This reasoning requires, however, that "teachers" understand why innovations are adaptive, and that they communicate this understanding to individuals who learn from them. If, instead, learners simply copy others without understanding why behavior is adaptive, evolutionary models show that there will be no increase in the ability of the population to adapt (Rogers 1988; Boyd and Richerson 1995). This surprising result emerges from processes that affect coevolution of the kinds of behaviors that are available to imitate and the psychology that controls learning and imitation. These evolutionary models of social learning rest on two assumptions. First, the propensities (a) to learn by individual experience and (b) to imitate the behavior of others are part of an evolved psychology shaped by natural selection. This means that the balance between learning and imitating will be governed by the relative fitness of the two modes of behavior; the average fitness of the population is irrelevant. When few individuals imitate, imitators will acquire the locally adaptive behavior with the same probability as individual learners. Because they do not pay the cost of learning, imitators have higher fitness, and the propensity to imitate spreads. As the number of imitators increases, some imitate individuals who imitated other individuals, who imitated other individuals, and so on until the chain is rooted in someone who extracted the information from the environment. As the fraction of imitators in the population increases, these chains extend further.

The second assumption is that the environment varies in time or space. This means that as chains of imitation get longer, there is a greater chance that the learner who roots the chain learned in an environment different from the current one, either because the environment has changed since then or because someone along the chain migrated from a different environment. Because they are imitating without understanding why the behavior is adaptive, imitators will be less likely to acquire the locally adaptive behavior than learners. The propensity to imitate will continue to increase until this reduction in fitness exactly balances the benefit of avoiding the costs of learning. At evolutionary equilibrium, the population has the same average fitness as a population without any imitation. There will be no increase in the ability to adapt to varying environments, and cumulative cultural adaptation will not occur.

Although this treatment is very simple, the basic result holds in more realistic models. The primary insight that emerges from these models is that imitation is a form of free riding where imitators scrounge information without producing anything of value. Free riders increase until they destroy the benefits of free riding. Realistic levels of relatedness among models and imitators do not qualitatively change the result (Lehmann et al. 2010). To understand the evolution of social learning psychology you have to know what is available to learn and this, in turn, is affected by the nature of the learning psychology. If imitators are simply information scroungers, then they will spread until selection no longer favors imitation.

Two lines of evidence suggest that many important cultural adaptations are not understood by the populations that use them. First, the anthropological literature on child development indicates that children and adolescents acquire most of their cultural information by learning from older individuals who typically discourage questions by young learners and rarely provide causal explanations of their behavior (Lancy 1996, 2009, 2010). Kids practice adult behaviors, often using toy versions of adult tools, during mixed-age play, and little experimentation is observed, except that which is necessary to master the adult repertoire (Hewlett et al. 2012; MacDonald 2007). Second, there is direct evidence of practices that are adaptive but not understood. Fijian food taboos are an interesting example. Many marine species in the Fijian diet contain toxins that are dangerous for pregnant women and nursing infants. Food taboos prohibit pregnant and lactating women from eating these species. Although women in these communities all share the same food taboos, they offer quite different causal explanations for them, and little information is exchanged among women save for the taboos themselves (Henrich and Henrich 2010). The taboos are learned and are not related to pregnancy sickness aversions. The transmission pathways for these taboos suggest the adaptive pattern is sustained by selective learning from prestigious women. Many other examples could be cited.

Social Learning Increases Adaptability When It Makes Individual Learning Less Costly or More Accurate

A number of models explore how cultural transmission can increase the ability of populations to adapt, even though individuals do not understand why behavior is adaptive. In these models, the propensity to imitate evolves because it is directly beneficial to the individual, but it nonetheless benefits the population as a side effect. There are at least two important ways this can happen. First, cultural learning can allow individuals to learn selectively, using environmental cues when they provide clear guidance and learning from others when they do not. Second, cultural learning allows the gradual accumulation of small improvements, and if small improvements are cheaper than big ones, cultural learning can reduce the cost of learning.

The ability to learn selectively is advantageous because opportunities to learn from experience or by observation of the world vary. For example, a rare chance observation might allow a hunter to associate a particular animal track with a wounded polar bear, or to link the color and texture of ice with its stability on windy days just after a thaw. Such rare cues allow accurate low-cost inferences about the environment. However, most individuals will not observe these cues, and thus making the same inference will be much more difficult for them. Organisms that cannot imitate must rely on individual learning, even when it is difficult and error prone. Such organisms are stuck with whatever information nature offers. In contrast, an organism capable of cultural learning can afford to be choosy, learning individually when it is cheap and accurate and relying on cultural learning when environmental information is costly or inaccurate.

Boyd and Richerson (1995) and Perreault et al. (2012) have shown that selection can lead to a psychology that causes most individuals to rely on cultural learning most of the time, and that this also simultaneously increases the average fitness of the population compared to a population that does not rely on cultural information. These models assume that our learning psychology has a genetically heritable *information quality threshold* that governs whether an individual relies on inferences from environmental cues or learns from others. Individuals with a low information quality threshold rely on even poor cues, whereas individuals with a high threshold usually imitate. As the mean information quality threshold in the population increases, the fitness of learners increases because they are more likely to make accurate or low cost inferences. At the same time, the frequency of imitators also increases. As a consequence, the population does not keep up with environmental changes as well as a population of individual learners. Eventually, equilibrium emerges in which individuals deploy individual and cultural learning in an optimal mix.

At this equilibrium, the average fitness of the population is higher than in an ancestral population without cultural learning. When most individuals in the population observe accurate environmental cues, the equilibrium threshold is low, individual learning predominates, and culture plays little role. When it is usually difficult for individuals to learn on their own, however, the equilibrium threshold is high and most people imitate, even when the environmental cues that they do observe indicate a different behavior than the one they acquire by cultural learning. We take the evidence on hunter-gatherer adaptations as indicating that most of the problems faced by the people in even the simplest societies were far too difficult for most individuals to solve (Boyd et al. 2011). As a result, we interpret this logic as predicting that selection should have favored a psychology that causes individuals to rely heavily on cultural learning.

The ability to learn culturally can also raise the average fitness of a population by allowing acquired improvements to accumulate from one generation to the next. Many kinds of traits admit successive improvements toward some optimum. Bows, for example, vary in many dimensions that affect performance such as length, width, cross section, taper, and degree of recurve.

It is typically more difficult to make large improvements by trial and error than small ones for the same reasons that Fisher (1930) identified in his "geometric model" of genetic adaptation. In a small neighborhood in design space, the performance surface is approximately flat, so that even if small changes are made at random, half of them will increase the payoff (unless the design is already at the optimum). Large changes will improve things only if they are in the small cone that includes the distant optimum. Thus, we expect it to be much harder to design a useful bow from scratch than to tinker with the dimensions of a reasonably good bow. Now, imagine that the environment varies so that different bows are optimal in different environments, perhaps because the kind of wood available varies. Sometimes a long bow with a round cross section is best, whereas at other times a short flat wide bow is best. Organisms which cannot imitate must start with whatever initial guess is provided by their genotype. Over their lifetimes, they may be able to learn and improve their bow. However, when they die, these improvements will disappear with them, leaving their offspring to begin again at the genetically inherited initial guess. In contrast, cultural species can learn how to make bows from others after these bows have been improved by experience. Therefore, cultural learners start their search closer to the best design than pure individual learners and can invest in further improvements. Then, they can transmit *those* improvements to the next generation, a process that will continue down through the generations until quite sophisticated artifacts evolve.

Adaptive Social Learning Mechanisms Will Cause Individuals to Adopt Maladaptive Beliefs and Behaviors

Both of the models discussed in the previous section predict that an adaptive evolved psychology will often cause individuals to acquire the behaviors they observe used by others, even when inferences based on environmental cues suggest that alternative behaviors would be better. In a species capable of acquiring behavior through teaching or imitation, individuals are exposed to two different kinds of cues that they can use to solve local adaptive problems. Like any other organism, they can make inferences based on cues from the environment. However, they can also observe the behaviors of a sample of their population. When most individuals solve the adaptive problem using environmental cues alone, the models predict that an optimal learning psychology will result and that social learning will play a significant but relatively modest role. Many people will rely on their own inferences, but some will copy to avoid learning costs. However, often only a minority will be able to solve the adaptive problem based on environmental cues alone because the appropriate environmental cues are rare or the adaptive problem is too complex. Then, if the environment is not too variable, an adaptive psychology will evolve in which most people ignore environmental cues and adopt behaviors that are common in the sample of the population they observe. They modify these behaviors rarely, or only at the margin, and as a result local adaptations evolve gradually, often over many generations. Many times individuals will have no idea why certain elements are included in a design, nor any notion of whether alternative designs would be better. We expect cultural learners to first acquire the local practices and occasionally experiment or modify them.

This picture of human cognition is consistent with recent experiments on imitation (Nielsen and Tomaselli 2010; Lyons et al. 2007). In these experiments, an adult performs a behavior such as opening a complex puzzle box to get a reward. The adult's behavior includes both necessary and unnecessary actions. A subject, either a child or a chimpanzee, observes the behavior. Children's performance on such tasks in both Western and small-scale societies differs in important ways from that of chimpanzees. Children accurately copy all steps, including steps that direct visual inspection would suggest are unnecessary. Children seem to assume implicitly that if the model performed an action, it was probably important, even if they do not understand why. Chimpanzees, however, do not seem to make this assumption; most of the time they skip the unnecessary steps, leading them to develop more efficient repertoires than children (Whiten et al. 2009) in these experimental settings.

Prepared Does Not Mean Perfect

From an engineer's point of view, prepared learning is an evolutionarily designed feedback mechanism by which humans and other animals adaptively control their actions. The wisdom of engineering tells us that feedback control is a subtle issue. Even in simple problems of regulation and control, the adaptive performance of feedback mechanisms can suffer from overshooting, oscillations, and all the other odd properties for which nonlinear dynamical systems are known.

Boom and Bust

If learning takes place in a social environment, peculiar problems can arise through interdependence of information processing. The behavior of human investors, for example, is often guided by imitating the actions of other investors who are particularly successful. This behavior is well known, even in the popular press. In 2011, *The Economist* reported that investors place their stakes disproportionately on funds whose recent returns were exceptional. At first glance, this decision heuristic seems quite reasonable, but as *The Economist* rightly pointed out, such decisions also entail "the foolishness of the crowd." At the aggregate level, imitating the successful can easily lead to a *boom and bust phenomenon*. Such economic bubbles have been observed for centuries. Consider, for example, the "tulip mania," which swept through Holland during the Dutch Golden Age, where newly introduced tulip bulbs were so high in demand (the boom) that prices escalated to unrealistically high levels, creating a grossly inflated market value. At its peak in 1637, a single tulip bulb is reported to have sold for more than 10 times the annual income of a skilled craftsman. As any rational decision maker would predict, demand for the bulbs collapsed suddenly (the bust), destroying in its wake the fortunes of many people.

Yet despite historical knowledge and experience with the growing and bursting of bubbles, they continue to occur. Consider the recent housing bubble in the United States, where the price of a typical house more than doubled within a relatively short time span, peaking in 2006. As the bubble burst, the relationship between mortgage debts and property value was severely thrown off balance due to the number of people who had participated in the boom. Just as in the tulip mania, investors imitated, and thus were susceptible to pitfalls, because our evolved learning mechanisms have their undisputable drawbacks. The small scale of ancestral human societies may have limited the scope for bubble-like phenomena, and thus evolution has not provided us with the means to avoid them.

Bossan et al. (in preparation) have shown, however, that dramatic boom-like phenomena can occur under very simple circumstances during the evolution of social learning. Investigating the evolutionary competition between individual and social learning strategies, their model has individuals choose repeatedly between one of two actions, A and B. In every period (during an individual's life), a payoff unit will be received with probabilities p_A and p_B for the two choices. These probabilities fluctuate independently and create the uncertainty that calls for continuous learning.

In Bossan et al.'s model (Figure 19.1), the learning strategies reproduce from generation to generation at rates proportional to the payoff collected

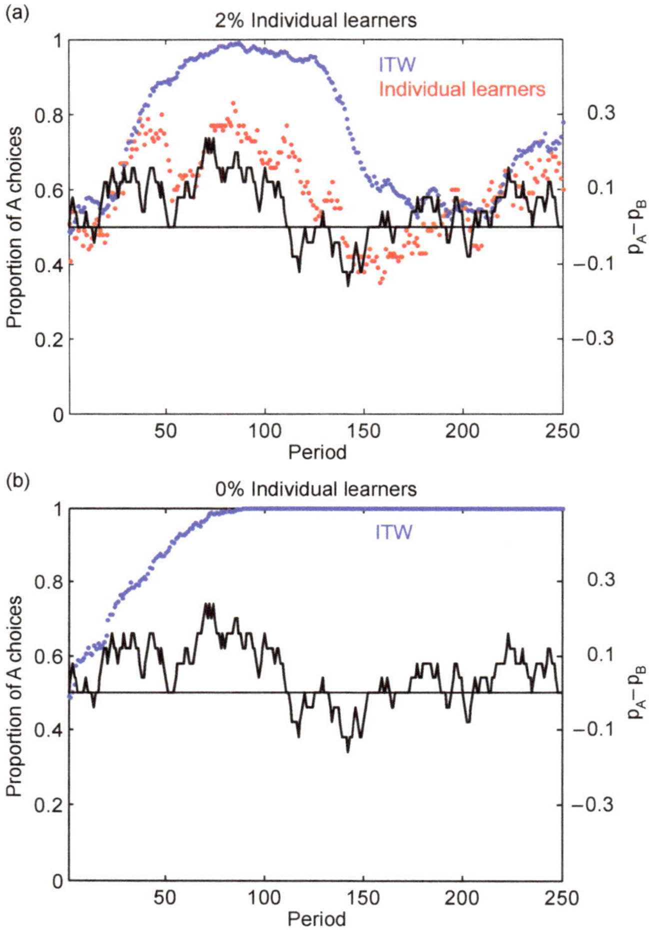

Figure 19.1 Behavior of individual and social learners in the model by Bossan et al. (in preparation). The black curve describes fluctuating differences in reward probabilities for choices *A* and *B*. These fluctuations are tracked reasonably well by the average proportion of *A* choices made by individual learners (red dots). (a) Social learners (blue dots) who imitate the wealthiest lag behind individual learners in their responses to environmental change. For substantial periods of time, social learners are very good at making the right choice and do not waste activities for exploration. Evolution has already reduced the proportion of individual learners to 2%. (b) Individual learners are extinct. Due to positive feedback, the social learners' choices now become totally independent of the environment. Complete informational breakdown occurs.

over a lifetime. Two of them are particularly interesting. One strategy is *reinforcement learning* based entirely on individual experience; the other is to investigate a small sample of other individuals, compare their cumulative payoff and *imitate the last choice of the wealthiest*. Individual learning is free of exogenous costs and thus there are no trivial incentives to free ride on the efforts of individual learners. Yet, despite the absence of such trivial incentives, social learners drive the fraction of individual learners in a population down to quantities so small that in finite populations the extinction of individual learning is likely to occur. Extinction actually does occur in the agent-based simulations, which take place, of course, in a finite population world.

The advantage of the social learners hinges on the following: by comparing the wealth of others, social learners indirectly base their decision on a larger set of data points than the individual learners, who only make use of self-collected information. Ultimately, however, when individual learners are extinct, the positive feedback among social learners will sooner or later produce an extremely high degree of coordination so that all individuals converge to make the same choice, say *A*. The social learners have now generated a social environment in which their learning efforts fail to provide any information about the real world, and a disastrous informational breakdown has occurred. Other learning strategies, such as the *critical social learning* suggested by Enquist et al. (2007), cannot prevent this breakdown. Two more theoretical studies have found informational breakdown, but only under more restrictive conditions (Rendell et al. 2010; Whitehead and Richerson 2009).

The theoretical insight from this simple evolutionary model may catch some of the flavor of the financial crises worrying central bankers. Theoreticians in economics use the term "herding effect" to describe similar insights gained from different models, where assumptions are made about the thoughtful formation of expectations. However, these models may sometimes impose too much economic reasoning on the cognitive mechanisms they are trying to depict. To illustrate this point, consider the following: Instead of forming more or less rational expectations about the market, investors may simply be driven by fear of punishment from a higher authority (e.g., a boss or a spouse). By following the decisions of whoever is most successful, investors do not expose themselves openly to attack. Even when a bubble bursts, they can defend themselves with the argument that other smart people encountered the same problem. This simple explanation may explain why imitation of the wealthiest occurs, even if we allow some reasoning to modulate the course of learning.

It is easy to understand why, in the world of economists, thoughtful behavior gives rise to the herding effect. For example, if there are two restaurants in the neighborhood, and the last time we visited both we found restaurant *A* to be better, we might want to visit *A* again. But if we see that restaurant *A* is empty, whereas *B* has many guests, we might decide to go to *B* instead. Our choice would then be guided by the following herding effect: the first guests entered *B* through happenstance but their presence precipitated an "informational cascade," which made all subsequent people visit *B* as well, even though *A* has the better chef. Too bad, for by basing our decision this way, we would miss a more delightful culinary experience. An important difference between this herding model and that of Bossan et al. is as follows: In the herding model, learning is at best implicit. The herding model only looks at one decision, whereas in Bossan et al.'s model the agents can make several decisions per lifetime and thus have the possibility to learn. Therefore, the Bossan et al. model gives a more complete picture of how decisions vary over time and how environmental trends are tracked.

The Archer's Style of Learning

Let us now return to the general problems associated with feedback loops in social learning. As we just saw, reasoning can influence these loops. Although it seems smart to integrate learning with higher forms of cognition, this integration involves numerous risks of developing maladaptive behavior. Before presenting an important example of maladaptive learning, let us discuss the phenomenon of the *winner's curse* (Thaler 1988). This phenomenon was originally found in real-life auctions of oil fields (Capen et al. 1971; Levinson 1987). Bidders who overestimated the value of an oilfield were the ones who won the auction and received the oilfield in the end. As a result, many of the bidders ended up with an unprofitable deal: the winner's curse.

Consider the following simple experiment in which a similar phenomenon occurs even though there is only one bidder who plays against a computer. A person has a car that she wants to sell. She knows exactly what the car is worth to her, a value denoted by x. Another person, the buyer (bidder), only knows that x is uniformly distributed between 0 and 1. The buyer knows, in addition, that the value of the car is 50% higher to herself than to the seller. The buyer's net payoff then is $1.5x - y$, where y is her bid and x the unknown value to the seller. The buyer is allowed only one bid in the experiment, where a computer determines x in a "random move" and also plays the seller. The computer will sell the car unless $y < x$.

What if subjects in this experiment asked us secretly for advice? We would strongly discourage them from offering any positive amount of money, since for every such bid the expected net payoff would be negative because the expected value of x conditional upon the seller's acceptance of a bid y is as small as $y/2$. Buyers would then on average only receive three-quarters in value of what they paid. Of course, subjects are not allowed to ask for advice, and bids near 0.5 were frequently observed in this experiment (Ball et al. 1991); the winner's curse hits most people under the appropriate circumstances. Obviously, humans are not prepared for computing conditional expectations. But aren't we supposedly prepared for learning?

Selten et al. (2005) ran winner's curse experiments similar to the one described above but with one primary difference: subjects had to make a hundred such bidding decisions before they could leave the laboratory (Figure 19.2). A minor difference was that the minimum bid could not be zero. The idea was, of course, to permit the subject to benefit from learning. Unfortunately,

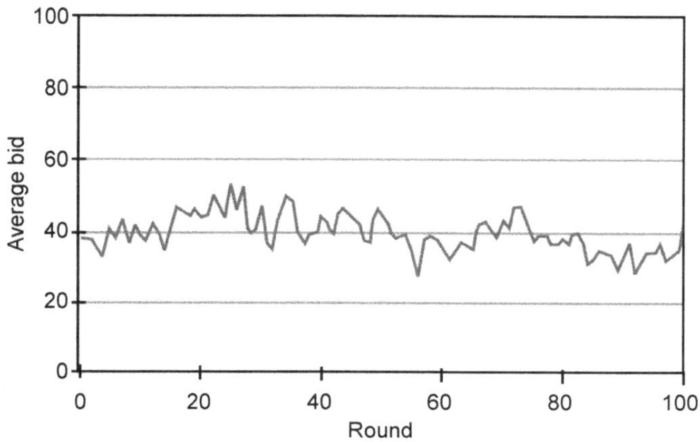

Figure 19.2 In the winner's curse learning experiment by Selten et al. (2005), average bids remained high, indicating a lack of learning throughout the 100 rounds of the game.

learning failed to transpire, as substantial amounts of money were still being offered on average toward the end of the learning period.

Why did learning fail to solve the winner's curse problem? Selten (1998) had previously suggested an answer based on what he calls *learning direction theory*. This concept was inspired by thinking of an archer who wants to hit the trunk of a tree with a bow and arrow (Figure 19.3). If the arrow misses the tree on the right (left) side, the archer will attempt to correct the error by aiming a little further to the left (right), if a behavioral change is made at all. Selten emphasizes that this example is less trivial than one might think at first glance. In standard models of reinforcement learning (Bush and Mosteller 1955; Roth and Erev 1995) the degree of reinforcement for an action depends on the size of the payoff obtained in the last period. What matters in learning direction theory is the additional payoff that might have been gained by other actions: had I aimed more to the left, I would have hit the tree. Obviously this is counterfactual reasoning about the past and demonstrates that in Selten's archer the feedback loop necessary for learning contains an element of "higher" cognition.

This cognitive element seems to be perfect for guiding the archer and may have helped our ancestors in hitting a mammoth or other prey, but it creates an impediment to learning during the winner's curse experiment. Why? When I get the car for my bid, I know that I could only have been better off by offering less. When the seller refuses to sell the car, I know that I could only have gotten the car by offering more. Being Selten's archer, I will have a propensity to decrease my bid in the first case and to increase it in the second. By offering only very low bids, I will almost never get the car; with very high bids, I will almost always get it. This way my archer's mentality will always push me

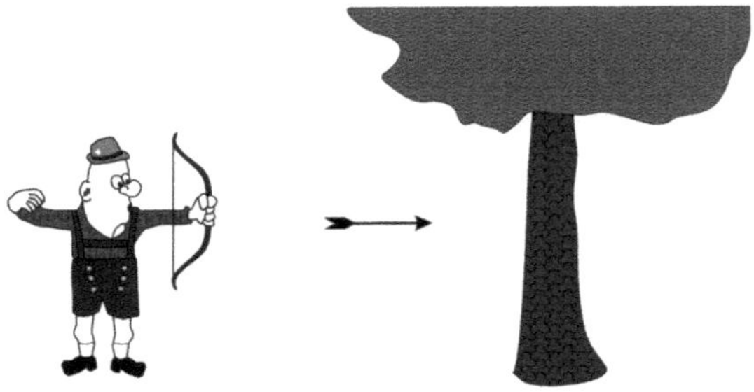

Figure 19.3 Learning direction theory (Selten 1998) compares the learner to an archer who attempts to hit a tree with his arrow. The archer's learning procedure includes counterfactual reasoning of the following kind: "had I aimed a little more to the right I would have hit the tree." This procedure is different from reinforcement learning and seems to explain several phenomena found in experimental games, where humans do not follow the logic of conventional game theory. One such phenomenon is the (nonrational) cooperation observed in repeated prisoner's dilemma games with a fixed number of rounds; another example is the great difficulty subjects seem to have with learning to avoid the winner's curse.

toward an intermediate bid; the learning disaster depicted in Figure 19.2 looks like a consequence of this mentality.

The key idea of learning direction theory was born when Selten and Stoecker (1986) studied the results of experiments in which a population of subjects had a chance of learning how to play repeated prisoner's dilemma games with a fixed number of rounds. Subjects played many of these repeated games and had to change partners whenever a new repeated game began. Narratively speaking, the following happened in this experiment. During exposure to the first repeated game, several of the subjects were confused but the community of subjects learned rather quickly to enjoy the benefits of sustained cooperation. It would be difficult to capture the exact cognitive process that led them to this state. The subjects' learning is easier to understand, however, once they also discovered the so-called *end effect* (in the last round it always pays to defect). A conventional game theorist would now expect learning to mimic backward induction. If it does not pay to cooperate in the last round, it will not pay in the second last round either, and so forth. However, here the archer comes into play again. Not being an academic game theorist, the subject's main worry will be when to switch from cooperation to noncooperation. If, in the last repeated game, I was the first to switch, then I could only have gained more by remaining cooperative for longer. If, on the other hand, my partner defected first, I could only have gained more by defecting earlier myself. Responding behaviorally to these simple insights in the subsequent repeated game, my onset of defection will fluctuate over time and the population will remain longer in a state were defection only begins "toward the end." The archers do not play the perfect equilibrium, but they are better off than academic theoreticians who would be stuck in a perfect equilibrium with low equilibrium payoff. In other words, it can be beneficial not to be perfect, and sociality may sometimes be protected by imperfections of prepared learning.

Emotions Are Promoters and Inhibitors of Sociality

For many years, economics and other rational choice disciplines paid scant attention to the role of emotions in shaping social life, and while there was much study of emotions in psychology, little of it was focused on how emotions gave rise to behavior (Elster 1998). Over the last decade or two, interest in emotions and other "visceral factors" has grown, as has the appreciation of their importance for understanding human sociality. Here we provide a brief sampler of this research.

The Logic and Illogic of Deterrence

Let us begin with a problem of great interest: nuclear deterrence. During the Cold War, NATO and Warsaw Pact forces faced each other from opposite sides of the Iron Curtain. Warsaw Pact forces greatly outnumbered NATO forces, such that most military observers believed that if the Soviets launched a conventional attack (picture a flood of T72 tanks pouring through the Fulda gap), NATO forces would be unable to halt their advance. NATO doctrine held that such an invasion would be met with a massive nuclear attack on the Soviet Union. Such an attack, NATO planners conceded, would lead to nuclear retaliation by the Soviets and most likely the end of civilized life, and perhaps all life, on the planet. Faced with a choice between nuclear war and maintaining the status quo, even the masters of Kremlin would—so the argument ran—choose the status quo.

Beginning with the seminal work of Thomas Schelling (1960), game theorists pointed to a flaw in the logic of deterrence; namely, that the decisions made by the Soviets and NATO would occur in sequence. First the Soviets would decide whether or not to invade; thereafter, based on this first decision, NATO would choose its action. If the Soviets decided to attack, NATO member states, most notably the United States, would face the decision of whether to launch a nuclear attack. So, suppose the Soviets invade. What decision would the American President take: lose Europe or end the world? Europe is nice, but not that nice, so any rational American leader should choose not to retaliate. This decision is predictable, and thus anticipating this, Soviet leaders should have discounted the threat of nuclear war as an "incredible threat" and launched a conventional attack as soon as their advantage in conventional forces was sufficient to guarantee victory. According to game theory, deterrence should not work because of the sequential nature of decisions: once the Soviets make the first move, the second player, NATO, is left with an unpalatable second choice.

The logic of deterrence occurs in less dramatic forms in everyday social and economic interactions. One obvious case takes the form of threats meant to deter undesired behavior: "If you hire that misguided dimwit, I'll quit" or "If you take that job, our relationship is over" are familiar examples. Similar patterns occur in many economic interactions, especially in bargaining situations, when there is some surplus to be divided but each party is tempted to hold out for a bigger share.

The *ultimatum game* captures the logic of bargaining in an especially stark form. Here, the first player proposes a division of an endowment. If the second player accepts, she gets what she was offered; if not, both players get nothing. Second players may threaten to reject an offer, but even if they receive an unfair offer, it is irrational to reject it in a one-shot anonymous setting.

The second player can solve this problem if she commits to retaliating; that is, if the first player attacks (or hires the dimwit or takes the job or makes a low offer), she won't evaluate the outcomes and make a choice, but will instead retaliate automatically. Such a commitment device has the effect of reversing the order of moves. In the example of nuclear deterrence, the Soviets (the former first player) now have a choice between the status quo and ending the world. By binding herself to a retaliatory response, the second player avoids having to retaliate and achieves a better outcome. The same holds for the ultimatum game. If the second players are able to bind themselves to reject unfair offers, they would only receive fair offers.

Nesse (2001a) distinguishes "external" from "internal" commitment devices. The second player creates an external commitment by changing the environment in a way that constrains her to retaliate. The presence of American troops in Germany was perceived to be an external device because, it was argued, no American President could avoid retaliating after the troops had been overrun. In a famous and more plausible example, the Spanish conquistador Hernán Cortés ordered all but one of his ships scuttled after landing at Veracruz in order to force his troops to move against Tenochtitlan.

Anger and Commitment

Jack Hirshleifer (1987; see also Frank 1998) argued that emotions can act as an "internal" commitment device and that anger, in particular, may have evolved because it commits actors to carry out threats. The intuition is simple: anger creates an unthinking, automatic response. A potential aggressor, who knows that aggression will anger his opponent, knows that his opponent will retaliate even though at that point retaliation does not pay. Thus an irrational opponent may do better than a rational one. According to one biographer, Richard Nixon intentionally cultivated a reputation for erratic emotionality to convince the leaders of the Soviet Union that he would respond irrationally to any attack (Isaacson 1992, pp. 163–162, 181–182). Recent events in the U.S. Congress may provide another example.

Everyday experience tells us that angry people do not rationally consider the future costs and benefits of their choices; they just act. Road rage is a familiar example; people who experience minor traffic insults retaliate in ways that make no sense in terms of self interest. For example, after being cut off by a Mercedes, an enraged Jack Nicholson got out of his car at the next stop light and pummeled the offender's windshield with a golf club. There are more than a thousand such incidents each year in the U. S., and sometimes they lead to serious injury or death (Mizell 1997). In these cases, anger at transgression gives rise to absurdly costly retaliation against strangers. This psychology is not limited to Western societies—the expression "blinded by anger" is used in several unrelated cultures (Fessler 2010).

This folk psychology is supported by a wide range of experimental evidence. Psychologists have studied how anger affects decision making (reviewed by Lerner and Tiedens 2006). In a typical experiment, decisions made by subjects who have been induced to experience anger are compared with control subjects. These experiments indicate that compared to control subjects, angry subjects are more likely to believe that their actions will turn out for the best and, as a result, take greater risks. Angry subjects are also more convinced about the cause of such events and are more likely to blame others. They also use heuristic reasoning based on superficial cues rather than careful in-depth reasoning.

Work in experimental economics is consistent with this picture. Subjects who receive low offers in the ultimatum game report experiencing anger and say that this motivated them to reject such offers. These findings are consistent with research in neurobiology, which indicates that computations underlying emotional decision making and those involved in "cooler" cognition occur in different brain regions, and that emotional cognition is more automatic and less available to conscious inspection (Ledoux 1998; Cohen 2005).

It is not clear whether anger evolved as an internal commitment device. Authors like Hirshleifer, Frank, and Nesse point to the fact that anger seems to function that way in contemporary societies. Data suggest that anger is focused on the punishment of transgression, and it reduces the salience of the costs of punishment. Anger has many visible effects on the phenotype. Most notably it generates an involuntary facial expression. These observable effects of anger may create a credible threat even in one-shot interactions because they signal a significant probability that the second player will retaliate. Fessler and Quintelier (2012) argue, however, that it is unlikely for selection to favor any internal commitment mechanisms. The problem in the end is that it does not pay to punish in one-shot interactions. Even in a world in which individuals honestly signal their willingness to punish, and as a result rarely need to punish, there will be rare occasions when they will need to carry through with their threat and, in these cases, selection will work against them. As a result, there should be selection in favor of bluffing—emitting signals associated with anger, but declining to punish if the bluff is called. Fessler and Quintelier argue instead that anger functions as part of an external commitment system based on reputation. People who, when push comes to shove, actually punish transgressors get a reputation for doing so, and this prevents future transgressions. Anger was favored by selection because it motivates individuals to punish. In current environments, anger causes people to punish irrationally, but in ancestral environments in which people lived in small groups, the reputational benefits were real. A possible problem with Fessler and Quintelier's argument is that it focuses solely on cases in which bluffs get called. Gale et al. (1995) point out that individuals who mean to bluff may occasionally carry through with their punishment by mistake (because they are really angry, perhaps). If the cost of being punished is high compared to the cost of punishing, they show that such errors can create strong selection against transgression even if the probability of such errors is small. Gale et al. point out that this is the case in the ultimatum game, where the rejection of unfair offers is much less costly to the second player than the first. Similar "error management" arguments (see also Nettle, this volume) have been used to explain cooperation in other seemingly anonymous one-shot games.

Whatever the evolutionary explanation for anger as a commitment device, it is clear that in the modern world its targets are shaped by cultural evolution. For example, Nisbett and Cohen (1996) have found that in the U.S., Southerners value personal honor more strongly than Northerners and, as a result, transgressions which lead to minor scuffles in the North can lead to lethal violence in the South. This hypothesis is consistent with differences in statistical patterns of violence as well as with differences in what Northerners and Southerners say about violence. Most impressive in the current context, behavior in the psychology laboratory also supports their hypothesis. Cohen and Nisbett recruited subjects from Northern and Southern backgrounds. During the experiment, an experimenters' confederate bumped some subjects and muttered "asshole" at them. After being bumped, subjects encountered the confederate walking toward them down the middle of a narrow hall, setting up a little game of chicken. The confederate, a six-foot-three 250-pound linebacker from the university football squad, was instructed to keep walking until the subject stepped aside to let him pass, even if a collision was imminent. Northerners stepped aside when the confederate was six feet away, regardless of whether they had been insulted or not. Southerners who had not been insulted stepped aside when they were 9 feet away from the confederate; Southerners who had been insulted, however, continued walking until they were just 3 feet away. These behavioral differences have physiological correlates. In a similar experiment, Nisbett and Cohen measured subject's cortisol and testosterone levels before and after insults. Insulted Southerners showed much larger jumps in cortisol and testosterone than non-insulted Southerners or insulted Northerners.

Moral Outrage and Cooperation

Willingness to punish can also sustain cooperation, even in large groups. Humans are unique among mammals in that they have a capacity for "moral outrage" (Fessler and Haley 2003; Fessler 2010). The propensity to punish others who transgress against the self is very common in nature (Clutton-Brock and Parker 1995), and this deters future transgression, even in one-shot interactions. For the reasons given above, anger may facilitate this function by serving as a commitment device. However, people get angry when third parties violate culturally transmitted moral norms, and these violations have only a small or no effect on the angered person. Much experimental work supports this picture. The clearest comes from the third-party punishment game (Fehr and Fischbacher 2004): There are three subjects. The first player receives an endowment and can transfer any fraction of the endowment to a second player, who is a passive recipient. The third player receives an endowment equal to half of the first player. She observes the first player's decision and can then remove money from the first player's fund, but must pay one monetary unit for each three removed. The first player knows that the third player will observe his choice and that she can remove money from his fund. In Fehr and Fischbacher's experiments on Swiss subjects, third players punished first players who made unequal offers; the amount of punishment was such that the first players who made unfair offers ended up with almost as much money as first players who made fair offers. Similar results were obtained in experiments in sample of 12 nonwestern societies (Henrich et al. 2006).

The role of punishment in maintaining cooperation outside of the laboratory is controversial. After reviewing ethnographic data, Guala (2012) argued that punishment occurs in state-level societies through the police, courts, and other coercive institutions, but that there are no good examples of costly punishment in small-scale societies. However, a recent study by Mathew and Boyd (2011) documents costly punishment among the Turkana, an East African pastoral group without centralized political institutions. The Turkana engage in frequent warfare with neighboring ethnic groups. Based on a representative sample of 88 recent raids, Mathew and Boyd show that the Turkana sustains costly cooperation in combat, at a remarkably large scale, through the punishment of free riders. Raiding parties comprise several hundred warriors, and participants are not kin or day-to-day interactants. Warriors incur substantial risk of death and produce collective benefits, for example, increased grazing land that benefits all. Cowardice and desertions occur and are punished by sanctions, including both fines and severe corporal punishment. The decision to impose sanctions is reached through informal community consensus and administered by community members, many of whom did not participate in the raid, and thus were not harmed by the cowardice or desertion.

Laboratory evidence suggests that emotions play an important role in motivating third-party punishment. Nelissen and Zeelenberg (2009) studied the role of anger in motivating punishment in a modified version of the third-party punishment game in which first-player allocations were affected by random noise (low offers sometimes increased and fair offers decreased). The second player was unaware of these changes, but the third player—the potential punisher—was. Nelissen and Zeelenberg reasoned that intentionally low offers would elicit greater anger on the part of third players, and this would result in more costly punishment; this seems to be what happened in their experiment. Self-reported anger was substantially higher when low offers were intentional compared to when low offers were unintentional, and when angry third players punished first players more severely.

The fact that moralistic anger can motivate punishment of norm violators is not by itself sufficient to explain human cooperation. Many moral norms act to create social order: norms against murder, theft, and dishonesty, at least toward in-group members, are nearly universal. However, many moral norms have nothing to do with cooperation or any other form of mutual benefit. Rules about diet, appropriate dress, and mode of speech are maintained by punishment but do not seem to have much to do with cooperation. Moreover, there are many examples of norms that are deleterious. A famous example is the system of mortuary cannibalism practiced by the Fore of Papua New Guinea. Prescribed funerary ritual required eating the bodies of dead relatives, and this led to the spread of *kuru*, a severe prion-based spongiform encephalopathy similar to mad cow disease (Whitfield et al. 2008). Many other such examples exist (Edgerton 2008). The diversity of moral systems is consistent with models of moralistic punishment (e.g., Boyd and Richerson 1992) which indicate that moralistic punishment can stabilize cooperation but also anything else. This means that sanctions are necessary, but a sufficient account must also include a mechanism that explains why moral norms tend to support cooperative and other mutually beneficial outcomes. Two possibilities suggest themselves: As Darwin argued in the Descent of Man (Darwin 1871, chapter 5), it could be that the moral systems of societies vary and that societies with systems that lead to more effective cooperation spread. The models of moralistic punishment explain why such between-group variation can persist. Another alternative is that humans evolved in small social groups with sufficient between-group genetic variation, and that selection led to the evolution of a psychology biased in favor of group beneficial moral systems (e.g., Bowles 2006).

Empathy and Cooperation

It has also been argued that emotions other than anger play an important role in explaining why humans are so much more cooperative than other mammals (Frank 1998). There is much experimental evidence that a variety of emotions affect the propensity to cooperate. Studies of empathy illustrate this well. For example, an important series of experiments by Batson (1991) indicate that empathy motivates helping. In a typical experiment, subjects were divided into experimental and control groups. Experimenters elicited an empathetic response from the experimental group (e.g., by asking them to take the point of view of the experiment's victim), while controls were asked to view the situation objectively. Then, the experimental conditions were manipulated to test whether subjects in the empathy condition were more likely to provide aid. For example, "Elaine," the sham victim in one experiment, was purported to suffer a series of moderately painful shocks. The experimental subjects were told at the beginning of the experiment that Elaine is unusually sensitive to shocks due to a traumatic childhood experience and that she finds them exceedingly uncomfortable. The experimenter expresses concern about this and offers the real subjects the chance to continue the "experiment" in place of Elaine. In the control condition, only about 20% of the subjects offered to help; in the empathy condition, nearly everyone offered to help.

Two recent neuroimaging experiments by Singer et al. (2004, 2006) back the conclusion that empathy supports cooperation. In the first experiment, one member of a couple was placed in an fMRI scanner where he or she could observe his or her partner, whose hand was subject to painful simulation by an electrode. The fMRI images indicated that brain regions associated with the experience of pain activated in the subject in the scanner. Interestingly, regions associated with the sensorimotor cortex did not light up, suggesting that people feel the pain of loved ones but do not experience it. The second experiment was similar except that it involved strangers who had previously played a prisoner's dilemma game. Here results differed for male and female subjects: fMRI imaging data showed that male subjects, who had previously been helped by the person experiencing the pain, felt pain themselves. Female subjects, however, did not.

It is important to understand that there is a crucial difference between commitment in deterrence and cooperative contexts. In the deterrence context, anger can signal a willingness to retaliate irrationally. If there is even a small probability that this willingness is real, then potential attackers will be deterred and the cost of retaliation will rarely need to be paid. In contrast, the individuals who signal their willingness to be costly cooperators sustain cooperation only if they actually cooperate, and thus the cost of cooperation will always need to be paid. This means that the emotions that support cooperation act as preferences, not commitment devices.

There are a vast number of experiments which demonstrate that people cooperate in anonymous, one-shot interactions, clearly indicating that people are not purely self-interested. Both the proximate and ultimate explanations for this behavior are controversial. At the proximate level, many believe that these results can be understood as resulting from prosocial preferences for fairness and equity, which are the result of emotions like empathy and shame. These interact with culturally evolved norms to generate observed cooperative behavior. An alternative view holds that people are truly self-interested. However, because defectors risk detection and this can result in severe reputational losses, we possess psychological mechanisms that make us behave as if anonymous interactions were actually public. At the ultimate level, the central question is: Did human psychology evolve within band-scale groups where mechanisms which in the modern world lead to cooperation among large groups of strangers were favored by kin selection and small-scale reciprocity? Or, alternatively, did human psychology evolve in tribal-scale groups, numbering hundreds or thousands of individuals, as the result of gene-culture coevolution? Notice that in both cases, the evolved psychology misfires in contemporary environments, leading to cooperation on very large scales.

Conclusion

The behavioral sciences confront us with a plethora of empirical examples in which learning and cognitive abilities seem to be strongly biased, constrained, and sometimes simply misleading, as in the case of the winner's curse. This may almost appear like a slap in the face for researchers who rely on the idea of adaptation through natural selection. What seems maladaptive at one level of analysis, however, turns out to be adaptive at a deeper level. Here we have demonstrated (as have many other contributions to this book; see also Hammerstein 2012) that the mechanisms of decision making have to be biased and constrained if they are to generate adaptive behavior efficiently in a given social and ecological context.

It is these biases that reflect particularly well the action of natural selection. When used out of the appropriate context, evolved mechanisms may, of course, generate maladaptive behavior. As we have seen, for example, the same cognitive mechanism that enables an archer to hit a mammoth with his arrow can cause bidders in auction experiments to engage repeatedly in unprofitable trades, thereby falling victim to a winner's curse. One could say that archers are not prepared for auctions. In general, however, human learning is superbly prepared for life in social environments. We learn from others to do "the right things," often without understanding why we do them. Even our teachers share this ignorance to a large extent and may teach us truly adaptive practices using absurd explanations for why these practices work well.

In a sense, the special preparedness of human social learning makes it possible for cultural knowledge to accumulate far beyond our individual understanding. The evolved mechanisms of social learning are, therefore, the most important key to understanding human culture and sociality. In combination with the study of emotions, we learn how these evolved mechanisms guide our decision making in directions that often promote sociality but sometimes inhibit it.

Acknowledgment

We benefited from many discussions with Benjamin Bossan, who enriched the chapter with his model of the evolution of social learning. We are also grateful to Arnulf Koehncke, David Stephens, Jeff Stevens, and Julia Lupp for their many useful suggestions and comments. PH received support from the Deutsche Forschungsgemeinschaft (SFB 618).

First column (top to bottom): Rob Boyd, Tom Griffiths, Keith Jensen, Felix Warneken, Simon Gächter, Peter Hammerstein, Tom Griffiths
Second column: Peter Hammerstein, Rosemarie Nagel, Benjamin Bossan, Rob Boyd, Keith Jensen
Third column: Thomas Bugnyar, Thomas Mussweiler, Simon Gächter, Felix Warneken, Rosemarie Nagel, Thomas Bugnyar, Thomas Mussweiler

20

Evolutionary Perspectives on Social Cognition

Thomas Bugnyar, Robert Boyd,
Benjamin Bossan, Simon Gächter, Thomas Griffiths,
Peter Hammerstein, Keith Jensen, Thomas Mussweiler,
Rosemarie Nagel, and Felix Warneken

Abstract

Game theory provides a useful framework for conceptualizing social decisions in which one person's behavior affects outcomes that matter to other individuals. Game theory can also help us understand the computational problems inherent in social decision making. To explain human adaptive success, culture plays an important role. Models from population biology are used to explain (a) culture as a dynamic process, (b) the role of psychological mechanisms in enabling cumulative cultural evolution, and (c) teaching as an evolved phenomenon. How game theoretic approaches can be integrated with algorithmic-level processes remains, however, to be resolved. Difficulties lie in how cognitive and social psychologists approach the study of cognition as well as in how existing data are used to interpret human (and nonhuman) decision making.

Introduction

During the winter, the Inuit of the central Arctic survive by hunting seals. When the sea ice begins to freeze over during the fall, seals claw a number breathing holes within their home range. As the ice thickens, the seals continuously maintain these openings until they become conical chambers under the ice. Inuit hunt in groups spread out so that there is one hunter for each hole. The goal is to cover as many holes in an area as possible. The hunters use a multipart harpoon: on the butt end is a sharpened point of hard bear bone, used as an ice pick; the other end has a detachable toggle-harpoon head connected to a heavy-duty braided sinew line. The hunter opens the hard snow that covers the hole using the ice pick, smells the interior to make sure it is still in use, and then investigates the shape of the chamber using a long, thin curved piece of caribou antler with a knob on one end. This helps him plan his thrust. Then he covers the hole with snow and places a bit of down tethered to a thread over the small opening that remains. The hunter then waits motionless in the frigid darkness, sometimes for hours. When the appearance of the seal disturbs the bit of down, the hunter strikes downward with all his weight, drops the harpoon shaft, but hangs on to the connected line. Once the seal tires, it is hauled onto the ice and killed. Successful hunters load their kill onto a dog sled and return to the encampment, where the seal is butchered and shared among all the members of the group (Balikci 1970).

This example highlights two important, distinguishing features of social decisions. First, the outcomes that people care about depend on *decisions made by others*. Hunters increase the group's chance of success by spreading out, one man at each breathing hole. This is an example of what game theorists call a coordination game; while there may be conflicts of interest, because some holes are better than others, each hunter improves his chances by choosing a different location. Like virtually all foraging peoples, the Inuit share large kills, a practice that ensures the group against variation in hunting success. However, this means that each individual depends on the willingness of successful hunters to share with the rest of the group, a payoff structure that game theorists label a public goods game. Second, many human decisions depend on *information acquired from others*. Hunters must know about the habits of the seal, hunting methods, and how to construct complex tools like toggle-harpoons and dog sleds. This complex body of knowledge is not invented by individual hunters, but is instead accumulated over generations by a population of individuals. Similarly, food sharing is regulated by culturally transmitted norms that specify who should get what, how disputes are to be adjudicated, and what kinds of sanctions are legitimate.

In this chapter, we focus on these two social dimensions of human decision making and summarize how social cognition informs models of social behavior. In both economics and biology, game theory is the main theoretical framework used to think about social life. Game theory rests, however, on assumptions that do not accurately reflect people's cognitive makeup because it neglects important problems, such as how people construe social settings and how they categorize people and outcomes. We begin by debating the pros and cons of game theoretical approaches for analyzing social behavior and evolution. Thereafter we explore ways of bridging different levels of analysis, approaching the topic from top-down and bottom-up perspectives and focusing on the concepts of beliefs and preferences, selfish and other-regarding behaviors, and social learning. We conclude with a discussion on cultural transmission and evolution.

The Adequacy of Game Theory

There are several versions of game theory (Table 20.1) which share some important assumptions but also differ in important ways. In all versions, there are a number of actors (players) and it is specified how their individual success depends on the actions of others. A player can use one of several (often many) alternative strategies, and each strategy specifies the actor's behavior in a variety of situations that may occur in the game. While all this provides a common framework for game-theoretic analysis, different versions of game theory incorporate different assumptions about how strategies are chosen or generated. Classical (or "eductive") game theory attributes strategy choices to reasoning, whereas evolutionary game theory attributes it to processes in populations. Depending on the timescales considered, these processes are either natural selection acting on genotypes or social learning.

Much of *classical game theory* is explicitly or implicitly based on the assumption that players are not only rational but also know that all other players are rational. Players may even know that they all share the knowledge about everybody's rationality. This assumption implies that, for example, the individual Inuit solves the coordination and public goods games by thinking about the payoffs and what the others in the group are likely to do. These assumptions sound absurd, given that real human players often fail to show rational behavior in experiments conducted in the behavioral sciences. One could argue, however, that if *all* players in a game were educated by game theorists, they would follow the academic wisdom of their teachers. Why should this be so? The central claim made by classical game theory is that actors should direct their attention to so-called Nash equilibria. A Nash equilbrium occurs when no player can do better by deviating from his/her choice of strategy. From the perspective of classical game theory, a rational player's reasoning about the pros and cons of strategies should thus typically come to a halt at a Nash equilibrium. Many game experiments have been conducted in which the subjects

Table 20.1 Summary of the different approaches to game theory. Note that different versions of game theory posit different equilibrating mechanisms.

Approach	Description
Classical or "eductive" game theory	All players are rational and know that every other player is also rational. A Nash equilibrium is a profile containing a strategy for each player from which no player is able to benefit by deviating as long as all other players follow the strategy assigned by the profile.
Evolutionary game theory	Strategies are assumed to be heritable. Individuals are sampled in some way from a larger population; they then interact according to their strategies and reproduce at a rate that is proportional to their payoff in the game. This process is iterated until the large population reaches a stable dynamic equilibrium.
Learning game theory	Players are assumed to interact and update their strategies according to some plausible learning process. This process can converge to a game theoretic equilibrium.

actually played such an equilibrium. This partly explains the success of Nash's mathematical concept, but it does not justify its foundation on strong rationality assumptions. From the cognitive science perspective, various processes can lead players to a Nash equilibrium.

Evolutionary game theory, as originally practiced in biology, assumes that strategies are heritable. In this Darwinian framework, strategy choices are made by natural selection and not by the individual. Maynard Smith and Price (1973) laid the foundation of this theory. In their mathematical framework, individuals are sampled in some way from a larger population and interact according to their strategies in game-like situations. Individuals then reproduce at a rate that is proportional to their game payoff. The evolutionary dynamics is defined by iteration of this process. Special attention is paid to cases where this process reaches a stable dynamic equilibrium. Under many circumstances, this equilibrium can be characterized as a Nash equilibrium. For human populations, this assumption means that cultural evolution gradually finds solutions as people learn social norms and then adjust their behavior at the margin. Thus, for example, whole populations of Inuit evolve solutions over generations.

A later version of evolutionary game theory replaces natural selection by *learning*. Here the assumption is that the players interact and use information from these interactions to update their strategies according to some plausible learning process. Game theoretic equilibrium can occur through convergence of this learning process. The idea is that individual Inuits adjust their behaviors over the course of months or years through some process of relatively myopic learning. Obviously, these two branches of evolutionary game theory study similar adaptive processes. Learning does not always mirror evolution on shorter timescales, however, even though a standard equation used in evolutionary game theory can be interpreted both as natural selection and social learning (see Hammerstein and Boyd, this volume).

A Critical Evaluation of the Approaches

Different versions of game theory are useful for different purposes. Classical game theory provides a tool for analyzing simple, multi-stage games, like the ultimatum game and the trust game, and these analyses provide insight into more complicated but structurally similar interactions in the real world. Evolutionary game theory is useful for understanding the genetic and cultural evolution of social interactions over longer timescales, and learning game theory provides predictions about behavior in experimental games.

Game theory, however, requires more information than is available in real-life situations. Specifically, classical game theory is based on strong assumptions about human social cognition:

1. People understand the game.
2. They know who the other actors are and what outcomes are possible.

3. They have consistent preferences over these outcomes and are also aware (at least in a probabilistic sense) of the preferences of other players.
4. People hold consistent rational beliefs about how other players will behave that are formed using Bayesian inference.

All these assumptions are contentious. In many situations, players neither know about all of the other players involved nor are they fully aware of the strategic situation. People often have vague preferences, which might also be intransitive. Beliefs are often not formed rationally but are influenced, for example, by contextual cues, stereotypes, and superstition. In addition, people often make cognitive errors and do not analyze the strategic nature of their interactions as deeply as fully rational players would. They also might want to contribute to the public good if everyone else does (i.e., conditional contribution).

Despite these shortcomings, game theory captures an essential feature of social decision making: people have goals, but these goals conflict. As a result, game theory provides a scientifically fruitful framework for studying strategic interactions of real people. The key problem is how to link game theory to other behavioral sciences, especially psychology. How do people actually construe the strategic situation in which they find themselves? How do people form preferences over possible outcomes? How are beliefs formed? These questions are fundamentally psychological in nature and thus game theoretic research can benefit from insights provided by cognitive and social psychology. These disciplines can help us understand how people form preferences and beliefs, a process which might depend on cognitive information processing and social cognition.

How Should We Think about Social Cognition (Bottom-Up or Top-Down) and How Does Our Thinking Account for Challenges with Game Theory?

If game theory is useful for structuring the strategy or payoff space of social interactions, but not necessarily for describing or predicting what humans actually do, it would be advantageous to find both a way to link solution concepts to underlying reasoning processes and to improve the fit between predictions and behavior. Marr (1982) identified three levels of analysis that can be useful in studying information-processing systems. The first level, the *computational level*, focuses on understanding the problem that the system is attempting to solve as well as on what constitutes a good solution to that problem. Next, the *algorithmic level* identifies which representations and algorithms can be used to execute that solution. Finally, the *implementation level* is concerned with how those algorithms are physically implemented. We consider these levels of analysis to be highly relevant as we lay out our goals and develop a theory of social decision making.

One possibility is that the different forms of game theory give us a good answer to the computational-level question of what the problem is that people are solving, and how it should be solved. In this case, differences between the predictions of the theory and human behavior are simply clues about the algorithmic level. We then need to work out how to incorporate cognitive constraints into game theory (working from the top down) or consider the psychological mechanisms that support social decision making (working from the bottom up) to develop an account that is closer to human behavior.

Identifying the different levels of analysis involved in social decision making sets up a new problem: If we believe that game theory is inadequate as an account of human behavior, and want to incorporate ideas about cognitive mechanisms, we need to work out how to bridge the computational and algorithmic levels. There are two strategies for doing this. A top-down strategy maintains the idea of identifying rational or optimal strategies but introduces additional cognitive constraints to the optimization problem which must be solved. A bottom-up strategy starts with the cognitive mechanisms and then explores the consequences of these mechanisms in the context of social decision making.

Bridging the Gap: Top Down

A number of game theoretical approaches, models of bounded rational behavior, attempt to fill some of the gap from the top down (for a detailed survey, see Camerer 2003). Most are based on experimental data, but they differ with varying degree from game theoretic approaches in the following ways:

1. The *quantal response equilibrium model* (error model, errors on beliefs) was proposed by McKelvey and Palfrey (1995, 1998) to explain differences in behavior from theory and remains close to the game theoretic solution concept. The model adds that players can make mistakes when choosing a strategy, assuming common knowledge of making mistakes. Otherwise, this model has not been modified as much as the others described below. Importantly, the result of the quantal response equilibrium model is that players behave in a homogeneous way by choosing an action given a common probabilistic function.
2. *Social preference models* remain within the game theoretic framework as well by calculating equilibrium strategies. These models extend the utility function, which can be done in many ways, for example, by introducing inequity aversion, intentions, reciprocity, and guilt aversion (Fehr and Schmidt 1999; Falk et al. 2008; Charness and Dufwenberg 2006). Here the differences between players can be very large, denoted by different parameter values for different intensities between players,

varying from self-interested players to social preference players, or different motivations, etc.

3. *Level-k models* or *cognitive hierarchy models* (see Camerer et al. 2004) are cognitive reasoning models which relax the assumption of common knowledge of rationality or the request to find equilibria. Here it is only important that a player believes what other players are choosing or believes the beliefs and actions of other players and gives an iterated best reply to those beliefs. The existing empirical evidence provided by lab experiments and first field experiments is that the majority of subjects choose levels between zero (random choice or intuitive choice), level 1 (best response to level 0 players), or level 2 (best response to level 1 or best response to a probability distribution of lower-level players). Thus a player might just think what a player of a one-step, less-sophisticated type thinks or what a player thinks that the others are distributed across lower-level types. The biggest challenge is to understand the level 0 players' process, because the recursive system starts there. The starting point can be interpreted according to the behavior of a player who looks at the problem in a "naïve" way and chooses randomly, or uses a focal point. However, in many cases, this is still a black box, which needs to be understood through the information process a player uses seeing the problem the first time, or incorporating his knowledge of the situation. Players can behave heterogeneously because they can have different starting points or different ways of interpreting the behavior of other players.
4. *Learning models* (reinforcement models, fictitious play, Bayesian learning model) try to track behavior (typically average behavior) over time to discover whether behavior converges to equilibrium in the long term (Erev and Haruvy 2012). Experimental economics is typically interested in behavior on short- or medium-term timescales. In contrast to the models described above, which require a very precise understanding of a game's parameters, the simplest learning model—the reinforcement model—needs very little information about the underlying game. Learning models are mechanistic, transforming payoff streams of different strategies from the past into probabilities of each possible strategy according to the principle, what was good in the past will most likely be chosen in the future. Bayesian learning models, which are used often in theoretical economics, update beliefs according to Bayes's rule. They are forward-looking models. Experienced-based attraction models encompass these different models in a unifying way. Learning direction theory, a cognitive learning model, assumes that a player shifts behavior in the direction that would have provided a higher payoff in the previous period. It thus assumes counterfactual reasoning about the past (see also Hammerstein and Boyd, this volume). Cognitive learning models are typically used to identify whether

a subject behaves in accordance to the model, unlike other learning models, which only check whether the models predict the aggregate behavior.

5. *Aspiration adaptation models* (satisficing models), are based on goal formation and originate from ideas put forth by Simon (1956). Thus far they have received the least attention in experimental economics (for new experiments, see Selten et al. 2011). The models are especially useful in complex decision problems, when a subject has to adjust many variables at the same time, or in cases where alternatives have to be constructed. A subject has only vague ideas about the underlying decision situation and its parameterization. The basic principles are that a subject has to choose which variables he wants to adjust and how those are adjusted from one period to the next. Goal formation directs these choices.

The quantal response equilibrium model as well as social preference and cognitive hierarchy models are constructed to address the interaction of players, whereas learning and aspiration adaptation models can also be used in a noninteractive context. For the latter, there are many models that describe specific situations, although the models of interaction try to be very parsimonious (Camerer 1995). Both learning and aspiration adaptation models require very little information about the underlying structure of the situation.

The distinction between models, theoretic solutions, and assumptions is critical and sometimes gets lost in the discussion; that is, we often speak of game theory in general when we are really talking about just solutions. These above-mentioned models were developed based on experimental data. With the exception of aspiration adaptation models, experiments are typically set up according to game theoretical models (which specify the strategy space, actions, time frame, etc.) that have well-known theoretic solutions. In most cases, these solutions serve as benchmarks or as a way of structuring the strategy space. The primary emphasis is not on testing whether the theory is right or wrong but rather on the usefulness of game theoretic models in answering an economic question, such as: Are people fair? This is done by providing an interactive situation with precise parameters (e.g., rules of the games, number of players).

Bridging the Gap: Bottom Up

The way in which humans and nonhuman animals process incoming information critically shapes how they construe a given situation, how they interact with others, and how they make decisions. Thus, social behavior, in general, and decision making, specifically, depend on what information is attended to and how this information is processed. Nevertheless, information processing is often treated as a "black box"—and consequently ignored—in non-psychological decision theories. The problem with this approach is that this black box is needed to arrive at a complete understanding of social decision making. In fact, a stronger focus on what is going on in this black box may allow for a particularly parsimonious account of decision making.

In humans, information processing relies on a number of core mechanisms that are applied to information processing across different contexts. For example, whenever humans process incoming information, they rely on mechanisms of categorization and comparison. Paying close attention to such core information-processing mechanisms, which play out in any decision-making context, allows us to examine the common cognitive architecture that contributes to any decision-making situation. This suggests that a bottom-up approach to bridging the algorithmic and computational levels, described by Marr (1982), may be a fruitful path toward a more complete understanding of human social behavior. In a first step, such a strategy would start at the algorithmic level and identify the basic cognitive mechanisms that underlie social behavior. In a second step, moving to the computational level, the strategy would examine the adaptive advantages of the mechanisms identified at the algorithmic level. If the first step, for example, identifies comparison mechanisms as one of the most fundamental elements of human information processing, the second step would then ask whether the ubiquitous use of comparison has adaptive advantages. In fact, recent research has demonstrated that comparative information processing allows humans to make judgments more efficiently (see Mussweiler et al., this volume). That is, decision makers who are induced to rely more heavily on comparison mechanisms make judgments more quickly and use less cognitive resources to do so. Importantly, these processing advantages are achieved without a loss in accuracy.

Conceptualizing Social Cognition in Terms of Beliefs and Preferences

Beliefs and preferences represent nice examples of how different levels of analysis could be integrated. *Beliefs* constitute a crucial concept in classical game theory (but not in most evolutionary game theory). They are formed on the basis of available information and involve, most importantly, expectations about how others will behave. Theory of mind is one way that people form beliefs; by inferring what others might think, believe, or desire, people can better predict the actions of others. Another mechanism involves taking the emotional perspective of others into account. Empathy (discussed further below) is one way of *feeling into* others.

Preferences are hypothetical constructs that are inferred from choices between different actions. In a nonsocial context, a person might prefer food over sleep, but such a preference could change circumstantially (e.g., when one has been sleep-deprived or has overeaten). *Social* or *other-regarding preferences* are motivated choices between actions that lead to outcomes for individuals other than the self. Thus, if given a choice between hoarding a seal and sharing it with others, the latter would indicate another-regarding preference. How much of the seal is shared, if any, depends on the internal drive state (hunger) of the individual, the perceived need of the others, the possibility that the hunter is being watched, and many other factors.

Can We Apply the Concept of Beliefs and Preferences to Daily Life Situations?

Any social decision-making task depends on how humans perceive the other individuals with whom they interact. For example, whether an individual decides to trust another person in a trust game critically depends on the perceived trustworthiness of this other person. How are such judgments about the characteristics of others made? One mechanism that may contribute to these judgments is egocentric comparison. When judging the trustworthiness of others, humans often use themselves as a reference point for comparison and thus use their perceived own level of trustworthiness as a basis for judging how trustworthy the other is. If I intend to share the seal should I be successful, then it is likely that others will as well. Importantly, egocentric comparison does not necessarily lead to similar judgments for self and other. Only if judges focus on similarities between themselves and another person are judgments about the other assimilated toward the self. If judges focus on self-other differences during comparison (e.g., as is the case for out-group members), judgments about the other are contrasted away from the self. If I am hunting with members of another band, perhaps they will not share. To predict how egocentric comparisons influence perceptions of others, we must therefore take into account the precise comparison mechanism that is engaged.

In cases of repeated interactions, such judgments about the qualities of others can be simply based on general learning mechanisms (e.g., following a rule like "I was nice to them previously and it was beneficial for me—thus I will be nice again"). In this scenario, conspecifics may be treated similarly as any other environmental cue (see Kacelnik, this volume); differences may be found just in the degree of salience. However, a crucial feature that sets the social domain apart from any ecological problem is the autonomous nature of living beings (Jolly 1966; Humphrey 1976). As intentional beings with their own goals and beliefs, conspecifics may respond differently from situation to situation, making it difficult to generalize their behavior across contexts based on learned contingencies only (e.g., Byrne and Whiten 1988; Tomasello and Call 1997).

When humans represent the behavior of other social agents, they often attribute psychological states as important causes of the behavior. One can distinguish three types of psychological (intentional) states, namely affective states (i.e., what others feel), conative states (i.e., what others desire, want, intend to do), and epistemic states (i.e., what others believe and know about the world). One goal of research in social cognition is to determine which (if any) psychological states humans and other animals are able to represent when predicting the behaviors of other agents. Human adults represent other people's actions not only in terms of behavioral regularities, but often understand and evaluate actions in terms of the underlying psychological states. As a matter of fact, this is often an automatic process and people cannot help but describe actions in mentalistic terms, as illustrated most famously in the seminal work by Heider and Simmel (1944), in which people describe the spatial movements of geometric shapes in mentalistic terms such as a triangles being angry or one triangle trying to catch another. A large body of work in developmental psychology has demonstrated that this tendency to represent behavior in terms of psychological states is already apparent in very young children, indicating that this capacity is foundational for the ontogeny of human decision making (see Warneken and Rosati, this volume).

Recent research in nonhuman primates has revealed precursor steps for representing mental states, such as perspective taking and knower-guesser differentiation in our closest living relatives (Hare et al. 2000; 2001; Flombaum and Santos 2005). Interestingly, some species of the bird family *Corvidae* show similar skills in competition for hidden food (Emery and Clayton 2001a; Bugnyar and Heinrich 2005; Dally et al. 2006; Bugnyar 2011), indicating that sophisticated socio-cognitive abilities may evolve convergently in phylogenetically distant taxonomic groups (Emery and Clayton 2004). Still, despite the ability to predict behavior based on certain psychological states such as intentions, to our knowledge, attribution of mental states in the form of beliefs has not been demonstrated in any nonhuman species yet (Call and Tomasello 2008).

Simulating Others' Mental States versus Using a Theory

One possible strategy for exploring the adequacy of simulation-based accounts of social cognition is to develop computational models that clearly quantify the predictions that result from simulation. An example of a setting in which this might be possible is inferring the preferences of others from their behavior. If you observe the choices that another person makes, how do you infer their utility function? This is a problem that has been studied extensively in economic theory and econometrics but is also something that people do whenever they observe another person making choices. Data from developmental psychology suggest that even young children are capable of this kind of inference; they are able to infer which objects a puppet would like based on the choices made by the puppet and the set of available alternatives (Kushnir et al. 2010).

The problem of inferring utilities from choices can be cast as one of Bayesian inference, where the observed data are the choices and the hypotheses to be evaluated concern different options for the utility function. To connect data and hypotheses, we need to define a "forward model" of choice that indicates how likely it is that people make certain choices if they have a certain utility function. This forward model defines the likelihood function used in Bayesian inference. Research in econometrics suggests a natural likelihood function—the mixed multinomial logit model—which assumes that people maximize a noisy representation of expected utility (McFadden and Train 2000).

The forward model used by Lucas et al. (2009) assumes that people use expected utility as the basis for interpreting the choices of others. It was able to explain the inferences children make about the preferences of others, such as those demonstrated by Kushnir et al. (2010). However, expected utility has been shown to be inconsistent with the choices that people actually make in cases where multiple attributes need to be considered, raising the tantalizing possibility that people assume others are more rational than they are themselves. This would provide evidence against simulation, since if people were inferring preferences by simulating choices, they would be using the same model for interpreting the choices of others as for making their own choices.

Preferences: Selfish versus Other-Regarding

Assumptions of Classical Models

A long-standing assumption in economics and other behavioral sciences has been that people are primarily self-interested. As a consequence, social outcomes needed to be explained by self-interest. Assuming selfishness and rationality is a defensible methodology in the absence of empirical evidence that shows systematic deviations. The development of experimental tools to investigate the selfishness assumption has changed the situation. Starting with the invention of the ultimatum game (Güth et al. 1982), numerous experiments have cast doubt on the assumption of selfishness and provide evidence for the importance of social preferences.

People are said to have social or other-regarding preferences if their choices also consider the consequences for the well-being of others. People are self-regarding if their choices are solely motivated by their own well-being. Consider the following example: An individual has to choose between two payoff allocations, (8 for me, 2 for the other person) or (5, 5). A selfish individual will choose (8, 2) because it gives the individual a higher payoff than (5, 5). A person with other-regarding preferences might prefer (5, 5) over (8, 2) because (5, 5) is equitable. This person may even choose (0, 0) over (2, 8) to avoid being better off than the other person.

There are several proximate sources of social preferences, all derived from a variety of carefully designed experiments that control for the material and social incentives people face. These sources include equity concerns, altruism, reciprocity, and guilt aversion. By now there is ample empirical evidence for these motivations (Camerer 2003), but there is also a high degree of variation: a majority of people typically makes choices that are consistent with social preferences, whereas a sizeable minority is best characterized by selfishness. For example, many people reject low offers in bargaining games and contribute to public goods, but some never contribute and are willing to accept low outcomes. Some people are willing to punish free riders, which gives the free riders an incentive to cooperate. Importantly, evidence for the relevance of social preferences is not constrained to the lab but has been achieved in naturally occurring situations (Gintis et al. 2005).

To what extent social preferences are individual traits that are stable across situations is an open research question. Some evidence, mostly from dictator games, suggests that social preferences are situation specific and can be influenced by situational cues (e.g., Haley and Fessler 2005), whereas other evidence (from public goods games) suggests that preferences are more stable (see Blanco et al. 2011). This is an important question because current theories of social preferences assume that these are stable and obey important consistency axioms, like transitivity (see below).

Other-Regarding Preferences

If other-regarding preferences are exhibited by individuals, two questions arise: How do they develop and how did they evolve? One point of caution is that similar phenomena can arise from very different mechanisms. Other-regarding concerns—emotions that are aligned or misaligned with the welfare of others—are likely to be an important motivating force for human other-regarding preferences, but prosocial and antisocial behaviors need not always be driven by these concerns in humans, and they may not exist at all in other animals (see Jensen, this volume). Evidence for the emergence of positive other-regarding concerns comes from studies on helping and sharing in young children. For example, infants as young as 14–18 months lend a hand when a person is struggling to access a desired object (Warneken and Tomasello 2009), based on an identification with the other person's failure to achieve an action goal. Toddlers as young as 18 months are more likely to give up a balloon to an experimenter who had been wronged than one who had not, even in the absence of emotional cues (Vaish et al. 2009). This indicates that empathy may be driving prosociality in children. Even very young infants (6 months old) appear to differentiate helpful individuals (in this case, a geometric shape with eyes) from those who hinder others and show a preference for nice actors, raising the possibility that the ability to evaluate the social intentions of others is innate (Hamlin et al. 2007).

Other animals also perform prosocial acts such as comforting others in distress and sharing food. These might be examples of positive other-regarding preferences in the sense that another individual benefits from the actor's actions, but they might not be driven by positive other-regarding concerns. For instance, in chimpanzees, *Pan troglodytes*, bystanders observing a conflict might affiliate with one of the distressed individuals. This affiliation might serve to console the recipient (de Waal and van Roosmalen 1979; Fraser et al. 2008), but it might also be a self-defense strategy in that the "consoler" experiences less redirected aggression (Koski et al. 2007; Koski and Sterck 2007). In ravens, *Corvus corax*, the function of bystander affiliation depends on who initiates the interaction: when bystanders offer affiliation to valuable partners after the partners received intense aggression, they may reduce the others' distress; however, when victims solicit affiliation from others, they may try to protect themselves from renewed aggression (Fraser and Bugnyar 2010). Likewise, food sharing in primates may be based on different motivations, such as "payment for labor" (de Waal and Berger 2000), and prosocial behaviors are apparently often merely the response to harassment and begging (Gilby 2006). Notably, in dictator-style games, where individuals pay no cost to choose outcomes that can benefit others, chimpanzees are indifferent to the well-being of others, including close affiliates and kin (Silk et al. 2005; Jensen et al. 2006; Vonk et al. 2008). Interestingly, these results do not hold for primates in general, who show other-regarding behavior in the same context (Burkart et al. 2007; Lakshminarayanan and Santos 2008; Massen et al. 2010). Moreover, chimpanzees may also behave prosocially in other experimental paradigms and contexts; namely, when another individual requires instrumental help to access, for example, an out-of-reach object or enter through a door to obtain food (Warneken and Tomasello 2006, 2007; Yamamoto et al. 2009). Taken together, nonhuman animals perform a variety of prosocial behaviors, although the breadth of these behaviors remains unclear, raising questions about the underlying motivations of these behaviors.

(When) Should People Be Selfish/Other-Regarding? An Evolutionary Perspective

There is much evidence that the solution of collective action problems plays an important role in even the simplest foraging groups of humans. In every well-studied foraging group, food sharing acts to ensure individuals against the vagaries of hunting success. Warfare is very common; historical data from a large sample of western North American foraging societies indicate that the average group engaged in four wars per year (Jorgensen 1980). In both warfare and food sharing, each individual has an incentive to free ride because the marginal effect of his or her behavior on the outcome is small but the effect on his or her own payoff is large. If viewed as a one-shot interaction, food sharing and warfare should be modeled as an *n*-person prisoner's dilemma. Therefore, if social groups were formed randomly, selection would not favor psychological mechanisms that give rise to other-regarding preferences.

Evolutionists differ about whether natural selection should be expected to produce other-regarding preferences. Cooperation is defined as behavior that increases the fitness of other individuals. Cooperation is altruistic if the behavior also reduces the fitness of the actor. Altruism can only be favored by selection if interactions are structured in such a way that altruists (or their genes) are more likely to receive the benefits of altruism by others than are non-altruists. This can occur when cues of recent common descent allow individuals to direct altruism toward kin, or when population structure creates relatedness within local groups. This process, called kin selection, could favor genuinely prosocial preferences toward family or other group members. Contemporary human groups are far too large for this mechanism to be important. Some believe that ancestral environments of humans led to significant relatedness in social groups and this explains other-regarding behavior observed in economic experiments (Hagen and Hammerstein 2005), but others disagree (Boyd and Richerson 2005). Cooperation can also be stabilized when individuals interact repeatedly, and behavioral strategies reward cooperators or punish non-cooperators. Once such strategies are common, cooperation is sustained but does not need to be interpreted as altruistic. These models do not, by themselves, explain the evolution of cooperation because many different behaviors can be stabilized in this way. A complete explanation requires some mechanism like cultural group selection, which selects cooperative equilibria over non-cooperative ones. However, once such behaviors are common, selection could favor prosocial preferences because they help individuals avoid punishment over the long run. In some contexts, these preferences lead to altruism, for example in economic experiments (Richerson and Boyd 2005, chapter 6), or to leveling mechanisms like food sharing, which reduces the cost of altruism (Bowles 2006).

What Proximate Mechanisms Underlie Self-Regard and Other-Regard?

The standard, gene-centric account of evolution predicts that individuals should act in ways to maximize their inclusive fitness. However, it is mute about the proximate mechanisms that give rise to this behavior. The care that a mother, for example, provides to her offspring is easily explained at the ultimate level by kin selection. At the proximate level, however, her motives are likely to be genuinely altruistic (i.e., positively other-regarding). It may be that humans have genuinely prosocial preferences, even toward distantly related individuals.

Indeed, there is much evidence that acts which benefit (and harm) others can have as their underlying causes positive (or negative) other-regarding concerns in most normal human beings. These include empathy, symhedonia (shared joy), schadenfreude (pleasure in the misfortunes of others), and envy (see Jensen, this volume). Shame may motivate people to avoid violating mutually beneficial social norms, and moral outrage may lead to the punishment of those who deviate from these norms (Fessler 2010).

However, some economists claim that many acts that seem to be driven by social preferences are indeed selfishly motivated. The argument, at least in a crude form, is that people help others out of strategic concerns; namely, out of an expectation of future reciprocity or to gain a reputation so that others will be helpful to the actor (indirect reciprocity). So, for example, members of an Inuit band may share their kills only because they calculate (or have learned) that this pays in the long run. If this is true, then it is not clear whether other-regarding concerns, such as empathy, belong in the human psychological toolkit. Perhaps, as the argument goes, other-regarding concerns motivate the actor to pursue selfish ends by acting as a short-term bridge to long-term gains. This is a rather dismal view of other-regarding concerns and, as is argued by Jensen (this volume), may not be true.

Related to this, a large part of our discussion concerned the sensitivity of prosocial emotions or selfishness to context. In a series of studies, images of eyes—from eyespots to robotic "eyes"—were found to elicit more generous donations in the dictator game and in a public goods situation, such as paying for milk in a coffee room (Haley and Fessler 2005; Bateson et al. 2006; Burnham and Hare 2007). One interpretation of these studies is that people act as if they are being watched, and because people are selfishly and strategically concerned about their reputations, they act nice. There are, however, at least two problems with this account. First, it does not always work. For instance, being observed by "real eyes" does not affect ultimatum game behavior (Lamba and Mace 2010) and stylized eyes had no effect in a trust game (Fehr and Schneider 2010). Second, it may be that the eyes prime other-regarding concerns. Rather than giving people the illusion that they are being watched, the eyes can subtly make people think of others and not just themselves. One way that this might be investigated is to see whether different kinds of eyes elicit different responses: Do sad or kind eyes prime generous behavior? Do angry or demanding expressions diminish generosity but prime punishment? Other primes can do this. For instance, priming religious people with words associated with their faith increases the amount of prosociality (Shariff and Norenzayan 2007).

The fact that generosity can be primed in no way diminishes the possibility of other-regarding preferences. These preferences, motivated out of concern for the welfare of others, are sensitive to various cues in the same way as hunger and lust. The fact that a picture of food can make people feel hungrier, and a picture of a rotting carcass will reduce their appetitive drive, does not mean that food preferences do not exist. Framing effects change the threshold for other-regarding preferences; they do not disprove them. The altruist may receive emotional benefits, such as reduction in empathic distress and guilt as well as an increase in symhedonia and pride at having helped others, but these are not the *primary* reasons for helping. The emotional benefits (or utilities) are by-products, not the causes (Clavien 2012). The improvement (in the case of positive other-regard) of the welfare of others is the primary goal of the altruist, and any personal benefits are secondary.

Do Preferences and Utilities Exist at All?

Some have expressed skepticism concerning "preferences" (in the sense of "a transitive ordering of options") and utilities (in the sense of "a one-dimensional scale representing an ordering of preferences"), and ask whether they exist at all (Vlaev et al. 2011). First, in experiments on simultaneous binary choices, it turns out regularly that the choices made do not satisfy the transitivity requirement. In other words, a subject chooses *A* when having the choice between *A* and *B*, *B* when having the choice between *B* and *C*, and *C* when having the choice between *A* and *C*. This happens not only in humans, but also in nonhuman animals, and even in situations that are highly relevant for fitness (e.g., Shafir 1994). Although under certain conditions this may make perfect sense, it undermines preference/utility theory in a fundamental way. Of course, one might save the theory by arguing that preferences changed from one moment to the next. This, however, would make the whole concept of preferences next to useless, since it is their stability that might give them some predictive power. Others are not convinced by this argument (see Gintis 2007). Most hungry Inuit want to eat, and so prefer seal meat to no seal meat. That's all you need to understand the interactions described in the introduction. The fact that individuals do not want any seal meat when they are sick or satiated is immaterial.

A second and related point is the dependence of real-world choices on "irrelevant" (according to the axioms) alternatives. It regularly happens that the presence or absence of an option *X* that is never chosen strongly affects the choice among items *A* and *B* (e.g., Tversky and Simonson 1993). Again, this is not only observed regularly in humans, but also in various species of nonhuman animals (e.g., ants choosing a new nest site). A third observation challenges the notion, common among economists, that at least in the case of "simple" choices, the monetary value of the various options is a reasonable proxy for fitness. Some members in our group argued that simple real-world experiments challenge this notion. For example, in recent pilot experiments in a food market of a small village, where virtually all customers know each other well, various types of fruit (apples, oranges, bananas) of comparable qualities were offered at three neighboring fruit stands at three different prices (Franjo Weissing, pers. comm.). According to these group members, economic theory predicts that customers will buy each type of fruit at the stand with the lowest price. This, however, was not the case. Some customers (about 25%) did indeed follow this strategy, but others (about 20%) consistently went for the medium price, while some (about 5%) consistently went for the highest price. The rest of the customers followed another strategy, like always buying at the same stand, irrespective of the price. Others in the group argued that this experiment is not an argument against preferences in general, but only against a particular claim that all people care about is the price and the quality of the fruit. These results can be easily handled by social preferences ("I feel loyalty and affection for a certain vendor") or by incorporating signaling theory ("I am rich and buy expensive things to signal my wealth"). Here it was argued that these are plausible accounts of people's real motivation, not wacky ad hoc fixes.

How Does Social Learning Affect Decision Making?

Success of Humans: Culture

According to many measures, humans are an exceptional outlier among the world's species. As of 10,000 years ago, human hunter gatherers had occupied every terrestrial habitat except Antarctica and the remotest islands of Oceania. This global expansion occurred because humans were able to acquire a range of habitat-specific adaptations that greatly exceed that of other species. For example, chimpanzees, one of the most sophisticated tool-using mammals, depend only on a handful of tools. By contrast, human foragers make use of a vast range of tool types: snares, weapons, clothing, shelters, etc. For example, Inuit seal hunters depend on complex multipart harpoons, dog sleds, and clothing sufficient to prevent hypothermia standing motionless in the Arctic cold. They also depend on a virtual encyclopedia of knowledge of the habits of seals, the properties of sea ice, the cues of oncoming blizzards, and so on. While these adaptations are complex and functionally integrated, they are mainly cultural adaptations, not genetic ones.

As a result of cultural adaptations, humans have come to dominate the world's biota. Contemporary humans, for example, account for about eight times as much biomass as do all other terrestrial wild vertebrates combined, and even late Pleistocene foragers are estimated to have had a total biomass greater than that of any other large vertebrate (Hill et al. 2009). Human foragers also cooperated in larger groups than any other mammal, and this has led to social complexity on an unparalleled scale. Hammerstein and Boyd (this volume) emphasize that although humans may be smarter than other mammals, much evidence suggests that increased cognitive ability *alone* does not account for human adaptability. Instead, human success depends on our unique ability to learn from each other, a capacity that enables humans to accumulate information gradually across generations and to develop well-adapted tools, beliefs, and practices which no individual could invent on their own.

Models of gene-culture coevolution shed some light on how culture can improve human adaptability. If the only benefit of social learning is to avoid costs associated with individual learning, then these models show that there will be no increase in the ability of the population to adapt. Instead, social learning must increase the quality of information available to the population. This can happen in at least two ways:

1. Cultural learning allows individuals to learn selectively, using environmental cues when they provide clear guidance and learning from others when they do not.
2. Cultural learning allows the gradual accumulation of small improvements; if small improvements are cheaper than big ones, cultural learning can reduce the cost of learning.

Both predict that an adaptive evolved psychology will often cause individuals to acquire the behaviors they observe used by others, even though inferences based on environmental cues suggest that alternative behaviors would be better. When nonsocial cues are unreliable and environments do not change too quickly, an adaptive psychology will evolve in which most people ignore environmental cues and adopt behaviors that are common in the sample of the population they observe. They modify these behaviors rarely, or only at the margin, and as a result local adaptations evolve gradually often over many generations. Often, individuals have no idea why certain elements are included in a design, nor any notion of whether alternative designs would be better. We expect cultural learners to first acquire the local practices, before they experiment or modify them.

Cognitive Prerequisites for Cumulative Cultural Change

In addition to asking about the circumstances under which cumulative cultural evolution can evolve, we can ask what the cognitive prerequisites for this ability might be. One way to explore this question is by exploring the conditions under which cumulative cultural evolution can be obtained in simple models of cultural transmission.

Beppu and Griffiths (2009) analyzed a model of cultural transmission in which a sequence of Bayesian agents each receive data generated by the world, together with some kind of message from the previous agent. The question was what properties such a message must have in order for the agents to converge on the hypothesis about how the world worked that most closely resembled the truth. Beppu and Griffiths (2009) considered two kinds of messages. In the first case, analogous to observational learning, the message was simply more data—a set of predictions generated by the previous agent, based on that agent's conception of how the world worked. This was shown not to be sufficient to produce cumulative cultural evolution. In the second case, the agents received the full posterior distribution of the previous agent and could use this as their own prior distribution. In this case, the sequence of agents could accumulate knowledge over time and ultimately converge on the most accurate hypothesis. Beppu and Griffiths (2009) then showed that the predictions of these two accounts are borne out in laboratory experiments simulating cultural transmission with human learners. People were shown data sampled from a complex function relating two variables and had to infer the function. When they were provided with a mixture of samples from the function and the predictions produced by the previous learner, there was no improvement in performance across generations. When they saw the samples from the function together with a message that the previous learner had typed into a box describing their current beliefs about the nature of the function, they gained a more accurate estimate of the function over successive generations.

This combination of theoretical and empirical results suggests one kind of mechanism that can support cumulative cultural evolution: the ability to identify the beliefs that others entertain about the world. A variety of cognitive capacities could support such a mechanism, the most obvious example being language, which provides a way to communicate one's beliefs directly to another person. However, in general, a capacity for certain kinds of "theory of mind" inferences might be sufficient. Observing the intentional acts of another, knowing that they are intentional, or engaging in pedagogical interactions might be sufficient to allow a learner to infer the hypotheses and degrees of belief of another person, and then to use those in interpreting new data and transferring the resulting knowledge to subsequent generations.

Rationality?

Rational interpretations of social learning situations may be unique to humans. Yet some sense of selectivity in copying the behavior of others may also be found in nonhuman animals under certain conditions. Domestic dogs, *Canis familiaris*, for instance, have been shown to copy a peculiar action of another dog (pull down a handle with its paw to get food) when there was no obvious "reason" for the demonstrator dog not to use the "default" method (grab handle with mouth). When the demonstrator was carrying a ball in its mouth, however, they did not copy the paw action, possibly because they could infer that the "default" option was not possible for the demonstrator on this particular occasion (Range et al. 2007). A similar effect has been demonstrated in chimpanzees (Buttelmann et al. 2007). Despite these interesting studies, it has to be emphasized that many forms of social learning do not require any reasoning and rational inferences.

An illustrative example may be the behavior of humans as well as nonhuman animals when faced with a foraging decision: Which of two restaurants/food patches (*A* and *B*, close to each other) should be selected? Person 1 receives a private signal (e.g., good smell) which, with reliability greater than 50%, informs her that *A* is the better option. However, she also observes that two other individuals have decided to go to *B*. In this case, Person 1 is most likely to choose *B*. On the proximate level, such behavior is parsimonously explained by a socially triggered attention shift (enhancement) toward *B*, which may reflect a phylogenetically evolved and/or ontogenetically acquired predisposition to do what others are doing. Any higher forms of cognition are not necessary.

From a game-theoretical perspective, herding behavior can be interpreted as rational on the individual level, but suboptimal on the group level. Assuming that the two other individuals also received private signals that had the same reliability as Person 1, Person 1 may compute that her private signal is likely more incorrect than the signals of the other two individuals. Thus, her best response is to ignore her private information and also opt for *B* (Banerjee 1992; Bikhchandani et al. 1992). Now imagine that a fourth individual, Person 4, arrives. Regardless of his private signal, Person 4 should choose *B*. However, observing three individuals in *B* is not more informative than one person seeing two conspecifics in *B*; Person 4 knows that the third individual just followed the two individuals. So, even though it is in Person 4's interest to follow the herd, his decision is not more informed, and his choice will, in turn, not be informative for subsequent customers.

In the case described, the probability that an informational cascade, which results from people making inferences from the actions of others, will lead to the choice of the better restaurant can be brought arbitrarily close to 50% if the signal reliability converges to 50%. This is because only the first two customers chose based on their private signal, so the signals that the other customers received are discarded. However, if one could somehow force the first ten customers to choose based on their private signal, more information would actually be used. Therefore, the choice of the eleventh customer would be much more informed and the probability that the cascade ends up with the better option is highly increased. The problem, of course, is that it is not in the personal interest of the first ten customers to ignore social information, as that would only benefit others but not themselves. Thus the end result will be that there are fewer individual learners than would be optimal for the population. This mirrors findings from gene-culture coevolution (Rogers 1988; see also Hammerstein and Boyd, this volume).

The question of whether people are likely to follow the behavior of others more than they should as a consequence of rational deliberation brings to mind another phenomenon. Recent work has suggested that 3- to 5-year-old children are likely to engage in "over-imitation," copying the behavior of an adult model exactly even when common sense might suggest otherwise (Lyons et al. 2007). This seems like an instance where an instinct to copy might overwhelm deliberative learning mechanisms. However, recent work by Buchsbaum et al. (2011) suggests that over-imitation might just be part of a more general rational interpretation of the actions of others by children. Buchsbaum et al. (2011) showed that imitation of the exact sequences of actions that an adult took to activate a toy could be reduced by providing further examples of sequences that activated the toy, all of which shared a common subsequence. For example, the child might see that action *A*, followed by action *B*, and then action *C* activated the toy, but then see that the sequences *DBC* and *EBC* also activated the toy. Children would then produce the critical subsequence, *BC*, rather than just copying *ABC*, *DBC*, or *EBC*. This suggests that copying adult behavior is defeasible (i.e., it can be overwhelmed by another source of evidence about the correct actions), and might simply be a consequence of believing that the intentional actions of adults are carefully selected to achieve their goals, with violations of common sense being the consequence of privileged knowledge on the part of the adults. This interpretation of over-imitation is supported by a second experiment conducted by Buchsbaum et al. (2011), in which children behaved differently when adults took an explicitly pedagogical stance in showing the child how the toy worked.

Role of Demonstrators/Teachers: Cooperative Environment

Naturally, the focus in social learning is on the side of the learner. Yet, it might be important to consider also the individuals providing information, generally referred to as demonstrators (when interest in information is on the side of the observer) or teachers (when both individuals have an interest in the transmission of information). Notably, providing information may come with costs. For instance, in our foraging example, the amount of good food may be limited and thus when a greater number of individuals choose *B*, individuals may have to compete to get enough pieces, either through scramble competition (eat fast) or through interference competition (monopolize/defend the food from being taken).

The critical role of social learning for cumulated culture in humans suggests that it has been in the interest of knowledgeable individuals (demonstrators, teachers) to allow and encourage transmission of information. Such a scenario requires a social environment with high levels of cooperation, in which knowledgeable persons could benefit from information transfer to others, for example, via kin selection, reciprocity and/or by-product mutualism. Indeed, hunter-gather societies are characterized by a high level of cooperation, including foraging, group defense, warfare, and child rearing. This fits to the comparative data from nonhuman animals, where simple forms of teaching have evolved in cooperatively breeding species such as meerkats, *Suricata suricatta* (Thornton and McAuliffe 2006) and ants, *Temnothorax albipennis* (Franks and Richardson 2006). In either species, knowledgeable individuals may modify their behavior in a way that is costly to themselves but allows naïve individuals to learn a given task (Caro and Hauser 1992). Note that the underlying cognitive mechanisms can be relatively simple (Thornton and McAuliffe 2006).

In human societies, the issue of teaching quickly becomes complex. If an Inuit wants to teach a skill, such as seal hunting, to his son, it costs nothing if others from his group also learn by watching. This is a nonrival service. However, not everyone in the group is equally as important to the teacher. For instance, if prestige came with being a good hunter (as it often does) and this prestige led to better mating prospects, it would be in the teacher's interests for his son to learn to be a better hunter than the son of someone else in the same group.

Groups can be defined at many levels. A group can be the tribe in which one lives and, in times of war, it can include neighboring tribes that form a coalition. Within the tribe, however, one can belong to a group of hunters, a family, a group of same-aged peers, a group of drinking buddies, and so on. In periods of strong selection, such as civil war or ethnic unrest, friends and neighbors can become enemies as they resort to their traditional ethnic boundaries; a hunter might leave his tribe with his kin if there is friction within the group. The interesting implication is that groups are malleable; group identities shift easily under some circumstances but remain fixed in others, the result of which is that people should be sensitive to numerous group markers. Thus, the minimal group paradigm of social psychology, whereby people can be easily manipulated to form in-groups and out-groups on the basis of trivial cues such as painting preferences (Tajfel et al. 1971), can be explained by cultural group selection.

Conclusion

Developing a complete account of social cognition in humans and other species will require an understanding of the formal structure of social decision making and the mechanisms that support it. Game theory provides a starting point for characterizing the problem that agents face in making decisions in a social context. It needs, however, to be modified and extended to take into account constraints from cognitive mechanisms, and to address the challenges posed by problems such as coordinating behavior within groups. Behavioral experiments and cross-species comparisons provide hints as to the mechanisms that are relevant to social decision making, but much remains to be done before we have a full understanding of what these mechanisms are, how they work, and why they might exist. Evolutionary thinking can inform both of these approaches to understanding social decision making. It can guide us in developing a formal theory of how agents compete, cooperate, and learn from one another, and can provide a framework in which to understand the nature and origins of the proximate mechanisms that underlie these behaviors.

Bibliography

Note: Numbers in square brackets denote the chapter in which an entry is cited.

Abdellaoui, M., A. Baillon, L. Placido, and P. Wakker. 2011. The rich domain of uncertainty: Source functions and their experimental implementation. *Am. Econ. Rev.* **101**:695–723. [15]

Adams, C. D., and A. Dickinson. 1981. Instrumental responding following reinforcer devaluation. *Q. J. Exp. Psychol* **33B**:109–121. [9]

Adams, R. B., and R. E. Kleck. 2003. Perceived gaze direction and the processing of facial displays of emotion. *Psychol. Sci.* **14**:644–647. [16]

Ainslie, G. 1974. Impulse control in pigeons. *J. Exp. Anal. Behav.* **21**:485–489. [7]

———. 1975. Specious reward: A behavioral theory of impulsiveness and impulse control. *Psychol. Bull.* **82**:463–496. [7]

Ajzen, I. 1991. The theory of planned behavior. *Org. Behav. Hum. Decis. Process.* **50**:179–211. [16]

Aktipis, C. A. 2004. Know when to walk away: Contingent movement and the evolution of cooperation. *J. Theor. Biol.* **231**:249–260. [10]

Albert, R., H. Jeong, and A. Barabási. 2000. Error and attack tolerance of complex networks. *Nature* **406**:378–382. [8]

Alcock, J. 2001. Animal Behavior: An Evolutionary Approach, 7th edition. Sunderland, MA: Sinauer. [11]

Allender, E., S. Arora, M. Kearns, C. Moore, and A. Russell. 2003. A note on the representational incompatibility of function approximation and factored dynamics. In: Advances in Neural Information Processing Systems 15, p. 447. Cambridge, MA: MIT Press. [9]

Alloy, L. B., and N. Tabachnik. 1984. Assessment of covariation by humans and animals: The joint influence of prior expectations and current situational information. *Psychol. Rev.* **91**:112–149. [10]

Alon, U., M. G. Surette, N. Barkai, and S. Leibler. 1999. Robustness in bacterial chemotaxis. *Nature* **397**:168–171. [8]

Ancel, L. W., and W. Fontana. 2000. Plasticity, evolvability, and modularity in RNA. *J. Exp. Zool.* **288**:242–283. [8]

Anderson, B., and J. B. Moore. 1979. Optimal Filtering, vol. 1. Englewood Cliffs: Prentice Hall. [6]

Anderson, J., S. Burks, C. DeYoung, and A. Rustichini. 2011. Toward the integration of personality theory and decision theory in the explanation of economic behavior. Unpublished manuscript available at http://www.cogsci.umn.edu/colloquia/colloquia_S11/Aldo%20Rustichini.pdf. (accessed 5 June 2012). [14]

Anderson, J. R. 1990. The Adaptive Character of Thought. Hillsdale, NJ: Lawrence Erlbaum. [4, 7]

Anderson, R. B., M. E. Doherty, N. D. Berg, and J. C. Friedrich. 2005. Sample size and the detection of correlation—a signal detection account: Comment on Kareev (2000) and Juslin and Olsson (2005). *Psychol. Rev.* **112**:268–279. [10]

Andreoni, J. 1990. Impure altruism and donations to public goods: A theory of warm-glow giving. *Econ. J.* **100**:464–477. [17, 18]

Armitage, C. J., and M. Conner. 2001. Efficacy of the theory of planned behaviour: A meta-analytic review. *Br. J. Soc. Psychol.* **40**:471–499. [16]

Arnold, T. W. 1992. The adaptive significance of eggshell removal by nesting birds: Testing the egg-capping hypothesis. *Condor* **94**:547–548. [2]

Arrow, K. J. 1950. A difficulty in the concept of social welfare. *J. Polit. Econ.* **58**:328–346. [7]

Arthur, L. B., ed. 1999. Religion, Dress and the Body. Oxford: Berg. [15]

Ashby, F. G., A. M. Isen, and U. Turken. 1999. A neuropsychological theory of positive affect and its influence on cognition. *Psychol. Rev.* **106**:529–550. [14]

Aston-Jones, G., and J. D. Cohen. 2005. An integrative theory of locus coeruleus-norepinephrine function: Adaptive gain and optimal performance. *Ann. Rev. Neurosci.* **28**:403–450. [6]

Auclair, Y., B. Konig, and A. Lindholm. 2012. A selfish genetic element influencing longevity correlates with reactive behavioural traits in female house mice (*Mus domesticus*). *Proc. R. Soc. B*, in preparation. [15]

Auer, P. 2003. Using confidence bounds for exploitation-exploration trade-offs. *J. Mach. Learn. Res.* **3**:397–422. [9, 12]

Aumann, R. J., and S. Sorin. 1989. Cooperation and bounded recall. *Games Econ. Behav.* **1**:5–39. [10, 14]

Aw, J. M., R. I. Holbrook, T. Burt de Perera, and A. Kacelnik. 2009. State-dependent valuation learning in fish: Banded tetras prefer stimuli associated with greater past deprivation. *Behav. Proc.* **81**:333–336. [7]

Aw, J. M., M. Vasconcelos, and A. Kacelnik. 2011. How costs affect preferences: Experiments on state dependence, hedonic state and within-trial contrast in starlings. *Anim. Behav.* **81**:1117–1128. [7]

Axelrod, R. 1984. The Evolution of Cooperation. New York: Basic Books. [12, 15]

———. 2006. The Evolution of Cooperation: Revised Edition. New York: Basic Books. [8]

Axelrod, R., and W. D. Hamilton. 1981. The evolution of cooperation. *Science* **211**:1390–1396. [9]

Axsom, D., S. Yates, and S. Chaiken. 1987. Audience response as a heuristic cue in persuasion. *J. Pers. Soc. Psychol.* **53**:30–40. [7]

Axtell, R. L., J. M. Epstein, J. S. Dean, et al. 2002. Population growth and collapse in a multiagent model of the Kayenta Anasazi in Long House Valley. *PNAS* **99**:7275–7279. [15]

Ay, N., J. C. Flack, and D. C. Krakauer. 2007. Robustness and complexity co-constructed in multimodal signaling networks. *Phil. Trans. R. Soc. B* **362**:441–447. [8]

Ay, N., and D. C. Krakauer. 2007. Geometric robustness theory and biological networks. *Theory Biosci.* **125**:93–121. [8]

Ay, N., and D. Polani. 2008. Information flows in causal networks. *Advances in Complex Systems* **11**:17–41. [8]

Azar, O. 2004. What sustains social norms and how they evolve? The case of tipping. *J. Econ. Behav. Org.* **54**:49–64. [15]

Backus, J. W. 1978. Can programming be liberated from the von Neumann style? A functional style and its algebra of programs (ACM Turing Award Lectures 1977). *Commun. ACM* **21**. [3]

Bacon, F. 1620/1905. Novum Organum, trans. R. L. Ellis and J. Spedding, ed. J. M. Robertson. London: George Routledge and Sons. [10]

Balikci, A. 1970. The Netsilik Eskimo. Garden City, NY: Natural History Press. [20]

Ball, S. B., M. H. Bazerman, and J. S. Carol. 1991. An evaluation of learning in the bilateral winner's curse. *Org. Behav. Hum. Decis. Process.* **48**:1–22. [19]

Balleine, B. W., C. Paredes-Olay, and A. Dickinson. 2005. Effects of outcome devaluation on the performance of a heterogeneous instrumental chain. *Int. J. Comp. Psychol.* **18**:257–272. [4]

Balsam, P., and C. R. Gallistel. 2009. Temporal maps and informativeness in associative learning. *Trends Neurosci.* **32**:73–78. [7]

Bandura, A., D. Ross, and S. A. Ross. 1961. Transmission of aggression through imitation of aggressive models. *J. Abnormal Soc. Psych.* **63**:575–582. [18]

Banerjee, A. 1992. A simple model of herd behavior. *Q. J. Econ.* **107**:797–817. [20]

Barabási, A., and R. Albert. 1999. Emergence of scaling in random networks. *Science* **286**:509. [9]

Barefoot, C., K. W. Maynard, J. C. Beckham, et al. 1998. Trust, health and longevity. *J. Behav. Med.* **6**: [15]

Bargh, J. A. 1982. Attention and automaticity in the processing of self-relevant information. *J. Pers. Soc. Psychol.* **43**:425–436. [16]

Barkan, C. P. L. 1990. A field test of risk-sensitive foraging in black-capped chickadees (*Parus atricapillus*). *Ecol. Lett.* **71**:391–400. [1]

Barraclough, D. J., M. L. Conroy, and D. Lee. 2004. Prefrontal cortex and decision making in a mixed-strategy game. *Nat. Neurosci.* **7**:404–410. [9]

Barrett, H. C. 2005. Enzymatic computation and cognitive modularity. *Mind Lang.* **20**:259–287. [11]

Barrett, H. C., L. Cosmides, and J. Tooby. 2007. The hominid entry into the cognitive niche. In: Evolution of Mind, Fundamental Questions and Controversies, ed. S. Gangestad and J. Simpson, pp. 241–248. New York: Guilford Press. [19]

Barrett, H. C., and R. Kurzban. 2006. Modularity in cognition: Framing the debate. *Psychol. Rev.* **113**:628–647. [11]

———. 2012. What are the functions of system 2 modules? A reply to Chiappe and Gardner. *Theory Psychol.*, in press. [11]

Barrett, H. C., P. M. Todd, G. F. Miller, and P. W. Blythe. 2005. Accurate judgments of intention from motion cues alone: A cross-cultural study. *Evol. Human Behav.* **26**:313–331. [5]

Bartal, I., J. Decety, and P. Mason. 2011. Empathy and pro-social behavior in rats. *Science* **224**:1427–1430. [18]

Bateson, M., S. Desire, S. E. Gartside, and G. A. Wright. 2011. Agitated honeybees exhibit pessimistic cognitive biases. *Curr. Biol.* **21**:1070–1073. [14]

Bateson, M., S. D. Healy, and T. A. Hurly. 2003. Context-dependent foraging decisions in rufous hummingbirds. *Proc. Biol. Sci.* **270**:1271–1276. [7]

Bateson, M., and A. Kacelnik. 1996. Rate currencies and the foraging starling: The fallacy of the averages revisited. *Behav. Ecol.* **7**:341. [6]

Bateson, M., D. Nettle, and G. Roberts. 2006. Cues of being watched enhance cooperation in a real-world setting. *Biol. Lett.* **2**:412–414. [20]

Bateson, P., D. Barker, T. Clutton-Brock, et al. 2004. Developmental plasticity and human health. *Nature* **430**:419–421. [15]

Batson, C. D. 1991. The Altruism Question: Toward a Social Psychological Answer. Hillsdale, NJ: Erlbaum [19]

Batson, C. D., B. D. Duncan, P. Ackerman, T. Buckley, and K. Birch. 1981. Is empathic emotion a source of altruistic motivation. *J. Pers. Soc. Psychol.* **40**:290–302. [18]

Bays, P. M., and M. Husain. 2008. Dynamic shifts of limited working memory resources in human vision. *Science* **321**:851–854. [9]

Bechara, A., and H. Damasio. 2000. Emotion, decision making and the orbitofrontal cortex. *Cereb. Cortex* **10**:295–307. [8]

Becker, G. M., M. H. Degroot, and J. Marschak. 1964. Measuring utility by a single-response sequential method. *Behav. Sci.* **9**:226–232. [14]

Bednekoff, P. A., and S. L. Lima. 1998. Randomness, chaos and confusion in the study of antipredator vigilance. *Trends Ecol. Evol.* **13**:284–287. [15]

Behne, T., M. Carpenter, J. Call, and M. Tomasello. 2005. Unwilling or unable? Infants' understanding of others' intentions. *Dev. Psychol.* **41**:328–337. [17]

Behrens, T. E. J., M. W. Woolrich, M. E. Walton, and M. F. S. Rushworth. 2007. Learning the value of information in an uncertain world. *Nat. Neurosci.* **10**:1214–1221. [9]

Beletsky, L. 1996. The Red-Winged Blackbird: The Biology of a Strongly Polygynous Songbird. London: Academic Press. [6]

Bell, A. M. 2005. Behavioral differences between individuals and two populations of stickleback (*Gasterosteus aculeatus*). *J. Evol. Biol.* **18**:464–473. [13, 15]

Bell, A. M., S. J. Hankison, and K. L. Laskowski. 2009. The repeatability of behaviour: A meta-analysis. *Anim. Behav.* **77**:771–783. [13, 15]

Bell, A. M., and A. Sih. 2007. Exposure to predation generates personality in three-spined sticklebacks (*Gasterosteus aculeatus*). *Ecol. Lett.* **10**:828–834. [15]

Bellemare, C., and S. Kröger. 2007. On representative social capital. *Eur. Econ. Rev.* **51**:183–202. [15]

Bendesky, A., M. Tsunozaki, M. V. Rockman, L. Kruglyak, and C. I. Bargmann. 2011. Catecholamine receptor polymorphisms affect decision-making in *C. elegans*. *Nature* **472**:313–318. [6]

Benedict, R. 1934. Patterns of Culture. New York: Houghton Mifflin. [8]

Bennett, S. M. 2002. Preference reversal and the estimation of indifference points using a fast-adjusting-delay procedure with rats. Dissertation, Dept. of Psychology, Univ. of Florida, Gainesville. [7]

Bentham, J. 1789. Introduction to the Principles of Morals and Legislation. Oxford: Clarendon Press. [7]

Benzion, U., A. Rapoport, and J. Yagil. 1989. Discount rates inferred from decisions: An experimental study. *Manag. Sci.* **35**:270–284. [7]

Beppu, A., and T. L. Griffiths. 2009. Iterated learning and the cultural ratchet. Proc. of the 31st Ann. Conf. of the Cognitive Science Society. http://cocosci.berkeley.edu/tom/papers/ratchet1.pdf. (accessed 27 Sept. 2011). [20]

Berg, N., and G. Gigerenzer. 2010. As-if behavioral economics: Neoclassical economics in disguise? *Hist. Econ. Ideas* **18**:133–165. [5]

Bergmann, G., and K. O. Donner. 1964. An analysis of the spring migration of the common scoter and the long-tailed duck in southern Finland. *Acta Zool. Fenn.* **105**:1–60. [12]

Bernardes, A. T., D. Stauffer, and J. Kertesz. 2002. Election results and the Sznajd model on Barabasi network. *Eur. Phys. J. B* **25**:123–127. [15]

Bernheim, B. D., and R. Thomadsen. 2005. Memory and anticipation. *Econ. J.* **115**:271–304. [14]

Bernoulli, D. 1738/1954. Exposition of a new theory on the measurement of risk. *Econometrica* **22**:23–36. [7]

Berridge, K. C., and T. E. Robinson. 1998. What is the role of dopamine in reward: Hedonic impact, reward learning, or incentive salience? *Brain Res. Rev.* **28**:309–369. [9]

Berry, D. A., and B. Fristedt. 1985. Bandit Problems: Sequential Allocation of Experiments. Heidelberg: Springer. [9]

Bikhchandani, S., D. Hirshleifer, and I. Welch. 1992. A theory of fads, fashion, custom, and cultural change as informational cascades. *J. Polit. Econ.* **100**:992–1026. [20]

Biro, P. A., and J. A. Stamps. 2010. Do consistent individual differences in metabolic rate promote consistent individual differences in behavior? *Trends Ecol. Evol.* **25**:653–659. [13]

Bisley, J. W., and M. E. Goldberg. 2010. Attention, intention, and priority in the parietal lobe. *Ann. Rev. Neurosci.* **33**:1–21. [6]

Blake, P. R., and K. McAuliffe. 2011. "I had so much it didn't seem fair": Eight-year-olds reject two forms of inequity. *Cognition* **120**:215–224. [18]

Blanco, M., D. Engelmann, and H. Normann. 2011. A within-subject analysis of other-regarding preferences. *Games Econ. Behav.* **72**:321–338. [15, 20]

Blavatskyy, P. R. 2011. A model of probabilistic choice satisfying first-order stochastic dominance. *Manag. Sci.* **57**:542–548. [14]

Blavatskyy, P. R., and G. Pogrebna. 2010. Models of stochastic choice and decision theories: Why both are important for analyzing decisions. *J. Appl. Econom.* **25**:963–986. [14]

Bodenhausen, G. V. 1990. Stereotypes as judgmental heuristics: Evidence of circadian variations in discrimination. *Psychol. Sci.* **1**:319–322. [16]

Bodenhausen, G. V., and K. Hugenberg. 2009. Attention, perception, and social cognition. In: Social Cognition: The Basis of Human Interaction, ed. F. Strack and J. Förster, pp. 1–22. Philadelphia: Psychology Press. [16]

Bodenhausen, G. V., and C. N. Macrae. 1998. Stereotype activation and inhibition. In: Stereotype Activation And Inhibition: Advances in Social Cognition, ed. R. S. Wyer, Jr., vol. 11, pp. 1–52. Hillsdale, NJ: Erlbaum. [16]

Boehm, C., and J. C. Flack. 2010. The emergence of simple and complex power structures through social niche construction. In: The Social Psychology of Power, ed. A. Guinote and T. K. Vescio, pp. 46–86. New York: Guilford Press. [8]

Boesch, C. 1992. New elements of a theory of mind in wild chimpanzees. *Behav. Brain Sci.* **15**:149–150. [18]

Bogacz, R., E. Brown, J. Moehlis, P. Holmes, and J. D. Cohen. 2006. The physics of optimal decision making: A formal analysis of models of performance in two-alternative forced choice tasks. *Psychol. Rev.* **113**:700–765. [7]

Bolhuis, J. J., and E. M. Macphail. 2001. A critique of the neuroecology of learning and memory. *Trends Cogn. Sci.* **4**:426–433. [1]

Bolles, R. C. 1970. Species-specific defense reactions and avoidance learning. *Psychol. Rev.* **77**:32–48. [9, 12]

Bolton, G. E., and A. Ockenfels. 2000. ERC: A theory of equity, reciprocity and competition. *Am. Econ. Rev.* **90**:166–193. [15]

Bolton, G. E., and R. Zwick. 1995. Anonymity versus punishment in ultimatum bargaining. *Games Econ. Behav.* **10**:95–121. [18]

Borghans, L., A. L. Duckworth, J. J. Heckman, and B. ter Weel. 2008. The economics and psychology of personality traits. *J. Human Res.* **43**:972–1059. [14]

Bosch-Domenech, A., R. Nagel, and J. V. Sanchez-Andres. 2010. Prosocial capabilities in Alzheimer's patients. *J. Geront. Ser. B* **65**:119–128. [14]

Both, C., N. J. Dingemanse, P. J. Drent, and J. M. Tinbergen. 2005. Pairs of extreme avian personalities have highest reproductive success. *J. Anim. Ecol.* **74**:667–674. [15]

Bouchard, T. J., and J. C. Loehlin. 2001. Genes, evolution, and personality. *Behav. Genet.* **31**:243–273. [13]

Boureau, Y.-L., and P. Dayan. 2011. Opponency revisited: Competition and cooperation between dopamine and serotonin. *Neuropsychopharm.* **36**:74–97. [9]

Bouret, S., and S. J. Sara. 2005. Network reset: A simplified overarching theory of locus coeruleus noradrenaline function. *Trends Neurosci.* **28**:574–582. [9]

Bourke, A. F. G. 2011. Principles of Social Evolution. New York: Oxford Univ. Press. [8]

Bovet, D., and D. A. Washburn. 2003. Rhesus macaques (*Macaca mulatta*) categorize unknown conspecifics according to their dominance relations. *J. Comp. Psychol.* **117**:400–405. [6]

Bowles, S. 2006. Group competition, reproductive leveling, and the evolution of human altruism. *Science* **314**:1569–1572. [19, 20]

Bowles, S., H. Gintis, and M. Osborne. 2001. Incentive-enhancing preferences: Personality, behavior, and earnings. *Am. Econ. Rev.* **91**:155–158. [14]

Box, S., C. Hale, and G. Andrews. 1988. Explaining fear of crime. *Br. J. Criminol.* **28**:340–356. [5]

Boyce, C. J., G. D. A. Brown, and S. C. Moore. 2010a. Money and happiness: Rank of income, not income, affects life satisfaction. *Psychol. Sci.* **21**:471–475. [14, 15]

Boyce, C. J., and A. M. Wood. 2011. Personality and the marginal utility of income: Personality interacts with increases in household income to determine life satisfaction. *J. Econ. Behav. Org.* **78**:183–191. [14]

Boyce, C. J., A. M. Wood, and G. D. A. Brown. 2010b. The dark side of conscientiousness: Conscientious people experience greater drops in life satisfaction following unemployment. *J. Res. Pers.* **44**:535–539. [14]

Boyd, R., and P. J. Richerson. 1985. Culture and the Evolutionary Process. Chicago: Univ. of Chicago Press. [15]

———. 1992. Punishment allows the evolution of cooperation (or anything else) in sizable groups. *Ethol. Sociobiol.* **13**:171–195. [18, 19]

———. 1995. Why does culture increase human adaptability. *Ethol. Sociobiol.* **16**:125–143. [19]

———. 2005. Solving the puzzle of human cooperation. In: Evolution and Culture, ed. S. Levinson, pp. 105–132. Cambridge MA: MIT Press. [18, 20]

Boyd, R., P. J. Richerson, and J. Henrich. 2011. The cultural niche: How social learning transformed human evolution. *PNAS* **108**:10,918–10,925. [19]

Brafman, R., and M. Tennenholtz. 2003. R-max: A general polynomial time algorithm for near-optimal reinforcement learning *J. Mach. Learn. Res.* **3**:213–231. [9]

Bräuer, J., J. Call, and M. Tomasello. 2006. Are apes really inequity averse? *Proc. Roy. Soc. B* **273**:3123–3128. [18]

Breland, K., and M. Breland. 1961. The misbehavior of organisms. *Am. Psychol.* **16**:681–684. [4, 9]

———. 1966. Animal Behavior. New York: Macmillan. [19]

Brereton, A. R. 1994. Return-benefit spite hypothesis: An explanation for sexual interference in stumptail macaques. *Primates* **35**:123–136. [18]

Brewer, M., and R. Kramer. 1986. Choice behavior in social dilemma: Effects of social identity, group size, and decision framing. *J. Pers. Soc. Psychol.* **71**:83–93. [11]

Brighton, H., and G. Gigerenzer. 2011. Towards competitive instead of biased testing of heuristics. *Top. Cogn. Sci.* **3**:197–205. [15]

Brighton, H., S. Kirby, and K. Smith. 2005. Language as an evolutionary system. *Phys. Life Rev.* **2**:177–226. [12]

Brockmann, H. J., A. Grafen, and R. Dawkins. 1979. Evolutionarily stable strategy in a digger wasp. *J. Theor. Biol.* **77**:473–496. [19]

Bröder, A. 2012. The quest for take-the-best: Insights and outlooks from experimental research. In: Ecological Rationality: Intelligence in the World, ed. P. M. Todd et al., pp. 216–240. New York: Oxford Univ. Press, in press. [15]

Brogan, W. L. 1985. Modern Control Theory. Englewood Cliffs, NJ: Prentice Hall. [6]

Bromberg-Martin, E. S., and O. Hikosaka. 2009. Midbrain dopamine neurons signal preference for advance information about upcoming rewards. *Neuron* **63**:119–126. [6]

Brooks, R. A. 1991. New approaches to robotics. *Science* **253**:1227–1232. [4]

Broome, J. 1991. Weighing Goods. Oxford: Blackwell. [4]

Brosnan, S. F., and F. B. M. de Waal. 2003. Monkeys reject unequal pay. *Nature* **425**:297–299. [15, 18]

Brown, A. E. 1879. Grief in the chimpanzee. *Am. Natural.* **13**:173–175. [18]

Brown, G. D. A., J. Gardner, A. J. Oswald, and J. Qian. 2008. Does wage rank affect employees' well-being? *Indust. Rel.* **47**:355–389. [14]

Browne, A., and J. Pilkington. 1994. Variable binding in a neural network using a distributed representation. In: European Symposium on Artificial Neural Networks (April 20–22), pp. 189–204. Brussels: IEE Colloquium. [3]

Browne, A., and R. Sun. 2001. Connectionist inference models. *Neural Netw.* **14**:1331–1355. [3]

Brownell, C., G. B. Ramani, and S. Zerwas. 2006. Becoming a social partner with peers: Cooperation and social understanding in one- to two-year-olds. *Child Devel.* **77**:803–821. [17]

Brownell, C. A., M. Svetlova, and S. Nichols. 2009. To share or not to share: When do toddlers respond to another's needs? *Infancy* **1**: [18]

Bshary, R., and A. S. Grutter. 2005. Punishment and partner switching cause cooperative behaviour in a cleaning mutualism. *Biol. Lett.* **1**:396–399. [18]

Bshary, R., and R. Noe. 2003. Biological markets: The ubiquitous influence of partner choice on the dynamics of cleaner fish-client reef fish interactions. In: Genetic and Cultural Evolution of Cooperation, ed. P. Hammerstein, pp. 167–184, Dahlem Workshop Reports, vol. 90, J. Lupp, series ed. Cambridge, MA: MIT Press. [6]

Buchsbaum, D., A. Gopnik, T. L. Griffiths, and P. Shafto. 2011. Children's imitation of causal action sequences is influenced by statistical and pedagogical evidence. *Cognition* **120**:331–340. [20]

Budzynski, C. A., F. C. Dyer, and V. P. Bingman. 2000. Partial experience with the arc of the sun is sufficient for all-day sun compass orientation in homing pigeons, *Columba livia*. *J. Exp. Biol.* **203**:2341–2348. [3]

Bugnyar, T. 2011. Knower-guesser differentiation in ravens: Others' viewpoints matter. *Proc. Roy. Soc. B* **278**:634–640. [20]

Bugnyar, T., and B. Heinrich. 2005. Food-storing ravens differentiate between knowlegeable and ignorant competitors. *Proc. Roy. Soc. B* **272**:1641–1646. [20]

Burkart, J. M., E. Fehr, C. Efferson, and C. P. van Schaik. 2007. Other-regarding preferences in a non-human primate: Common marmosets provision food altruistically. *PNAS* **104**:19,762–19,766. [18, 20]

Burks, S. V., J. P. Carpenter, L. Goette, and A. Rustichini. 2009. Cognitive skills affect economic preferences, strategic behavior, and job attachment. *PNAS* **106**:7745–7750. [14]

Burnham, T., and B. Hare. 2007. Engineering cooperation: Does involuntary neural activation increase public goods contributions? *Human Nature* **18**:88–108. [20]

Burnham, T., and J. Phelan, eds. 2000. Mean Genes. Cambridge, MA: Perseus. [11]

Busemeyer, J. R., and J. T. Townsend. 1993. Decision field theory: A dynamic-cognitive approach to decision making in an uncertain environment. *Psychol. Rev.* **100**:432–459. [15]

Bush, R., and F. Mosteller. 1955. Stochastic Models of Learning. New York: Wiley. [2, 19]

Buss, D. M. 2009. How can evolutionary psychology successfully explain personality and individual differences? *Perspect. Psychol. Sci.* **4**:359–366. [14]

Buss, D. M., and P. Hawley. 2010. The Evolution of Personality and Individual Differences. Oxford: Oxford Univ. Press. [13]

Bustamante, C. D., A. Fledel-Alon, S. Williamson, et al. 2005. Natural selection on protein coding genes in the human genome. *Nature* **437**:1153–1157. [15]

Butler, D., and G. Loomes. 2007. Imprecision as an account of the preference reversal phenomenon. *Am. Econ. Rev.* **97**:277–297. [15]

Buttelmann, D., M. Carpenter, J. Call, and M. Tomasello. 2007. Enculturated chimpanzees imitate rationally. *Dev. Sci.* **10**:31–38. [20]

Byrne, R. W., and A. Whiten, eds. 1988. Machiavellian Intelligence. Social Expertise and the Evolution of Intellect in Monkeys, Apes, and Humans. New York: Oxford Univ. Press. [17, 20]

Calabretta, R., S. Nole, D. Parisi, and G. Wagner. 1998. Emergence of functional modularity in robots. In: Artificial Life VI, ed. H. Kitano and C. Taylor, pp. 497–504. Boston: MIT Press. [8]

Call, J., B. Hare, M. Carpenter, and M. Tomasello. 2004. "Unwilling" versus "unable": Chimpanzees' understanding of human intentional action. *Dev. Sci.* **7**:488–498. [17]

Call, J., and M. Tomasello. 2008. Does the chimpanzee have a theory of mind? 30 years later. *Trends Cogn. Sci.* **12**:187–192. [4, 18, 20]

Camerer, C. 2003. Behavioral Game Theory. Princeton: Princeton Univ. Press. [11, 15, 18, 20]

———. 1995. Individual decision making. In: The Handbook of Experimental Economics, vol. 1, ed. J. H. Kagel and A. E. Roth, pp. 587–703. Princeton: Princeton Univ. Press. [20]

Camerer, C., T. H. Ho, and J. K. Chong. 2004. A cognitive hierarchy model of games. *Q. J. Econ.* **119**:861–898. [20]

Camerer, C., and R. H. Thaler. 1995. Anomalies: Ultimatums, Dictators and Manners. *J. Econ. Persp.* **9**:209–219. [18]

Campbell, D. T. 1958. Common fate, similarity, and other indices of the status of aggregates of persons as social entities. *Behav. Sci.* **3**:14–25. [12]

———. 1966. Pattern-matching as an essential in distal knowing. In: The Psychology of Egon Brunswik, ed. K. R. Hammond. New York: Holt Rinehart and Winston. [12]

———. 1974. Evolutionary epistemology. In: The Philosophy of Karl Popper, ed. P. A. Schlipp, pp. 413–463. LaSalle: Open Court. [5]

Campbell, D. T., and D. W. Fiske. 1959. Convergent and discriminant validation by the multi-trait multi-method matrix. *Psychol. Bull.* **56**:81–105. [12]

Capen, E. C., R. V. Clapp, and W. M. Campbell. 1971. Competitive bidding in high risk situations. *J. Petrol. Tech.* **23**:641–653. [19]

Carere, C., D. Caramaschi, and T. W. Fawcett. 2010. Covariation between personalities and individual differences in coping with stress: Converging evidence and hypotheses. *Curr. Zool.* **56**:728–740. [13]

Caro, T. M., and M. D. Hauser. 1992. Is there teaching in nonhuman animals? *Q. Rev. Biol.* **67**:151–174. [20]

Carpenter, G., and S. Grossberg. 1988. The ART of adaptive pattern recognition by a self-organizing neural network. *Computer* **21**:77–88. [9]

Carpenter, M., N. Akhtar, and M. Tomasello. 1998. Fourteen- through 18-month-old infants differentially imitate intentional and accidental actions. *Infant Behav. Devel.* **21**:315–330. [17]

Carpenter, M., M. Tomasello, and T. Striano. 2005. Role reversal imitation and language in typically-developing infants and children with autism. *Infancy* **8**:253–278. [17]

Case, R., D. M. Kurland, and J. Goldberg. 1982. Operational efficiency and the growth of short-term memory span. *J. Exp. Child Psychol.* **33**:386–404. [10]

Caspi, A., B. W. Roberts, and R. L. Shiner. 2005. Personality development: Stability and change. *Ann. Rev. Psychol.* **56**:453–484. [13]

Chaiken, S. 1980. Heuristic versus systematic information processing and the use of source versus message cues in persuasion. *J. Pers. Soc. Psychol.* **39**:752–766. [7]

Chamley, C. P. 2003. Rational Herds: Economic Models of Social Learning. Cambridge: Cambridge Univ. Press. [15]

Charness, G., and M. Dufwenberg. 2006. Promises and partnership. *Econometrica* **74**:1579–1601. [20]

Charness, G., and M. Rabin. 2002. Understanding social preferences with simple tests. *Q. J. Econ.* **117**:817–869. [15]

Charnov, E. L. 1976. Optimal foraging: The marginal value theorem. *Theor. Popul. Biol.* **9**:129–136. [1, 2, 6, 7]

Chase, W. G., and H. A. Simon. 1973. Perception in chess. *Cogn. Psychol.* **4**:55–81. [10]

Chater, N. 2009. Rational and mechanistic perspectives on reinforcement learning. *Cognition* **113**:350–364. [4]

Chater, N., and C. D. Manning. 2006. Probabilistic models of language processing and acquisition. *Trends Cogn. Sci.* **10**:335–344. [15]

Chater, N., J. B. Tenenbaum, and A. Yuille. 2006. Probabilistic models of cognition: Conceptual foundations. *Trends Cogn. Sci.* **10**:287–291. [4, 7]

Chen, Y. X., G. Iyer, and A. Pazgal. 2010. Limited memory, categorization, and competition. *Mark. Sci.* **29**:650–670. [14]

Chiappe, D. 2000. Metaphor, modularity, and the evolution of conceptual integration. *Metaphor Symbol* **15**:137–158. [11]

Chiappe, R., and R. Gardner. 2011. The modularity debate in evolutionary psychology. *Theory Psychol.* doi: 0959354311398703. [11]

Chittka, L., and J. Niven. 2009. Are bigger brains better? *Curr. Biol.* **19**:R995–R1008. [12]

Chomsky, N. 1981. Principles and Parameters in Syntactic Theory. London: Longman Publ. Group. [8]

Christakis, N. A., and J. H. Fowler. 2007. The spread of obesity in a large social network over 32 years. *New Eng. J. Med.* **357**:370–379. [15]

———. 2008. The collective dynamics of smoking in a large social network. *New Eng. J. Med.* **358**:2249–2258. [15]

Chung, S. H., and R. J. Herrnstein. 1967. Choice and delay of reinforcement. *J. Exp. Anal. Behav.* **10**:67–74. [7]

Clark, A. B., and T. J. Ehlinger. 1987. Pattern and adaptation in individual behavioral differences. In: Perspectives in Ethology, ed. P. P. G. Bateson and P. H. Klopfer, pp. 1–47. New York: Plenum. [13]

Clark, A. E., N. Kristensen, and N. Westergard-Nielsen. 2008. Economic satisfaction and income rank in small neighbourhoods. *J. Europ. Econ. Assn.* **7**:519–527. [14]

Clark, A. E., and A. J. Oswald. 1998. Comparison-concave utility and following behaviour in social and economic settings. *J. Pub. Econ.* **70**:133–155. [15]

Clark, C. W. 1994. Antipredator behavior and the asset-protection principle. *Behav. Ecol.* **5**:159–170. [15]

Clavien, C. 2012. Altruistic emotional motivation: An argument in favour of psychological altruism. In: Philosophy of Behavioral Biology, ed. K. Plaisance and T. Reydon, pp. 275–296, Boston Studies in Philosophy of Science. New York: Springer. [18, 20]

Clayton, N. S., T. J. Bussey, and A. Dickinson. 2003. Can animals recall the past and plan for the future? *Nat. Rev. Neurosci.* **4**:685–691. [3]

Clayton, N. S., and A. Dickinson. 1998. Episodic-like memory during cache recovery by scrub jays. *Nature* **395**:272–274. [3]

———. 1999a. Memory for the content of caches by scrub jays (*Aphelocoma coerulescens*). *J. Exp. Psychol. Anim. Behav. Proc.* **25**:82–91. [3]

———. 1999b. Scrub jays *(Aphelocoma coerulescens*) remember the relative time of caching as well as the location and content of their caches. *J. Comp. Physiol.* **113**:403–416. [3]

Clayton, N. S., N. Emery, and A. Dickinson. 2006. The rationality of animal memory: Complex caching strategies of western scrub jays. In: Rational Animals?, ed. M. Nuuds and S. Hurley, pp. 197–216. Oxford: Oxford Univ. Press. [3]

Clayton, N. S., D. P. Griffiths, N. Emery, and A. Dickinson. 2001a. Elements of episodic-like memory in animals. *Phil. Trans. R. Soc. B* **356**:1483–1491. [1]

Clayton, N. S., J. Russell, and A. Dickinson. 2009. Are animals stuck in time or are they chronesthetic creatures? *Top. Cogn. Sci.* **1**:59–72. [3]

Clayton, N. S., K. Yu, and A. Dickinson. 2001b. Scrub jays (*Aphelocoma coerulescens*) can form integrated memory for multiple features of caching episodes. *J. Exp. Psychol. Anim. Behav. Proc.* **27**:17–29. [3]

Clutton-Brock, T. H., and G. A. Parker. 1995. Punishment in animal societies. *Nature* **272**:209–216. [18, 19]

Cohen, D. 1966. Optimizing reproduction in randomly varying environments. *J. Theor. Biol.* **12**:119–129. [12]

Cohen, J. D. 2005. The vulcanization of the human brain: A neural perspective on interactions between cognition and emotion. *J. Econ. Persp.* **19**:3–24. [19]

Cohen, J. D., S. M. McClure, and A. J. Yu. 2007. Should I stay or should I go? How the human brain manages the trade-off between exploitation and exploration. *Phil. Trans. R. Soc. B* **362**:933. [6, 10]

Cohen, M., and A. Kohn. 2011. Measuring and interpreting neuronal correlations. *Nat. Neurosci.* **14**:811–819. [9]

Cohen, S., W. J. Doyle, R. B. Turner, C. M. Alper, and D. P. Skoner. 2003. Emotional style and susceptibility to the common cold. *Psychosom. Med.* **65**:652–657. [14]

Cohen-Cole, E., and J. M. Fletcher. 2008. Is obesity contagious? Social networks vs. environmental factors in the obesity epidemic. *J. Health Econ.* **27**:1382–1387. [15]

Collett, M. 2009. Spatial memories in insects. *Curr. Biol.* **19**:R1103–R1108. [3]

———. 2010. How desert ants use a visual landmark for guidance along a habitual route. *PNAS* **107**:11,638–11,643. [3]

Collett, T. S., and P. Graham. 2004. Animal navigation: Path integration, visual landmarks and cognitive maps. *Curr. Biol.* **14**:R475–R477. [7]

Cook, M., and S. Mineka. 1989. Observational conditioning of fear to fear-relevant versus fear-irrelevant stimuli in rhesus monkeys. *J. Abnorm. Psychol.* **98**:448–459. [19]

Cools, R., K. Nakamura, and N. D. Daw. 2011. Serotonin and dopamine: Unifying affective, activational, and decision functions. *Neuropsychopharm.* **36**:98–113. [9]

Cools, R., A. C. Roberts, and T. W. Robbins. 2008. Serotoninergic regulation of emotional and behavioural control processes. *Trends Cogn. Sci.* **12**:31–40. [9]

Cooter, R. D., and P. Rappoport. 1984. Were the ordinalists wrong about welfare economics? *J. Econ. Lit.* **22**:507–530. [4]

Correll, J., B. Park, C. M. Judd, and B. Wittenbrink. 2002. The police officer's dilemma: Using ethnicity to disambiguate potentially threatening individuals. *J. Pers. Soc. Psychol.* **83**:1314–1329. [16]

Cosmides, L., H. C. Barrett, and J. Tooby. 2010. Adaptive specializations, social exchange, and the evolution of human intelligence. *PNAS* **107**:9007–9014. [1, 11]

Cosmides, L., and J. Tooby. 1992. Cognitive adaptations for social exchange. In: The Adapted Mind: Evolutionary Psychology and the Generation of Culture, ed. J. Barkow et al., pp. 163–228. New York: Oxford Univ. Press. [11]

———. 1994. Beyond intuition and instinct blindness: Toward an evolutionarily rigorous cognitive science. *Cognition* **50**:41–77. [1]

———. 2008. Can a general deontic logic capture the facts of human moral reasoning? How the mind interprets social exchange rules and detects cheaters. In: Moral Psychology, ed. W. Sinnott-Armstrong, pp. 53–119. Cambridge, MA: MIT Press. [11]

Costa-Gomes, M., and V. Crawford. 2006. Cognition and behavior in two-person guessing games: An experimental study. *Am. Econ. Rev.* **96**:1737–1768. [15]

Cote, J., S. Fogarty, T. Brodin, K. Weinersmith, and A. Sih. 2011. Personality-dependent dispersal in the invasive mosquitofish: Group composition matters. *Proc. Roy. Soc. B* **278**:1670–1678. [15]

Couzin, I. D. 2009. Collective cognition in animal groups. *Trends Cogn. Sci.* **13**:36–43. [8]

Couzin, I. D., and J. Krause. 2003. Self-organisation and collective behaviour of vertebrates. *Adv. Study Behav.* **32**:1–67. [15]

Cowan, N. 2001. The magical number 4 in short-term memory: A reconsideration of mental storage capacity. *Behav. Brain Sci.* **24**:87–114. [10]

Cowen, T., and A. Glazer. 1996. More monitoring can induce less effort. *J. Econ. Behav. Org.* **30**:113–123. [10]

Crawford, V., and N. Iriberri. 2007. Level-k auctions: Can boundedly rational strategic thinking explain the winner's curse and overbidding in private-value auctions? *Econometrica* **75**:1721–1770. [15]

Cronin, K. A., K. K. E. Schroeder, E. S. Rothwell, J. B. Silk, and C. T. Snowdon. 2009. Cooperatively breeding cottontop tamarins (*Saguinus oedipus*) do not donate rewards to their long-term mates. *J. Comp. Psychol.* **123**:231–241. [18]

Cronin, K. A., K. K. E. Schroeder, and C. T. Snowdon. 2010. Prosocial behavior emerges independent of reciprocity in cottontop tamarins. *Proc. R. Soc. B* **277**:3845–3851. [18]

Cruse, H., and R. Wehner. 2011. No need for a cognitive map: Decentralized memory for insect navigation. *PloS Comp. Biol.* **7**:e1002009. [7]

Crusius, J., and T. Mussweiler. 2011. When people want what others have: The impulsive side of envious desire. *Emotion* **93**:1163–1184. [16]

Cuthill, I. C., and A. Kacelnik. 1990. Central place foraging: A re-appraisal of the "loading effect." *Anim. Behav.* **40**:1087–1101. [2]

Cuthill, I. C., A. Kacelnik, J. R. Krebs, P. Haccou, and Y. Iwasa. 1990. Starlings exploiting patches: The effect of recent experience on foraging decisions. *Anim. Behav.* **40**:625–640. [6]

D'Acremont, M., and P. Bossaerts. 2008. Neurobiological studies of risk assessment: a comparison of expected utility and mean-variance approaches. *Cogn. Affect. Behav. Neurosci.* **8**:363–374. [9]

Dall, S. R. X. 2004. Behavioural biology: Fortune favours bold and shy personalities. *Curr. Biol.* **14**:R470–R472. [14]

———. 2010. Managing risk: The perils of uncertainty. In: Evolutionary Behavioral Ecology, ed. D. F. Westneat and C. W. Fox, pp. 194–206. Oxford: Oxford Univ. Press. [15]

Dall, S. R. X., L.-A. Giraldeau, O. Olsson, J. M. McNamara, and D. W. Stephens. 2005. Information and its use by animals in evolutionary ecology. *Trends Ecol. Evol.* **20**:187–193. [15]

Dall, S. R. X., A. I. Houston, and J. M. McNamara. 2004. The behavioural ecology of personality: Consistent individual differences from an adaptive perspective. *Ecol. Lett.* **7**:734–739. [14, 15]

Dally, J. M., N. J. Emery, and N. S. Clayton. 2005. Cache protection strategies by western scrub-jays, *Aphelocoma californica*: Implications for social cognition. *Anim. Behav.* **70**:1251–1263. [3]

———. 2006. Food-caching western scrub-jays keep track of who was watching when. *Science* **312**:1662–1666. [3, 20]

Damasio, A. R. 1994. Descartes' Error: Emotion, Reason, and the Human Brain. New York: Putnam. [18]

Daniels, B. C., Y.-J. Chen, J. P. Sethna, R. N. Gutenkunst, and C. R. Myers. 2008. Sloppiness, robustness, and evolvability in systems biology. *Curr. Opin. Biotech.* **19**:389–395. [8]

Darwin, C. 1871. The Descent of Man, and Selection in Relation to Sex. London: John Murray. [18, 19]

———. 1872. The Expression of Emotion in Man and Animals, 1st ed. London: John Murray. [18]

Daston, L. 1995. Classical Probability in the Enlightenment. Princeton: Princeton Univ. Press. [1]

David, F. N. 1954. Tables of the Ordinates and Probabiliy Integral of the Distribution of the Correlation Coefficient in Small Samples Issued by the Biometrika Office University College, London. Cambridge: Cambridge Univ. Press. [10]

Davidson, E. H. 2010. Emerging properties of animal gene regulatory networks. *Nature* **468**:911–920. [8]

Daw, N. D., and K. Doya. 2006. The computational neurobiology of learning and reward. *Curr. Opin. Neurobiol.* **16**:199–204. [9]

Daw, N. D., S. J. Gershman, B. Seymour, P. Dayan, and R. J. Dolan. 2011. Model-based influences on humans' choices and striatal prediction errors. *Neuron* **69**:1204–1215. [9]

Daw, N. D., S. Kakade, and P. Dayan. 2002. Opponent interactions between serotonin and dopamine. *Neural Netw.* **6**:603–616. [9]

Daw, N. D., Y. Niv, and P. Dayan. 2005. Uncertainty-based competition between prefrontal and dorsolateral striatal systems for behavioral control. *Nat. Neurosci.* **8**:1704–1711. [9]

Dawkins, R. 1982. The Extended Phenotype. Oxford: W. H. Freeman. [8]

Dayan, P. 1994. Computational modelling. *Curr. Opin. Neurobiol.* **4**:212–217. [9]

———. 2008. The role of value systems in decision-making. In: Better than Conscious?: Decision Making, the Human Mind, and Implications For Institutions, ed. C. Engel and W. Singer, pp. 51–70, Strüngmann Forum Reports, vol. 1, J. Lupp, series ed. Cambridge, MA: MIT Press. [9]

Dayan, P., and L. Abbott. 2001. Theoretical Neuroscience: Computational and Mathematical Modeling of Neural Systems. Cambridge, MA: MIT Press. [4, 9, 12]

Dayan, P., and N. D. Daw. 2008. Decision theory, reinforcement learning, and the brain. *Cogn. Affect. Behav. Neurosci.* **8**:429–453. [9]

Dayan, P., S. Kakade, and P. R. Montague. 2000. Learning and selective attention. *Nat. Neurosci.* **3**:1218–1223. [9]

Dayan, P., Y. Niv, B. Seymour, and N. D. Daw. 2006. The misbehavior of value and the discipline of the will. *Neural Netw.* **19**:1153–1160. [4, 9]

Dayan, P., and T. Sejnowski. 1996. Exploration bonuses and dual control. *Mach. Learn.* **25**:5–22. [9]

Dayan, P., and A. J. Yu. 2006. Phasic norepinephrine: A neural interrupt signal for unexpected events. *Network* **17**:335–350. [9]

de Finetti, B. 1937/1980. La prévision: Ses lois logiques, ses sources subjectives. In: Studies in Subjective Probability, 2nd edition (translated and reprinted), ed. H. E. Kyburg and H. E. Smokler, pp. 53–118. New York: Robert Krieger. [15]

De Groot, A. D. 1965. Thought and Choice in Chess. The Hague: Mouton Publishers. [10]

de Jong, P. F., and E. A. Das-Small. 1995. Attention and intelligence: The validity of the Star Counting Test. *J. Educ. Psychol.* **87**:80–92. [10]

de Sousa, R. 2004. Emotions: What I know, what I'd like to think I know, and what I'd like to think. In: Thinking about Feeling: Contemporary Philosophers on Emotions ed. R. Solomon, pp. 61–75. Oxford: Oxford Univ. Press. [18]

———. 2008. Really, what else is there? Emotions, value and morality. *Crit. Q.* **50**:12–23. [18]

de Visser, J. A., J. Hermisson, G. P. Wagner, et al. 2003. Evolution and detection of genetic robustness. *Evolution* **57**:1959–1972. [8]

de Waal, F. B. M. 1982. Chimpanzee Politics. Power and Sex Among Apes. Baltimore: Johns Hopkins Univ. Press. [17]

———. 1993. Reconciliation among primates: A review of empirical evidence and unresolved issues. In: Primate Social Conflict, ed. W. A. Mason and S. Mendoza, pp. 111–144. Albany: SUNY Press. [8]

———. 1996. Good Natured. Cambridge, MA: Harvard Univ. Press. [18]

———. 2000. Primates: A natural heritage of conflict resolution. *Science* **289**:586–590. [8, 17]

de Waal, F. B. M., and M. L. Berger. 2000. Payment for labour in monkeys. *Nature* **404**:563. [20]

de Waal, F. B. M., and L. M. Luttrell. 1988. Mechanisms of social reciprocity in three primate species: Symmetrical relationship characteristics or cognition? *Ethol. Sociobiol.* **9**:101–118. [17]

de Waal, F. B. M., and A. van Roosmalen. 1979. Reconciliation and consolation among chimpanzees. *Behav. Ecol. Sociobiol.* **5**:55–66. [18, 20]

Deakin, J. F. W. 1983. Roles of brain serotonergic neurons in escape, avoidance and other behaviors. *J. Psychopharmacol.* **43**:563–577. [9]

Deakin, J. F. W., M. Aitken, T. Robbins, and B. J. Sahakian. 2004. Risk taking during decision-making in normal volunteers changes with age. *J. Int. Neuropsychol. Soc.* **10**:590–598. [14]

Deakin, J. F. W., and F. G. Graeff. 1991. 5-HT and mechanisms of defence. *J. Psychopharmacol.* **5**:305–316. [9]

Deaner, R. O., A. V. Khera, and M. L. Platt. 2005. Monkeys pay per view: Adaptive valuation of social images by rhesus macaques. *Curr. Biol.* **15**:543–548. [6]

Deaner, R. O., and M. L. Platt. 2003. Reflexive social attention in monkeys and humans. *Curr. Biol.* **13**:1609–1613. [6]

Dechêne, A., C. Stahl, J. Hansen, and M. Wänke. 2010. The truth about the truth: A meta-analytic review of the truth effect. *Pers. Soc. Psychol. Rev.* **14**:238–257. [7]

Dedeo, S., D. C. Krakauer, and J. C. Flack. 2010. Inductive game theory and the dynamics of animal conflict. *PLoS Comp. Biol.* **6**:e1000782. [8]

———. 2011. Evidence of strategic periodicities in collective conflict dynamics. **8**:1260–1273. [8]

Dehaene, S. 1997. The Number Sense. New York: Oxford Univ. Press. [4]

Delgado, M. R., R. H. Frank, and E. A. Phelps. 2005. Perceptions of moral character modulate the neural systems of reward during the trust game. *Nat. Neurosci.* **8**:1611–1618. [17]

DeNeve, K. M., and H. Cooper. 1998. The happy personality: A meta-analysis of 137 personality traits and subjective well-being. *Psychol. Bull.* **124**:197–229. [14]

DeScioli, P., and R. Kurzban. 2009. The alliance hypothesis for human friendship. *PLoS ONE* **4**:e5802. [11]

DeScioli, P., and B. J. Wilson. 2011. The territorial foundations of human property. *Evol. Human Behav.* **32**:297–304. [11]

DeWall, C. N., J. K. Maner, and D. A. Rouby. 2009. Social exclusion and early-stage interpersonal perception: Selective attention to signs of acceptance. *J. Pers. Soc. Psychol.* **96**:729–741. [16]

Diamond, D. M., M. Fleshner, N. Ingersoll, and G. Rose. 1996. Psychological stress impairs spatial working memory: Relevance to electrophysiological studies of hippocampal function. *Behav. Neurosci.* **110**:661–672. [10]

Dickinson, A. 1985. Actions and habits: The development of behavioural autonomy. *Phil. Trans. R. Soc. B* **308**:67–78. [9]

———. 1994. Instrumental conditioning. In: Animal Learning and Cognition, ed. N. Mackintosh, pp. 45–79. San Diego: Academic Press. [9]

Dickinson, A., and B. Balleine. 1994. Motivational control of goal-directed action. *Anim. Learn. Behav.* **22**:1–18. [9]

———. 2002. The role of learning in motivation. In: Stevens' Handbook of Experimental Psychology, ed. C. Gallistel, vol. 3, pp. 497–533. New York: Wiley. [9]

Dickinson, A., J. Smith, and J. Mirenowicz. 2000. Dissociation of Pavlovian and instrumental incentive learning under dopamine antagonists. *Behav. Neurosci.* **114**:468–483. [9]

Dickinson, A., A. Watt, and W. J. H. Griffiths. 1992. Free-operant acquisition with delayed reinforcement. *Q. J. Exp. Psychol. B* **45**:241–258. [2]

Dickinson, J., and F. Dyer. 1996. How insects learn about the sun's course: Alternative modeling approaches. In: From Animals to Animats, ed. S. Wilson and J. A. Meyer, vol. 4, pp. 193–203. Cambridge, MA: MIT Press. [3]

Diener, E., and M. Y. Chan. 2011. Happy people live longer: Subjective well-being contributes to health and longevity. *Appl. Psychol. Health Well-Being* **3**:1–43. [14]

Dingemanse, N. J., K. M. Bouwman, M. van de Pol, et al. 2012. Variation in personality and behavioral plasticity in four populations of the great tit *Parus major*. *J. Anim. Ecol.* **81**:116–126. [13]

Dingemanse, N. J., N. A. Dochtermann, and J. Wright. 2010a. A method for exploring the structure of behavioural syndromes to allow formal comparison within and between datasets. *Anim. Behav.* **79**:439–450. [15]

Dingemanse, N. J., A. J. N. Kazem, D. Réale, and J. Wright. 2010b. Behavioural reaction norms: Where animal personality meets individual plasticity. *Trends Ecol. Evol.* **25**:81–89. [13–15]

Dingemanse, N. J., and D. Réale. 2012. What is the evidence for natural selection maintaining animal personality variation? In: Animal Personalities: Behavior, Physiology, and Evolution, ed. C. Carere and D. Maestripieri. Chicago: Chicago Univ. Press, in press. [13]

Dingemanse, N. J., F. van der Plas, J. Wright, et al. 2009. Individual experience and evolutionary history of predation affect expression of heritable variation in fish personality and morphology. *Proc. Roy. Soc. B* **276**:1285–1293. [13, 15]

Dingemanse, N. J., and M. Wolf. 2010. Recent models for adaptive personality differences: A review. *Phil. Trans. R. Soc. B* **365**:3947–3958. [14, 15]

Dingemanse, N. J., J. Wright, A. J. N. Kazem, et al. 2007. Behavioural syndromes differ predictably between 12 populations of three-spined stickleback. *J. Anim. Ecol.* **76**:1128–1138. [13–15]

Dobzhansky, T. 1937. Genetics and the Origin of Species. New York: Columbia Univ. Press. [2]

Dochtermann, N. A. 2011. Testing Cheverud's conjecture for behavioral correlations and behavioral syndromes. *Evolution* **65**:1814–1820. [13]

Doherty, M. J. 2008. Theory of Mind: How Children Understand Others' Thoughts and Feelings. Hove: Psychology Press. [18]

Dorris, M. C., and P. W. Glimcher. 2004. Activity in posterior parietal cortex is correlated with the relative subjective desirability of action. *Neuron* **44**:365–378. [9]

Dougherty, M. R., and A. Sprenger. 2006. The influence of improper sets of information on judgment: How irrelevant information can bias judged probability. *J. Exp. Psychol. Gen.* **135**:262–281. [14, 15]

Dougherty, M. R. P., C. F. Gettys, and E. E. Ogden. 1999. MINERVA-DM: A memory processes model for judgments of likelihood. *Psychol. Rev.* **106**:180–209. [14, 15]

Doya, K., K. Samejima, K. Katagiri, and M. Kawato. 2002. Multiple model-based reinforcement learning. *Neural Comput.* **14**:1347–1369. [9]

Doyle, J., K. Glover, P. Khargonekar, and B. Francis. 1989. State-space solutions to standard H_2 and H_∞ control problems. *IEEE Trans. Automat. Contr.* **34**:831–847. [9, 12]

Dreber, A., D. G. Rand, D. Fudenberg, and M. Nowak. 2008. Winners don't punish. *Nature* **452**:348–351. [18]

Drent, P. J., K. van Oers, and A. J. van Noordwijk. 2003. Realized heritability of personalities in the great tit (*Parus major*). *Proc. Roy. Soc. B* **270**:45–51. [13]

Driver, P. M., and D. A. Humphries. 1988. Protean Behavior: The Biology of Unpredictability. Oxford: Oxford Univ. Press. [10]

Dubey, P., and O. Haimanko. 2003. Optimal scrutiny in multi-period promotion tournaments. *Games Econ. Behav.* **42**:1–24. [10]

Dubey, P., and C. Wu. 2001. Competitive prizes: When less scrutiny induces more effort. *J. Math. Econ.* **36**:311–336. [10]

Duffy, J., and R. Nagel. 1997. On the robustness of behaviour in experimental "beauty contest" games. *Econ. J.* **107**:1684–1700. [14]

Dunbar, R. I. M., and S. Schultz. 2007. Evolution in the social brain. *Science* **317**:1344. [17]

Dunfield, K. A., and V. A. Kuhlmeier. 2010. Intention mediated selective helping in human infants. *Psychol. Sci.* **21**:523–527. [17]

Dunlap, A. S., and D. W. Stevens. 2009. Components of change in the evolution of learning and unlearned preference. *Proc. R. Soc. B* **276**:3201–3208. [19]

Durkheim, E. 1895/1964. The Division of Labor in Society. New York: The Free Press. [8]

Dyer, F. C. 1991. Bees acquire route-based memories but not cognitive maps in a familiar landscape. *Anim. Behav.* **41**:239–246. [7]

Dyer, F. C., and J. A. Dickinson. 1994. Development of sun compensation by honeybees: How partially experienced bees estimate the sun's course. *PNAS* **91**:4471–4474. [3, 7]

Dyer, J. R. G., A. Johansson, D. Helbing, I. D. Couzin, and J. Krause. 2009. Leadership, consensus decision making and collective behaviour in humans. *Phil. Trans. R. Soc. B* **364**:781–789. [15]

Ecker, A. S., P. Berens, G. A. Keliris, et al. 2010. Decorrelated neuronal firing in cortical microcircuits. *Science* **327**:584–587. [9]

Eckerman, C. O., and K. Peterman. 2001. Peers and infant social/communicative development. In: Blackwell Handbook of Infant Development Handbooks of Developmental Psychology, ed. G. Bremner and A. Fogel, pp. 326–350. Malden: Blackwell. [17]

Edgar, J. L., J. C. Lowe, E. S. Paul, and C. J. Nicol. 2011. Avian maternal response to chick distress. *Proc. Roy. Soc. B* **278**:3129–3134. [18]

Edgerton, R. B. 2008. Sick Societies. New York: The Free Press. [19]

Ein-Dor, T., M. Mikulincer, G. Doron, and P. R. Shaver. 2010. The attachment paradox: How can so many of us (the insecure ones) have no adaptive advantages? *Perspect. Psychol. Sci.* **5**:123–141. [14]

Eisenberg, N., T. L. Spinrad, and A. Sadovsky. 2006. Empathy-related responding in children. In: Handbook of Moral Development ed. M. S. Killen, J. G., pp. 517–549. Mahwah, NJ: Lawrence Erlbaum. [18]

Elenkov, I. J., and G. P. Chrousos. 2006. Stress system-organization, physiology and immunoregulation. *Neuroimmun.* **13**:257–267. [9]

Elman, J. L. 1990. Finding structure in time. *Cogn. Sci.* **14**:179–211. [3]

———. 1993. Learning and development in neural networks: The importance of starting small. *Cognition* **48**:71–99. [10]

Elster, J. 1998. Emotions and Economic Theory. *J. Econ. Lit.* **36**:47–74. [19]

Emanuel, P. 1979. Introduction to Feedback Control Systems. New York: McGraw-Hill. [8]

Emery, N. J., and N. S. Clayton. 2001a. Effects of experience and social context on prospective caching strategies in scrub jays. *Nature* **414**:443–446. [3, 20]

———. 2001b. It takes a thief to know a thief: Effects of social context on prospective caching strategies in scrub jays. *Nature* **414**:443–446. [3]

———. 2004. The mentality of crows: convergent evolution of intelligence in corvids and apes. *Science* **306**:1903–1907. [20]

Emery, N. J., J. Dally, and N. S. Clayton. 2004. Western scrub-jays (*Aphelocoma californica*) use cognitive strategies to protect their caches from thieving conspecifics. *Anim. Cogn.* **7**:37–43. [3]

Engle, R. W. 2002. Working memory capacity as executive attention. *Curr. Dir. Psychol. Sci.* **11**:19–23. [10]

Engle, R. W., S. W. Tuholski, J. E. Laughlin, and A. R. A. Conway. 1999. Working memory, short-term memory, and general fluid intelligence: A latent-variable approach. *J. Exp. Psychol. Gen.* **128**:309–331. [10]

Enquist, M., K. Eriksson, and S. Ghirlanda. 2007. Critical social learning: A solution to Rogers's paradox. *Am. Anthropol.* **109**:727–734. [19]

Ens, B. J., and J. T. Cayford. 1996. Feeding with other oystercatchers. In: The Oystercatcher: From Individual to Population, ed. J. D. Goss-Custard, pp. 77–104. Oxford: Oxford Univ. Press. [15]

Epstein, J. M. 2001. Learning to be thoughtless: Social normals and individual computation. *Comp. Econ.* **18**:9–24. [15]

Epstein, S. 1979. Stability of behavior. 1. Predicting most of the people much of the time. *J. Pers. Soc. Psychol.* **37**:1097–1126. [14]

Erev, I., and E. Haruvy. 2012. Learning and the economics of small decisions. In: The Handbook of Experimental Economics, ed. J. H. Kagel and A. E. Roth. Princeton: Princeton Univ. Press, in press. [15, 20]

Falk, A., E. Fehr, and U. Fischbacher. 2003. On the nature of fair behavior. *Econ. Inq.* **41**:20–26. [17, 18]

———. 2008. Testing theories of fairness: Intentions matter. *Games Econ. Behav.* **62**:287–303. [20]

Falk, A., and U. Fischbacher. 2006. A theory of reciprocity. *Games Econ. Behav.* **54**:293–315. [17]

Fanselow, M. S. 1980. Conditioned and unconditional components of post-shock freezing. *Pavlov J. Biol. Sci.* **15**:177–182. [9]

Faure, A., S. M. Reynolds, J. M. Richard, and K. C. Berridge. 2008. Mesolimbic dopamine in desire and dread: Enabling motivation to be generated by localized glutamate disruptions in nucleus accumbens. *J. Neurosci.* **28**:7184–7192. [9]

Fawcett, T. W., S. Hamblin, and L.-A. Giraldeau. 2012. Exposing the behavioral gambit: the evolution of learning and decision rules. *Behav. Ecol.*, in press. [1]

Fawcett, T. W., J. M. McNamara, and A. I. Houston. 2011. When is it adaptive to be patient? a general framework for evaluating delayed rewards. *Behav. Proc.* **89**:128–136. [7]

Fehr, E., H. Bernhard, and B. Rockenbach. 2008a. Egalitarianism in young children. *Nature* **454**:1079–1083. [17, 18]

Fehr, E., and C. F. Camerer. 2007. Social neuroeconornics: The neural circuitry of social preferences. *Trends Cogn. Sci.* **11**:419–427. [17]

Fehr, E., and U. Fischbacher. 2003. The nature of human altruism. *Nature* **425**:785–791. [18]

———. 2004. Third-party punishment and social norms. *Evol. Human Behav.* **25**:63–87. [19]

Fehr, E., and S. Gächter. 2000. Cooperation and punishment in public goods experiments. *Am. Econ. Rev.* **90**:980–994. [18]

———. 2002. Altruistic punishment in humans. *Nature* **415**:137–140. [18]

Fehr, E., K. Hoff, and M. Kshetramade. 2008b. Spite and development. *Am. Econ. Rev.* **98**:494–499. [18]

Fehr, E., A. Klein, and K. Schmidt. 2007. Fairness and contract design. *Econometrica* **75**:121–154. [15]

Fehr, E., and K. M. Schmidt. 1999. A theory of fairness, competition, and cooperation. *Q. J. Econ.* **114**:817–868. [15, 17, 18, 20]

Fehr, E., and F. Schneider. 2010. Eyes are on us, but nobody cares: Are eye cues relevant for strong reciprocity? *Proc. Roy. Soc. B* **277**:1315–1323. [20]

Ferguson, T. 1973. A Bayesian analysis of some nonparametric problems. *Ann. Stat.* **1**:209–230. [9]

Ferrari, P. F., E. Kohler, L. Fogassi, and V. Gallese. 2000. The ability to follow eye gaze and its emergence during development in macaque monkeys. *PNAS* **97**:13,997–14,002. [6]

Fessler, D. M. T. 1999. Toward an understanding of the universality of second order emotions. In: Beyond Nature or Nurture: Biocultural Approaches to the Emotions, ed. A. Hinton, pp. 75–116. New York: Cambridge Univ. Press. [18]

———. 2010. Madmen: An evolutionary perspective on anger and men's violent responses to transgression. In: International Handbook of Anger, ed. M. Potegal et al., pp. 361–381. New York: Springer. [19, 20]

Fessler, D. M. T., and K. J. Haley. 2003. The strategy of affect: Emotions in human cooperation. In: Genetic and Cultural Evolution of Cooperation, ed. P. Hammerstein, pp. 7–36, Dahlem Workshop Reports, J. Lupp, series ed. Cambridge, MA: MIT Press. [18, 19]

Fessler, D. M. T., and K. Quintelier. 2012. Suicide, weddings, and prison tattoos. An evolutionary perspective on subjective commitment and objective commitment. In: Signaling, Commitment, and Emotion, ed. R. Joyce et al. Cambridge, MA: MIT Press, in press. [19]

Festinger, L. 1954. A theory of social comparison processes. *Human Relat.* **7**:117–140. [16, 18]

Feynman, R. 1967. The Character of Physical Law. Cambridge, MA: MIT Press. [12]

Fidler, A. E., K. van Oers, P. J. Drent, et al. 2007. Drd4 gene polymorphisms are associated with personality variation in a passerine bird. *Proc. Roy. Soc. B* **274**:1685–1691. [13]

Fiedler, K., and P. Juslin, eds. 2005. Information sampling and adaptive cognition. Cambridge: Cambridge Univ. Press. [14, 15]

Fiser, J., P. Berkes, G. Orbán, and M. Lengyel. 2010. Statistically optimal perception and learning: from behavior to neural representations. *Trends Cogn. Sci.* **14**:119–130. [9]

Fishbein, M., and I. Ajzen. 1975. Belief, Attitude, Intention, and Behavior: An Introduction to Theory and Research. Reading, MA: Addison-Wesley. [17]

Fishburn, P. C., and A. Rubinstein. 1982. Time preference. *Intl. Econ. Rev.* **23**:677–694. [7]

Fisher, R. 1930. The Genetical Theory of Natural Selection. Oxford: Oxford Univ. Press. [19]

Fiske, S. T., and S. E. Taylor. 2008. Social Cognition: From Brains to Culture. New York: McGraw-Hill. [16]

Fisman, R., and E. Miguel. 2006. Cultures of corruption: Evidence from diplomatic parking tickets, NBER Working Paper No. 12312. http://www.nber.org/papers/w12312. (accessed 12 Dec 2011). [15]

FitzGibbon, C. D. 1990. Mixed-species grouping in Thomson's and Grant's gazelles: The antipredator benefits. *Anim. Behav.* **39**:1116–1126. [15]

Flack, J. C., and F. B. M. de Waal. 2000. Any animal whatever: Darwinian building blocks of morality in monkeys and apes. *J. Conscious Stud.* **7**:1–29. [18]

Flack, J. C., F. B. M. de Waal, and D. C. Krakauer. 2005a. Social structure, robustness, and policing cost in a cognitively sophisticated species. *Am. Natural.* **165**:E126–139. [8]

Flack, J. C., M. Girvan, F. B. M. de Waal, and D. C. Krakauer. 2006. Policing stabilizes construction of social niches in primates. *Nature* **439**:426–429. [8, 18]

Flack, J. C., and D. C. Krakauer. 2006. Encoding power in communication networks. *Am. Natural.* **168**:E87–102. [8]

Flack, J. C., D. C. Krakauer, and F. B. M. d. Waal. 2005b. Robustness mechanisms in primate societies: A perturbation study. *Proc. R. Soc. B* **272**:1091–1099. [8]

Fleeson, W. 2001. Toward a structure- and process-integrated view of personality: Traits as density distributions of states. *J. Pers. Soc. Psychol.* **80**:1011–1027. [14, 15]

———. 2004. Moving personality beyond the person-situation debate: The challenge and the opportunity of within-person variability. *Curr. Dir. Psychol. Sci.* **13**:83–87. [13–15]

Flombaum, J. I., and L. R. Santos. 2005. Rhesus monkeys attribute perceptions to others. *Curr. Biol.* **15**:447–452. [20]

Fodor, J. 1983. The Modularity of Mind. Cambridge, MA: MIT Press. [1, 11]

Fogarty, S., J. Cote, and A. Sih. 2011. Social personality polymorphism and the spread of invasive species: A model. *Am. Natural.* **177**:273–287. [13, 15]

Forehand, M., J. Gastil, and M. A. Smith. 2004. Endorsements as voting cues: Heuristic and systematic processing in initiative elections. *J. Appl. Soc. Psychol.* **34**:2215–2233. [7]

Förster, J., and N. Liberman. 2007. Knowledge activation. In: Social Psychology: Handbook of Basic Principles, 2nd edition, ed. A. W. Kruglanski and E. T. Higgins, pp. 201–231. New York: Guilford Press. [16]

Foster, D. J., and M. A. Wilson. 2006. Reverse replay of behavioural sequences in hippocampal place cells during the awake state. *Nature* **440**:680–683. [9]

———. 2007. Hippocampal theta sequences. *Hippocampus* **17**:1093–1099. [9]

Foster, K. R., T. Wenseleers, and F. L. W. Ratnieks. 2001. Spite: Hamilton's unproven theory. *Ann. Zool. Fenn.* **38**:229–238. [18]

Foxe, J. J., and G. V. Simpson. 2002. Flow of activation from V1 to frontal cortex in humans: A framework for defining early visual processing. *Exp. Brain Res.* **142**:139–150. [9]

Franconeri, S. L., and D. J. Simons. 2003. Moving and looming stimuli capture attention. *Percept. Psychophys.* **65**:1–12. [16]

Frank, M. J., B. B. Doll, J. Oas-Terpstra, and F. Moreno. 2009. Prefrontal and striatal dopaminergic genes predict individual differences in exploration and exploitation. *Nat. Neurosci.* **12**:1062–1068. [6]

Frank, R. H. 1987. If *Homo economicus* could choose his own utility function, would he want one with a conscience? *Am. Econ. Rev.* **77**:593–604. [18]

———. 1988. Passions within Reason: The Strategic Role of the Emotions. New York: W. W. Norton. [18]

Frank, S. A. 1995. Mutual policing and repression of competition in the evolution of cooperative groups. *Nature* **377**:520–522. [18]

———. 1998. Foundations of Social Evolution. Princeton: Princeton Univ. Press. [8, 19]

Franks, N. R., F. X. Dechaume-Moncharmont, E. Hanmore, and J. K. Reynolds. 2009. Speed versus accuracy in decision making ants: Expediting politics and policy implementation. *Phil. Trans. R. Soc. B* **364**:845–852. [12]

Franks, N. R., and T. Richardson. 2006. Teaching in tandem-running ants. *Nature* **439**:153. [20]

Frasconia, P., M. Gori, F. Kurfessc, and A. Sperdutid. 2002. Special issue on integration of symbolic and connectionist systems. *Cogn. Sys. Res.* **3**:121–123. [3]

Fraser, O. N., and T. Bugnyar. 2010. Do ravens show consolation? Responses to distressed other. *PLoS One* **5**:e10605. [20]

Fraser, O. N., D. Stahl, and F. Aureli. 2008. Stress reduction through consolation in chimpanzees. *PNAS* **105**:8557–8562. [20]

Frederick, S., G. Loewenstein, and T. O'Donoghue. 2002. Time discounting and time preference: A critical review. *J. Econ. Lit.* **40**:351–401. [1, 7]

Friedberg, E., G. C. Walker, and W. Siede. 1995. DNA Repair and Mutagenesis. New York: W. H. Freeman. [8]

Funder, D. C., and D. J. Ozer. 1983. Behavior as a function of the situation. *J. Pers. Soc. Psychol.* **44**:107–112. [14]

Gadagkar, R. 1993. Can animals be spiteful? *Trends Ecol. Evol.* **8**:232–234. [18]

Gaissmaier, W., L. Schooler, and J. Rieskamp. 2006. Simple predictions fueled by capacity limitations: When are they successful? *J. Exp. Psychol. Learn. Mem. Cogn.* **32**:966–982. [10]

Gale, J., K. G. Binmore, and L. Samuelson. 1995. Learning to be imperfect: The ultimatum game. *Games Econ. Behav.* **8**:59–90. [19]

Galef, B. G. 1989. Enduring social enhancement of rats' preferences for the palatable and piquant. *Appetite* **13**:81–92. [19]

———. 1996. Food selection: Problems in understanding how we choose foods to eat. *Neurosci. Biobehav. Rev.* **20**:67–73. [9]

Galef, B. G., J. R. Mason, G. Preti, and N. J. Bean. 1988. Carbon disulfide: A semiochemical mediating socially-induced diet choice in rats. *Physiol. Behav.* **42**:119–124. [19]

Galef, B. G., and E. E. Whiskin. 2003. Socially transmitted food preferences can be used to study long-term memory in rats. *Learn. Behav.* **31**:160–164. [19]

Galef, B. G., and S. Wigmore. 1983. Transfer of information concerning distant foods: A laboratory investigation of the "information-centre" hypothesis. *Anim. Behav.* **31**:748–758. [9, 19]

Gallistel, C. R. 1990. The Organization of Learning. Cambridge, MA: MIT Press. [3]

———. 2000. The replacement of general-purpose learning models with adaptively specialized learning modules. In: The Cognitive Neurosciences, 2nd edition, ed. M. S. Gazzaniga, pp. 1179–1191. Cambridge, MA: MIT Press. [11]

———. 2001. Mental representations; psychology of. In: International Encyclopedia of the Social & Behavioral Sciences, ed. N. J. Smelser and P. B. Baltes, pp. 9691–9695. Oxford: Elsevier. [7]

———. 2011. Mental magnitudes. In: Space, Time and Number in the Brain: Searching for the Foundations of Mathematical Thought, ed. S. Dehaene and L. Brannon, pp. 3–12. New York: Elsevier. [7]

Gallistel, C. R., and A. E. Cramer. 1996. Computations on metric maps in mammals: Getting oriented and choosing a multi-destination route. *J. Exp. Biol.* **199**:211–217. [7]

Gallistel, C. R., and A. P. King. 2009. Memory and the Computational Brain: Why Cognitive Science Will Transform Neuroscience. New York: Wiley. [3]

Gallistel, C. R., A. P. King, D. Gottlieb, et al. 2007. Is matching innate? *J. Exp. Anal. Behav.* **87**:161–199. [3]

Gallistel, C. R., T. A. Mark, A. P. King, and P. Latham. 2001. The rat approximates an ideal detector of changes in rates of reward: Implications for the law of effect. *J. Exp. Psychol. Anim. Behav. Proc.* **27**:354–372. [3, 10]

Galton, F. 1907. Vox populi. *Nature* **75**:450–451. [12]

Garcia, J., W. G. Hankins, and K. W. Rusiniak. 1974. Behavioral regulation of the milieu interne in man and rat. *Science* **185**:824–831. [9, 19]

Garcia, J., D. J. Kimeldorf, and R. A. Koelling. 1955. Conditioned aversion to saccharin resulting from exposure to gamma radiation. *Science* **122**:157–158. [4]

Garcia, J., and R. A. Koelling. 1966. The relation of cue to consequence in avoidance learning. *Psychonom. Sci.* **4**:123–124. [7, 19]

Gardner, A., and S. A. West. 2004. Spite and the scale of competition. *J. Evol. Biol.* **17**:1195–1203. [18]

Geman, S., E. Bienenstock, and R. Doursat. 1992. Neural networks and the bias/variance dilemma. *Neural Comput.* **4**:1–58. [12]

Geman, S., and D. Geman. 1984. Stochastic relaxation, Gibbs distributions, and the Bayesian restoration of images. *IEEE-PAMI* **6**:721–741. [4]

Gergely, G., H. Bekkering, and I. Kiraly. 2002. Rational imitation in preverbal infants. *Nature* **415**:755. [17, 18]

Gergely, G., Z. Nadasdy, G. Csibra, and S. Biro. 1995. Taking the intentional stance at 12 months of age. *Cognition* **56**:165–193. [17]

Ghiselin, M. T. 1974. The Economy of Nature and the Evolution of Sex. Berkeley: Univ. of California Press. [18]

Gibbon, J. 1995. Dynamics of time matching: Arousal makes better seem worse. *Psychon. Bull. Rev.* **2**:208–215. [3]

Gick, M. L., and K. J. Holyoak. 1980. Analogical problem solving. *Cogn. Psychol.* **12**:306–355. [4]

Gigerenzer, G. 2000. Adaptive Thinking: Rationality in the Real World. Oxford: Oxford Univ. Press. [5]

———. 2008. Why heuristics work. *Perspect. Psychol. Sci.* **3**:20–29. [11]

Gigerenzer, G., and W. Gaissmaier. 2010. Heuristic decision making. *Ann. Rev. Psychol.* **62**:451–482. [1]

Gigerenzer, G., and D. G. Goldstein. 1996. Reasoning the fast and frugal way: Models of bounded rationality. *Psychol. Rev.* **103**:650–669. [4, 6, 15]

Gigerenzer, G., and K. Hug. 1992. Domain-specific reasoning: Social contracts, cheating, and perspective change. *Cognition* **43**:127–171. [11]

Gigerenzer, G., and R. Selten, eds. 2001. Bounded Rationality: The Adaptive Toolbox. Dahlem Workshop Reports, vol. 84, J. Lupp, series ed. Cambridge, MA: MIT Press. [4, 6]

Gigerenzer, G., P. M. Todd, and the ABC Research Group, eds. 1999. Simple Heuristics That Make Us Smart. New York: Oxford Univ. Press. [1, 2, 4, 7, 10, 15]

Gilbert, D. T., B. W. Pelham, and D. S. Krull. 1988. On cognitive business: When person perceivers meet persons perceived. *J. Pers. Soc. Psychol.* **54**:733–740. [10]

Gilby, I. C. 2006. Meat sharing among the Gombe chimpanzees: Harassment and reciprocal exchange. *Anim. Behav.* **71**:953–963. [18, 20]

Gilovich, T., D. Griffin, and D. Kahneman, eds. 2002. Heuristics and Biases: The Psychology of Intuitive Judgment. Cambridge: Cambridge Univ. Press. [10]

Gilovich, T., R. Vallone, and A. Tversky. 1985. The hot hand in basketball: On the misperception of random sequences. *Cogn. Psychol.* **17**:295–314. [11]

Gilroy, S., and A. Trewavas. 2001. Signal processing and transduction in plant cells: The end of the beginning? *Nat. Rev. Molec. Cell Biol.* **2**:307–314. [7]

Giner-Sorella, R., and S. Chaiken. 1997. Selective use of heuristic and systematic processing under defence motivation. *Pers. Soc. Psychol. Rev.* **23**:84–97. [7]

Gintis, H. 2007. A framework for the unification of the behavioral sciences. *Behav. Brain Sci.* **30**:1–61. [20]

Gintis, H., S. Bowles, R. Boyd, and E. Fehr. 2003. Explaining altruistic behavior in humans. *Evol. Human Behav.* **24**:153–172. [11]

Gintis, H., S. Bowles, R. Boyd, and E. Fehr.. 2005. Moral Sentiments and Material Interests. Princeton: Princeton Univ. Press. [20]

Gittins, J. C. 1989. Multi-Armed Bandit Allocation Indices. New York: Wiley [9]

Glimcher, P. W. 2004. Decisions, Uncertainty, and the Brain: The Science of Neuroeconomics. Cambridge, MA: MIT Press. [6]

Glimcher, P. W., M. C. Dorris, and H. M. Bayer. 2005. Physiological utlity theory and the neuroeconomics of choice. *Games Econ. Behav.* **52**:213–256. [6, 17]

Gold, J. I., and M. N. Shadlen. 2007. The neural basis of decision making. *Ann. Rev. Neurosci.* **30**:535–574. [6]

Goldberg, E. 1995. Rise and fall of modular orthodoxy. *J. Clin. Exp. Neuropsychol.* **17**:193–208. [8]

Goldberg, L. R. 1993. The structure of phenotypic personality-traits. *Am. Psychol.* **48**:26–34. [14]

Goldberg, M. E., J. W. Bisley, K. D. Powell, and J. Gottlieb. 2006. Saccades, salience and attention: The role of the lateral intraparietal area in visual behavior. *Prog. Brain Res.* **155**:157–175. [6]

Goldstein, D. G., and G. Gigerenzer. 2009. Fast and frugal forecasting. *Intl. J. Forecasting* **25**:760–772. [12]

Goodall, J. 1986. The Chimpanzees of Gombe: Patterns of Behaviour. Cambridge, MA: Belknap Press. [18]

Goodman, N. 1970. Seven strictures on similarity. In: Experience and Theory, ed. L. Foster and J. W. Swanson, pp. 19–29. Amherst: Univ. of Massachusetts Press. [4]

Gopnik, A., A. N. Meltzoff, and P. K. Kuhl. 1999. The Scientist in the Crib. New York: William Morris & Company. [4]

Gosling, S. 2001. From mice to men: What can we learn about personality from animal research? *Psychol. Bull.* **127**:45–86. [13]

———. 2008. Personality in non-human animals. *Soc. Personal. Psychol. Compass* **2**:985–1001. [13]

Gosling, S., and O. P. John. 1999. Personality dimensions in non-human animals: A cross-species review. *Curr. Dir. Psychol. Sci.* **8**:69–75. [13–15]

Goss-Custard, J. D. 1996. The Oystercatcher: From Individual to Population. Oxford: Oxford Univ. Press. [15]

Gould, E. 2007. How widespread is adult neurogenesis in mammals? *Nat. Rev. Neurosci.* **8**:481–488. [9]

Gould, J. L. 1986. The locale map of honey bees: Do insects have cognitive maps? *Science* **232**:861–863. [7]

Gould, S. 2002. The Structure of Evolutionary Theory. Cambridge, MA: Belknap Press. [8]

Grace, R. C. 1993. Violations of transitivity: Implications for a theory of contextual choice. *J. Exp. Anal. Behav.* **60**:185–201. [7]

Gradinaru, V., K. R. Thompson, F. Zhang, et al. 2007. Targeting and readout strategies for fast optical neural control *in vitro* and *in vivo*. *J. Neurosci.* **27**:14,231–14,238. [9]

Graeff, F. G., F. S. Guimaraes, T. G. C. S. De Andrade, and J. F. W. Deakin. 1998. Role of 5HT in stress, anxiety and depression. *Pharm. Biochem. Behav.* **54**:129–141. [9]

Grafen, A. 1984. Natural selection, kin selection and group selection. In: Behavioral Ecology: An Evolutionary Approach, 2nd, ed. J. R. Krebs and N. B. Davies, pp. 62–84. Oxford: Blackwell. [2]

Gräfenhain, M., T. Behne, M. Carpenter, and M. Tomasello. 2009. Young children's understanding of joint commitments. *Dev. Psychol.* **45**:1430–1443. [17]

Green, D. M., and J. A. Swets. 1966. Signal Detection Theory and Psychophysics. New York: Wiley. [5]

Green, L., E. B. Fisher, S. Perlow, and L. Sherman. 1981. Preference reversal and self-control: Choice as a function of reward amount and delay. *Behav. Anal. Lett.* **1**:43–51. [7]

Green, L., N. Fristoe, and J. Myerson. 1994. Temporal discounting and preference reversals in choice between delayed outcomes. *Psychon. Bull. Rev.* **1**:383–389. [7]

Green, L., J. Myerson, and E. McFadden. 1997. Rate of temporal discounting decreases with amount of reward. *Mem. Cogn.* **25**:715–723. [7]

Green, R. F. 1984. Stopping rules for optimal foragers. *Am. Natural.* **123**:30–43. [2]

Greenberg, J. R., K. Hamann, F. Warneken, and M. Tomasello. 2010. Chimpanzee helping in collaborative and noncollaborative contexts. *Anim. Behav.* **80**:873–880. [17]

Griffiths, T. L., N. Chater, C. Kemp, A. Perfors, and J. B. Tenenbaum. 2010. Probabilistic models of cognition: Exploring representations and inductive biases. *Trends Cogn. Sci.* **14**:357–364. [3]

Grinnell, J., C. Packer, and A. E. Pusey. 1995. Cooperation in male lions: Kinship, reciprocity or mutualism? *Anim. Behav.* **49**:95–105. [15]

Groothuis, T. G. G., and C. Carere. 2005. Avian personalities: Characterization and epigenesis. *Neurosci. Biobehav. Rev.* **29**:137–150. [13]

Groothuis, T. G. G., and F. Trillmich. 2011. Unfolding personalities: The importance of studying ontogeny. *Dev. Psychobiol.* **53**:641–655. [15]

Gross, M. R. 1996. Alternative reproductive strategies and tactics: Diversity within sexes. *Trends Ecol. Evol.* **11**:92–98. [14]

Gross, R., A. I. Houston, E. J. Collins, et al. 2008. Simple learning rules to cope with changing environments. *J. R. Soc. Interface* **5**:1193–1202. [10]

Guala, F. 2012. Reciprocity: weak or strong? What punishment experiments do (and do not) demonstrate. *Behav. Brain Sci.*, in press. [19]

Gualtiero, P. 2008. Some neural networks compute, others don't. *Neural Netw.* **21**:311–321. [3]

Gueron, S., S. A. Levin, and D. I. Rubenstein. 1996. The dynamics of herds: From individuals to aggregations. *J. Theor. Biol.* **182**:85–98. [15]

Güroglu, B., W. van den Bos, and E. A. Crone. 2009. Fairness considerations: Increasing understanding of intentionality during adolescence. *J. Exp. Child Psychol.* **104**:398–409. [18]

Güth, W., R. Schmittberger, and B. Schwarze. 1982. An experimental analysis of ultimatum bargaining. *J. Econ. Behav. Org.* **3**:367–388. [18, 20]

Guthrie, S. 2001. Why gods? A cognitive theory. In: Religion in Mind: Cognitive Perspectives on Religious Belief, Ritual and Experience, ed. J. R. Anderson, pp. 94–112. Cambridge: Cambridge Univ. Press. [5]

Hagen, E. H., and P. Hammerstein. 2005. Evolutionary biology and the strategic view of ontogeny: Genetic strategies provide robustness and flexibility in the life course. *Res. Human Devel.* **2**:83–97. [7, 20]

———. 2006. Game theory and human evolution: A critique of some recent interpretations of experimental games. *Theor. Popul. Biol.* **69**:339–348. [7, 18]

Hagerty, M. R. 2000. Social comparisons of income in one's community: Evidence from national surveys of income and happiness. *J. Pers. Soc. Psychol.* **78**:764–771. [14]

Hahn, U., N. Chater, and L. B. C. Richardson. 2003. Similarity as transformation. *Cognition* **87**:1–32. [4]

Haldane, J. B. S. 1957. The cost of natural selection. *J. Genetics* **55**:511–524. [15]

Halder, G., P. Callaerts, and W. J. Gehring. 1995. Induction of ectopic eyes by target expression of the *eyeless* gene in *Drosophila*. *Science* **267**:1788–1792. [3]

Haley, K. J., and D. l. M. Fessler. 2005. Nobody's watching? Subtle cues affect generosity in an anonymous economic game. *Evol. Human Behav.* **26**:245–256. [20]

Hamann, K., F. Warneken, J. Greenberg, and M. Tomasello. 2011. Collaboration encourages equal sharing in children but not chimpanzees. *Nature* **476**:328–331. [17]

Hamann, K., F. Warneken, and M. Tomasello. 2012. Children's developing commitment to joint goals. *Child Devel.* **83**:137–145. [17]

Hamilton, W. D. 1964. The genetical evolution of social behaviour. I & II. *J. Theor. Biol.* **7**:1–52. [18]

Hamlin, J. K., K. Wynn, and P. Bloom. 2007. Social evaluation by preverbal infants. *Nature* **450**:557–559. [17, 18, 20]

———. 2010. Three-month-olds show a negativity bias in their social evaluations. *Devel. Sci.* **13**:923–929. [18]

Hammerstein, P. 2012. Toward a Darwinian theory of decision making: Games and the biological roots of behavior. In: Evolution and Rationality, ed. K. Binmore and S. Okasha, pp. 7–22. Cambridge: Cambridge Univ. Press. [19]

Hammerstein, P., and E. H. Hagen. 2005. The second wave of evolutionary economics in biology. *Trends Ecol. Evol.* **20**:604–609. [7, 17]

Hammerstein, P., E. H. Hagen, A. V. M. Herz, and H. Herzel. 2006. Robustness: A key to evolutionary design. *Biol. Theory* **1**:90–93. [7, 8, 12]

Hammerstein, P., and S. E. Riechert. 1988. Payoffs and strategies in territorial contests: ESS analyses of two ecotypes of the spider *Agelenopsis aperta*. *Evolutionary Ecology* **2**:115–138. [19]

Hammond, K. R. 2007. Beyond Rationality: The Search for Wisdom in a Troubled Time. Oxford: Oxford Univ. Press. [4]

Hampton, A. N., P. Bossaerts, and J. P. O'Doherty. 2008. Neural correlates of mentalizing-related computations during strategic interactions in humans. *PNAS* **105**:6741–6746. [9]

Hansen, C. H., and R. D. Hansen. 1988. Finding the face in the crowd: An anger superiority effect. *J. Pers. Soc. Psychol.* **54**:917–924. [16]

Hanski, I. 2001. Spatially realistic theory of metapopulation ecology. *Naturwiss.* **88**:372–381. [8]

Hare, B., J. Call, B. Agnetta, and M. Tomasello. 2000. Chimpanzees know what conspecifics do and do not see. *Anim. Behav.* **59**:771–785. [20]

Hare, B., J. Call, and M. Tomasello. 2001. Do chimpanzees know what conspecifics know? *Anim. Behav.* **61**:771–785. [20]

Hare, B., A. P. Melis, V. Woods, S. Hastings, and R. Wrangham. 2007. Tolerance allows bonobos to outperform chimpanzees in a cooperative task. *Curr. Biol.* **17**:619–623. [17]

Harkness, R. D., and N. G. Maroudas. 1985. Central place foraging by an ant (*Cataglyphis bicolor Fab.*): A model of searching. *Anim. Behav.* **33**:916–928. [3]

Hart, S. 2005. Adaptive heuristics. *Econometrica* **73**:1401–1430. [10]

Hart, S., and A. Mas-Colell. 2000. A simple adaptive procedure leading to correlated equilibrium. *Econometrica* **68**:1127–1150. [10]

Hartup, W. W. 1996. The company they keep: Friendships and their developmental significance. *Child Devel.* **67**:1–13. [17]

Haselton, M. G., D. A. Bryant, A. Wilke, et al. 2009. Adaptive rationality: An evolutionary perspective on cognitive bias. *Social Cogn.* **27**:733–763. [5]

Haselton, M. G., and D. M. Buss. 2000. Error management theory: A new perspective on biases in cross-sex mind-reading. *J. Pers. Soc. Psychol.* **78**:81–91. [1, 5, 10]

Haselton, M. G., and D. Nettle. 2006. The paranoid optimist: An integrative evolutionary model of cognitive biases. *Pers. Soc. Psychol. Bull.* **10**:47–66. [5, 10]

Hastorf, S., and J. Polefka. 1970. Person Perception. Reading, MA: Addison-Wesley. [18]

Hauser, M., K. McAuliffe, and P. R. Blake. 2009. Evolving the ingredients for reciprocity and spite. *Phil. Trans. R. Soc. B* **364**:3255–3266. [18]

Hayden, B. Y., J. M. Pearson, and M. L. Platt. 2011. Neuronal basis of sequential foraging decisions in a patchy environment. *Nat. Neurosci.* **14**:933–939. [6]

Hays, W. L. 1963. Statistics for Psychologists. New York: Holt Rinehart and Winston. [10]

Heath, C., and D. Heath. 2006. The curse of knowledge *Harvard Bus. Rev.* **84**:20–23. [4]

Hebb, D. O. 1946. Emotion in man and animal: An analysis of the intuitive processes of recognition. *Psychol. Rev.* **53**:88–106. [13]

Heider, F., and M. Simmel. 1944. An experimental study of apparent behavior. *Am. J. Psychol.* **57**:243–259. [20]

Heinsohn, R., and C. Packer. 1995. Complex cooperative strategies in group-territorial African lions. *Science* **269**:1260–1262. [15]

Henrich, J. 2004. Inequity aversion in capuchins? *Nature* **428**:139. [18]

Henrich, J., R. Boyd, S. Bowles, et al. 2001. Cooperation, reciprocity and punishment in fifteen small-scale societies. *Am. Econ. Rev.* **91**:73–78. [11]

———. 2004. Foundations of Human Sociality: Economic Experiments and Ethnographic Evidence in Fifteen Small-Scale Societies. Oxford: Oxford Univ. Press. [15]

Henrich, J., J. Ensminger, R. McElreath, et al. 2010. Markets, religion, community size, and the evolution of fairness and punishment. *Science* **327**:1480–1484. [18]

Henrich, J., and F. J. Gil-White. 2001. The evolution of prestige: Freely conferred status as a mechanism for enhancing the benefits of cultural transmission. *Evol. Human Behav.* **22**:165–196. [11]

Henrich, J., and N. Henrich. 2010. The evolution of cultural adaptations: Fijian food taboos protect against dangerous marine toxins. *Proc. R. Soc. B* **277**:3715–3724. [19]

Henrich, J., R. McElreath, A. Barr, et al. 2006. Costly punishment across human societies. *Science* **312**:1767–1770. [18, 19]

Hermalin, B. E., and A. M. Isen. 2008. A model of the effect of affect on economic decision making. *Q. Mark. Econ.* **6**:17–40. [14]

Herrmann, B., C. Thöni, and S. Gächter. 2008. Antisocial punishment across societies. *Science* **319**:1362–1367. [15, 18]

Herrnstein, R. J., and D. Prelec. 1991. Melioration: A theory of distributed choice. *J. Econ. Persp.* **5**:137–156. [3]

Herrnstein, R. J., and W. J. Vaughan. 1980. Melioration and behavioral allocation. In: Limits to Action: The Allocation of Individual Behavior, ed. J. E. R. Staddon, pp. 143–176. New York: Academic Press. [3]

Hershberger, W. A. 1986. An approach through the looking-glass. *Anim. Learn. Behav.* **14**:443–451. [9]

Hertwig, R., G. Barron, E. U. Weber, and I. Erev. 2004. Decisions from experience and the effect of rare events in risky choice. *Psychol. Sci.* **15**:534–539. [9]

Hertwig, R., U. Hoffrage, and ABC Research Group, eds. 2012. Simple Heuristics in a Social World. New York: Oxford Univ. Press. [12]

Hertwig, R., and T. J. Pleskac. 2010. Decisions from experience: Why small samples. *Cognition* **115**:225–237. [10]

Hertwig, R., and P. M. Todd. 2003. More is not always better: The benefits of cognitive limits. In: Thinking: Psychological Perspectives on Reasoning, Judgment and Decision Making, ed. D. Hardman and L. Macchi, pp. 213–231. New York: John Wiley & Sons. [10]

Hertz, J., A. Krogh, and R. G. Palmer. 1991. Introduction to the Theory of Neural Computation. Boston: Addison-Wesley. [8]

Hess, N. H., and E. H. Hagen. 2006. Psychological adaptations for assessing gossip veracity. *Human Nature* **17**:337–354. [7]

Hewlett, B. S., H. N. Fouts, A. H. Boyette, and B. L. Hewlett. 2012. *Phil. Trans. R. Soc. B*, in press. [19]

Hey, J. D., and C. Orme. 1994. Investigating generalizations of expected utility theory using experimental data. *Econometrica* **62**:1291–1326. [14]

Heyman, G. M. 1982. Is time allocation unconditioned behavior? In: Quantitative Analyses of Behavior, Vol. 2: Matching and Maximizing Accounts, ed. M. Commons et al., vol. 2, pp. 459–490. Cambridge, MA: Ballinger Press. [3]

Heyman, G. M., and R. D. Luce. 1979. Operant matching is not a logical consequence of maximizing reinforcement rate. *Anim. Learn. Behav.* **7**:133–140. [3]

Higgins, E. T. 1996. Knowledge activation: Accessibility, applicability, and salience. In: Social Psychology: Handbook of Basic Principles, ed. E. T. Higgins and A. W. Kruglanski, pp. 133–168. New York: Guilford Press. [16]

Hill, K., M. Barton, and A. M. Hurtado. 2009. The emergence of human uniqueness: Characters underlying behavioral modernity. *Evolutionary Anthropology* **18**:187–200. [18, 20]

Hills, T. T., P. M. Todd, and R. L. Goldstone. 2008. Search in external and internal spaces: Evidence for generalized cognitive search processes. *Psychol. Sci.* **19**:802–808. [1]

Hirshleifer, J. 1987. Economics from a biological viewpoint. *J. Law Econ.* **20**:1–52. [19]

Hockey, G. R. J., A. J. Maule, P. J. Clough, and L. Bdzola. 2000. Effects of negative mood states on risk in everyday decision making. *Cogn. Emot.* **14**:823–855. [14]

Hoffman, M. L. 1982. Development of prosocial motivation: Empathy and guilt. In: The Development of Prosocial Behavior, ed. N. Eisenberg, pp. 281–338. New York: Academic Press. [18]

Hofmann, W., M. Friese, and F. Strack. 2009. Impulse and self-control from a dual-systems perspective. *Perspect. Psychol. Sci.* **4**:162–176. [16]

Hogarth, R. M., and H. J. Einhorn. 1992. Order effects in belief updating: The belief-adjustment model. *Cogn. Psychol.* **24**:1–55. [10]

Holder, M. D., F. Bermudez-Rattoni, and J. Garcia. 1988. Taste potentiated noise-illness associations. *Behav. Neurosci.* **102**:363–370. [7]

Holland, P. C. 2004. Relations between Pavlovian-instrumental transfer and reinforcer devaluation. *J. Exp. Psychol. Anim. Behav. Proc.* **30**:104–117. [9]

Hölldobler, B., and E. O. Wilson. 2011. The Leafcutter Ants: Civilization by Instinct. New York: W. W. Norton. [8]

Hook, J., and T. Cook. 1979. Equity theory and the cognitive ability of children. *Psychol. Bull.* **86**:429–445. [18]

Horner, V., D. Carter, M. Suchak, and F. B. M. de Waal. 2011. Spontaneous prosocial choice by chimpanzees. *PNAS* **108**:13,847–13,851[18]

Horowitz, A. 2009. Disambiguating the "guilty look": Salient prompts to a familiar dog behaviour. *Behav. Proc.* **81**:447–452. [18]

Horrocks, J., and W. Hunte. 1981. "Spite": A constraint on optimal foraging in the vervet monkey, *Cercopithecus aethiops sabaues*, in Barbados. *Am. Zool.* **21**:939. [18]

Horvitz, J. C. 2000. Mesolimbocortical and nigrostriatal dopamine responses to salient non-reward events. *Neuroscience* **96**:651–656. [9]

Houston, A. I. 1991. Violations of stochastic transitivity on concurrent chains: Implications for theories of choice. *J. Exp. Anal. Behav.* **55**:1991. [7]

———. 1997. Natural selection and context-dependent values. *Proc. R. Soc. B* **264**:1539–1541. [7]

Houston, A. I., A. Kacelnik, and J. M. McNamara. 1982. Some learning rules for acquiring information. In: Functional Ontogeny, ed. D. J. McFarland, pp. 140–191. London: Pitman. [2, 9]

Houston, A. I., and J. M. McNamara. 1988. Fighting for food: A dynamic version of the hawk-dove game. *Evol. Ecol.* **2**:51–64. [15]

———. 1999. Models of Adaptive Behaviour: An Approach Based on State. Cambridge: Cambridge Univ. Press. [2, 4, 15]

Houston, A. I., J. M. McNamara, and M. D. Steer. 2007a. Do we expect natural selection to produce rational behaviour? *Phil. Trans. R. Soc. B* **362**:1531–1543. [5]

———. 2007b. Violations of transitivity under fitness maximization. *Biol. Lett.* **3**:365–367. [7]

Howard, R. A. 1960. Dynamic Programming and Markov Processes. Cambridge, MA: MIT Press. [9]

Howell, R. T., and C. J. Howell. 2008. The relation of economic status to subjective well-being in developing countries: A meta-analysis. *Psychol. Bull.* **134**:536–560. [14]

Hsee, C. K., and Y. Rottenstreich. 2004. Music, pandas and muggers: On the affective psychology of value. *J. Exp. Psychol.* **133**:23–30. [7]

Hsu, M. 2006. Three correlated essays on the neural foundations of economic decision-making. PhD thesis, California Institute of Technology. [9]

Hsu, Y. Y., R. L. Earley, and L. L. Wolf. 2006. Modulation of aggressive behaviour by fighting experience: Mechanisms and contest outcomes. *Biol. Rev.* **81**:33–74. [15]

Huang, Z. Y., E. Plettner, and G. E. Robinson. 1998. Effects of social environment and worker mandibular glands on endocrine-mediated behavioral development in honey bees. *J. Comp. Physiol. A* **183**:143–152. [12]

Huber, P. 1972. Robust statistics: A review. *Ann. Math. Statist.* **43**:1041–1067. [9]

Hume, D. 1739/2007. A Treatise of Human Nature. Oxford: Clarendon Press. [4]

Humphrey, N. 1976. The social function of intellect. In: Growing Points in Ethology: Proceedings, ed. P. P. G. Bateson and R. A. Hinde, pp. 303–317. London: Cambridge Univ. Press. [17, 20]

Humphreys, L. G. 1939. The effect of random alternation of reinforcement on the acquisition and extinction of conditioned eyelid reactions. *J. Exp. Psychol.* **25**:141–158. [15]

Humphries, D. A., and P. M. Driver. 1967. Erratic display as a device against predators. *Science* **156**:1767–1768. [10]

Humphries, D. A., and P. M. Driver. 1970. Protean defence by prey animals. *Oecologia* **5**:285–302. [10]

Huntingford, F. A. 1976. The relationship between anti-predator behavior and aggression among conspecifics in the three-spined stickleback, *Gasterosteus aculeatus*. *Anim. Behav.* **24**:245–260. [13]

Huntingford, F. A., P. J. Wright, and J. F. Tierney. 1994. Adaptive variation in antipredator behaviour in threespine stickleback. In: The Evolutionary Biology of the Threespine Stickleback ed. M. A. Bell and S. A. Foster, pp. 277–296. Oxford: Oxford Univ. Press. [15]

Hurley, S., and N. Chater, eds. 2005. Mechanisms of Imitation and Imitation in Animals. Perspectives on Imitation: From Neuroscience to Social Science, vol. 1, Cambridge, MA: MIT Press. [4]

Hutchinson, J. M. C., and G. Gigerenzer. 2005. Simple heuristics and rules of thumb: where psychologists and behavioural biologists might meet. *Behav. Proc.* **69**:97–124. [1, 12]

Hutchinson, J. M. C., A. Wilke, and P. M. Todd. 2008. Patch leaving in humans: Can a generalist adapt its rules to dispersal of items across patches? *Anim. Behav.* **75**:1331–1349. [1]

Huttenlocher, J., and D. Burke. 1976. Why does memory span increase with age? *Cogn. Psychol.* **8**:1–31. [10]

Huys, Q. J. M., and P. Dayan. 2009. A Bayesian formulation of behavioral control. *Cognition* **113**:314–328. [9]

Ihmels, J., S. R. Collins, M. Schuldiner, N. J. Krogan, and J. S. Weissman. 2007. Backup without redundancy: Genetic interactions reveal the cost of duplicate gene loss. *Mol. Sys. Biol.* **3**:86. [8]

Ikemoto, S., and J. Panksepp. 1999. The role of nucleus accumbens dopamine in motivated behavior: A unifying interpretation with special reference to reward-seeking. *Brain Res. Rev.* **31**:6–41. [9]

Ioannides, Y. M., and G. Topa. 2010. Neighborhood effects: Accomplishments and looking beyond them. *J. Reg. Sci.* **50**:343–362. [15]

Isaacson, W. 1992. Kissinger. New York: Simon and Schuster. [19]

Isen, A. M. 2001. An influence of positive affect on decision making in complex situations: Theoretical issues with practical implications. *J. Consum. Psychol.* **11**:75–85. [14]

Isen, A. M., and B. Means. 1983. The influence of positive affect on decision-making strategy. *Social Cogn.* **2**:18–31. [14]

Isen, A. M., and R. Patrick. 1983. The effect of positive feelings on risk-taking: When the chips are down. *Org. Behav. Hum. Perform.* **31**:194–202. [14]

Jacob, F. 1977. Evolution and tinkering. *Science* **196**:1161–1166. [7]

Jacquez, J. 1985. Compartmental Analysis in Biology and Medicine. Ann Arbor: Univ. of Michigan Press. [8]

Janis, I. L. 1972. Victims of Groupthink: A Psychological Study of Foreign-Policy Decisions and Fiascoes. Boston: Houghton Mifflin. [15]

Jensen, K. 2010. Punishment and spite, the dark side of cooperation. *Phil. Trans. R. Soc. B* **365**:2635–2650. [18]

———. 2012. Social regard: Evolving a psychology of cooperation. In: The Evolution of Primate Societies, ed. J. Mitani et al. Chicago: Chicago Univ. Press, in press. [18]

Jensen, K., J. Call, and M. Tomasello. 2007a. Chimpanzees are rational maximizers in an ultimatum game. *Science* **318**:107–109. [18]

———. 2007b. Chimpanzees are vengeful but not spiteful. *PNAS* **104**:13,046–13,050. [17, 18]

Jensen, K., B. A. Hare, J. Call, and M. Tomasello. 2006. What's in it for me? Self-regard precludes altruism and spite in chimpanzees. *Proc. Roy. Soc. B* **273**:1013–1021. [18, 20]

Jensen, K., J. B. Silk, K. Andrews, et al. 2011. Social knowledge. In: Animal Thinking: Contemporary Issues in Comparative Cognition, ed. R. Menzel and J. Fischer, pp. 265–289, Strüngmann Forum Reports, J. Lupp, series ed. Cambridge, MA: MIT Press. [18]

Jensen, K., and M. Tomasello. 2010. Punishment. In: Encyclopedia of Animal Behavior, ed. M. Breed and J. Moore, pp. 800–805. Oxford: Academic Press. [18]

Jessup, R. K., A. J. Bishara, and J. R. Busemeyer. 2008. Feedback produces divergence from prospect theory in descriptive choice. *Psychol. Sci.* **19**:1015–1022. [9]

Ji, D., and M. A. Wilson. 2007. Coordinated memory replay in the visual cortex and hippocampus during sleep. *Nat. Neurosci.* **10**:100–107. [9]

Johnson, A., and A. D. Redish. 2007. Neural ensembles in CA3 transiently encode paths forward of the animal at a decision point. *J. Neurosci.* **27**:12,176–12,189. [9]

Johnson, E. J., S. Gächter, and A. Herrmann. 2006. Exploring the nature of loss aversion. [14].

Johnson, J., and A. Sih. 2005. Pre-copulatory sexual cannibalism in fishing spiders (*Dolomedes triton*): A role for behavioral syndromes. *Behav. Ecol. Sociobiol.* **58**:390–396. [1, 15]

Johnson, T. R., D. V. Budescu, and T. S. Wallsten. 2001. Averaging probability judgments: Monte Carlo analyses of asymptotic diagnostic value. *J. Behav. Decis. Mak.* **14**:123–140. [10]

Johnson-Laird, P. N. 1983. Mental Models. Cambridge: Cambridge Univ. Press. [4]

Jolly, A. 1966. Lemur social behavior and primate intelligence. *Science* **153**:501–506. [17, 20]

Jorgensen, J. G. 1980. Western Indians: Comparative Environments, Languages, and Cultures of 172 Western American Indian Tribes. San Francisco: W. H. Freeman. [20]

Judge, T. A., and R. Ilies. 2002. Relationship of personality to performance motivation: A meta-analytic review. *J. Appl. Psychol.* **87**:797–807. [14]

Jurden, F. H. 1995. Individual differences in working memory and complex cognition. *J. Educ. Psychol.* **87**:93–102. [10]

Juslin, P., and H. Olsson. 2005. Capacity limitations and the detection of correlations: Comment on Kareev (2000). *Psychol. Rev.* **112**:256–267. [10]

Just, M. A., and P. A. Carpenter. 1992. A capacity theory of comprehension: Individual differences in working memory. *Psychol. Rev.* **99**:122–149. [10]

Kacelnik, A., and M. Bateson. 1996. Risky theories: The effects of variance on foraging decisions. *Am. Zool.* **36**:402–434. [2]

———. 1997. Risk-sensitivity: Cross-roads for theories of decision making. *Trends Cogn. Sci.* **1**:304–309. [2]

Kacelnik, A., and F. Brito e Abreu. 1998. Risky choice and Weber's Law. *J. Theor. Biol.* **194**:289–298. [2]

Kacelnik, A., and D. Brunner. 2002. Timing and foraging: Gibbon's scalar expectancy theory and optimal patch exploitation. *Learn. Motiv.* **33**:177–195. [2]

Kacelnik, A., C. Schuck-Paim, and L. Pompilio. 2006. Inconsistency in animal and human choice. In: Is There Value in Inconsistency?, ed. L. Daston and C. Engel, pp. 379–396. Baden-Baden: Nomos. [15]

Kacelnik, A., and I. A. Todd. 1992. Psychological mechanisms and the Marginal Value Theorem: Effect of variability in travel time on patch exploitation. *Anim. Behav.* **43**:313–322. [2]

Kacelnik, A., M. Vasconcelos, T. Monteiro, and J. Aw. 2011. Darwin's "tug-of-war" vs. Starlings' "horse-horse-racing": How adaptations for sequential encounters drive simultaneous choice. *Behav. Ecol. Sociobiol.* **65**:547–558. [2]

Kahneman, D., P. Slovic, and A. Tversky, eds. 1982. Judgment under Uncertainty: Heuristics and Biases. Cambridge: Cambridge Univ. Press. [4, 8, 10, 15]

Kahneman, D., and A. Tversky. 1979. Prospect theory: An analysis of decision under risk. *Econometrica* **47**:263–291. [1, 7, 14, 15]

———. 1981. The framing of decisions and the psychology of choice. *Science* **211**:453–458. [9]

———. 2000. Choices, Values, and Frames. New York: Russell Sage Foundation, Cambridge Univ. Press. [1]

Kakade, S., and P. Dayan. 2002. Dopamine: generalization and bonuses. *Neural Netw.* **15**:549–559. [9, 12]

Kalenscher, T., and C. Pennartz. 2010. Do intransitive choices reflect genuinely context-dependent preferences. In: Attention and Performance XIII: Decision Making, ed. M. R. Delgado et al., pp. 101–124. Oxford: Oxford Univ. Press. [7]

Kalenscher, T., P. N. Tobler, W. Huijbers, S. M. Daselaar, and C. Pennartz. 2010. Neural signatures of intransitive preferences. *Front. Human Neurosci.* **4**:1–14. [7]

Kamil, A. C., R. P. Balda, and D. J. Olson. 1994. Performance of four seed-caching corvid species in the radial-arm maze analog. *J. Comp. Psychol.* **108**:385–393. [1]

Kandori, M., G. T. Mailath, and R. Rob. 1993. Learning, mutation, and long run equilibria in games. *Econometrica* **61**:29–56. [10]

Kanwisher, N., J. McDermott, and M. M. Chun. 1997. The fusiform face area: A module in human extrastriate cortex specialized for face perception. *J. Neurosci.* **17**:4302–4311. [6]

Kareev, Y. 1995a. Positive bias in the perception of covariation. *Psychol. Rev.* **102**:490–502. [14]

———. 1995b. Through a narrow window: Working memory capacity and the detection of covariation. *Cognition* **56**:263–269. [10]

———. 2000. Seven (indeed, plus or minus two) and the perception of correlation. *Psychol. Rev.* **107**:397–402. [10]

———. 2005. And yet the small-sample effect does hold: Reply to Juslin and Olsson (2005) and Anderson, Doherty, Berg, and Friedrich (2005). *Psychol. Rev.* **112**:280–285. [10]

Kareev, Y., S. Arnon, and R. Horwitz-Zeliger. 2002. On the misperception of variability. *J. Exp. Psychol. Gen.* **131**:287–297. [10]

Kareev, Y., and J. Avrahami. 2007. Choosing between adaptive agents: Some unexpected implications of level of scrutiny. *Psychol. Sci.* **18**:636–641. [10]

Kareev, Y., and K. Fiedler. 2006. Non-proportional sampling and the amplification of correlations. *Psychol. Sci.* **17**:715–720. [10]

Kareev, Y., I. Lieberman, and M. Lev. 1997. Through a narrow window: Sample size and the perception of correlation. *J. Exp. Psychol. Gen.* **126**:278–287. [10]

Kareev, Y., and Y. Trope. 2011. Correct acceptance weighs more than correct rejection: A decision bias induced by question framing. *Psychon. Bull. Rev.* **18**:103–109. [10]

Kasser, T. 2002. The High Price of Materialism. Cambridge, MA: MIT Press. [14]

Kearns, M., and S. Singh. 2002. Near-optimal reinforcement learning in polynomial time. *Mach. Learn.* **49**:209–232. [9]

Keinan, G. 1987. Decision making under stress: Scanning of alternatives under controllable and uncontrollable threats. *J. Pers. Soc. Psychol.* **52**:639–644. [8]

Kennerley, S. W., A. F. Dahmubed, A. H. Lara, and J. D. Wallis. 2009. Neurons in the frontal lobe encode the value of multiple decision variables. *J. Cogn. Neurosci.* **21**:1162–1178. [9]

Killcross, S., and E. Coutureau. 2003. Coordination of actions and habits in the medial prefrontal conrtex of rats. *Cereb. Cortex* **13**:400–408. [9]

King-Casas, B., D. Tomlin, C. Anen, et al. 2005. Getting to know you: Reputation and trust in a two-person economic exchange. *Science* **308**:78–83. [9]

Kinnunen, U., and L. Pulkkinen. 2003. Childhood socioemotional characteristics as antecedents of marital stability and quality. *Eur. Psychol.* **8**:223–237. [15]

Kirby, K. N., and R. J. Herrnstein. 1995. Preference reversals due to myopic discounting of delayed reward. *Psychol. Sci.* **6**:83–89. [7]

Kirman, A. 1993. Ants, rationality, and recruitment. *Q. J. Econ.* **108**:137–156. [15]

Kitano, H. 2004. Biological robustness. *Nat. Rev. Gen.* **5**:826–837. [12]

Klein, G. A. 1998. Sources of Power: How People Make Decisions. Cambridge, MA: MIT Press. [4]

Klein, J. T., R. O. Deaner, and M. L. Platt. 2008. Neural correlates of social target value in macaque parietal cortex. *Curr. Biol.* **18**:419–424. [6]

Knill, D. C., and A. Pouget. 2004. The Bayesian brain: The role of uncertainty in neural coding and computation. *Trends Neurosci.* **27**:712–719. [7]

Koch, C. 1997. Computation and the single neuron. *Nature* **385**:207–210. [3]

———. 1999. Biophysics of Computation. Information processing in single neurons. Oxford: Oxford Univ. Press. [9]

Kocher, M., T. Cherry, S. Kroll, R. J. Netzer, and M. Sutter. 2008. Conditional cooperation on three continents. *Econ. Lett.* **101**:175–178. [15]

Kocher, M., S. Straus, and M. Sutter. 2006. Individual or team decision-making: Causes and consequences of self-selection. *Games Econ. Behav.* **56**:259–270. [15]

Kohn, A., and M. A. Smith. 2005. Stimulus dependence of neuronal correlation in primary visual cortex of the macaque. *J. Neurosci.* **25**:3661–3673. [9]

Kolodner, J. 1993. Case-Based Reasoning. San Mateo: Morgan Kaufmann. [4]

Koole, S. L., W. Jager, A. E. van den Berg, C. A. J. Vlek, and W. K. B. Hofstee. 2001. On the social nature of personality: Effects of extraversion, agreeableness, and feedback about collective resource use on cooperation in a resource dilemma. *Pers. Soc. Psychol. Bull.* **27**:289–301. [14]

Koolhaas, J. M., S. F. De Boer, C. M. Coppens, and B. Buwalda. 2010. Neuroendocrinology of coping styles: Towards understanding the biology of individual variation. *Front. Neuroendocrinol.* **31**:307–321. [13]

Koolhaas, J. M., S. M. Korte, S. F. de Boer, et al. 1999. Coping styles in animals: Current status in behavior and stress-physiology. *Neurosci. Biobehav. Rev.* **23**:925–935. [13]

Koopmans, T. C. 1960. Stationary ordinal utility and impatience. *Econometrica* **28**:287–309. [7]

Körding, K., and D. Wolpert. 2004. The loss function of sensorimotor learning. *PNAS* **101**:9839–9842. [9, 15]

Korsten, P., J. C. Mueller, C. Hermannstädter, et al. 2010. Association between DRD4 gene polymorphism and personality in great tits: A test across four wild populations. *Mol. Ecol.* **19**:832–843. [13]

Koski, S. E., H. M. H. de Vries, S. W. van den Tweel, and E. H. M. Sterck. 2007. What to do after a fight? The determinants and inter-dependency of post-conflict interactions in chimpanzees. *Behaviour* **144**:529–555. [20]

Koski, S. E., and E. H. M. Sterck. 2007. Triadic postconflict affiliation in captive chimpanzees: Does consolation console? *Anim. Behav.* **73**:133–142. [20]

———. 2010. Empathic chimpanzees: A proposal of the levels of emotional and cognitive processing in chimpanzee empathy. *Eur. J. Dev. Psychol.* 7:38–66. [18]

Koyama, N. F., C. Caws, and F. Aureli. 2006. Interchange of grooming and agonistic support in chimpanzees. *Intl. J. Primatol.* **72**:1293–1309. [17]

Krakauer, D. C. 2000. Evolving cell death in the virus-infected nervous system. *Trends Neurosci.* **23**:611–612. [8]

———. 2006. Robustness in biological systems: A provisional taxonomy. In: Complex Systems Science in Biomedicine, ed. T. Deisboek and J. Y. Kresh, pp. 183–205. New York: Springer. [8]

Krakauer, D. C., J. C. Flack, S. Dedeo, and D. Farmer. 2010. Intelligent data analysis of intelligent systems. *Lecture Notes Comp. Sci.* **6065**:8–17. [8]

Krakauer, D. C., and M. A. Nowak. 1999. Evolutionary preservation of redundant duplicated genes. *Sem. Cell Develop. Biol.* **10**:555–559. [8]

Krakauer, D. C., K. Page, and E. D. 2009. Diversity, dilemmas and monopolies of niche construction. *Am. Natural.* **173**:26–40. [8]

Krakauer, D. C., and J. Plotkin. 2004. Principles and parameters of molecular robustness. In: Robust Design: A Repertoire for Biology, Ecology, and Engineering, ed. E. Jen, pp. 71–103. Oxford: Oxford Univ. Press. [8]

Krause, A., and C. Guestrin. 2005. Optimal nonmyopic value of information in graphical models: Efficient algorithms and theoretical limits. In: Proc. 19th Intl. Joint Conf. on Artificial Intelligence, pp. 1339–1345. San Francisco: Morgan Kaufmann. [9]

Krause, J., and G. D. Ruxton. 2002. Living in Groups. Oxford: Oxford Univ. Press. [12, 15]

Krause, J., G. D. Ruxton, and S. Krause. 2010. Swarm intelligence in animals and humans. *Trends Ecol. Evol.* **25**:28–34. [12]

Krebs, J. R., and N. B. Davies. 1993. An Introduction to Behavioral Ecology, 3rd ed. Oxford: Blackwell. [2]

Krebs, J. R., A. Kacelnik, and P. Taylor. 1978. Test of optimal sampling by foraging great tits. *Nature* **275**:27–31. [2, 15]

Krebs, J. R., J. C. Ryan, and E. L. Charnov. 1974. Hunting by expectation or optimal foraging: Study of patch use by chickadees. *Anim. Behav.* **22**:953–956. [6]

Kreps, D., P. Milgrom, J. Roberts, and R. Wilson. 1982. Rational cooperation in the finitely repeated Prisoner's Dilemma. *J. Econ. Theory* **27**:245–252. [10]

Ku, S. P., A. S. Tolias, N. K. Logothetis, and J. Goense. 2011. fMRI of the face-processing network in the ventral temporal lobe of awake and anesthetized macaques. *Neuron* **70**:352–362. [6]

Kuhl, P. K. 1991. Human adults and human infants show a "perceptual magnet effect" for the prototypes of speech categories, monkeys do not. *Percept. Psychophys.* **50**:93–107. [15]

Kuhl, P. K., F.-M. Tsao, and H.-M. Liu. 2003. Foreign-language experience in infancy: Effects of short-term exposure and social interaction on phonetic learning. *PNAS* **100**:9096–9101. [15]

Kuhl, P. K., K. A. Williams, F. Lacerda, K. N. Stevens, and B. Lindblom. 1992. Linguistic experience alters phonetic perception in infants by 6 months of age. *Science* **255**:606–608. [15]

Kuhlmeier, V., K. Wynn, and P. Bloom. 2003. Attribution of dispositional states by 12-month-olds. *Psychol. Sci.* **14**:402–408. [17]

Kühn, R., and W. Wurst, eds. 2009. Gene Knockout Protocols, 2nd edition. Clifton, NJ: Humana Press. [8]

Kurzban, R. 2010. Why Everyone (Else) Is a Hypocrite: Evolution and the Modular Mind. Princeton: Princeton Univ. Press. [11]

Kushnir, T., F. Xu, and H. M. Wellman. 2010. Young children use statistical sampling to infer the preferences of others. *Psychol. Sci.* **21**:1134–1140. [20]

Laibson, D. I. 1997. Golden eggs and hyperbolic discounting. *Q. J. Econ.* **112**:443–477. [7]

Lakshminarayanan, V. 2005. Vision and the single photon. In: The Nature of Light: What Is a Photon?, ed. C. Roychoudhuri et al., vol. 5866, pp. 332–337. Bellingham: SPIE. [9]

Lakshminarayanan, V., and L. R. Santos. 2008. Capuchin monkeys are sensitive to others' welfare. *Curr. Biol.* **18**:R999–1000. [20]

Laland, K. N. 2004. Social learning strategies. *Learn. Behav.* **32**:4–14. [15]

Lamba, S., and R. Mace. 2010. People recognise when they are really anonymous in an economic game. *Evol. Human Behav.* **31**:271–278. [20]

Laming, D. R. 1997. The Measurement of Sensation. Oxford: Oxford Univ. Press. [4]

Lancaster, K. 1963. An axiomatic theory of consumer time preference. *Intl. Econ. Rev.* **4**:221–231. [7]

Lancy, D. 1996. Playing on Mother Ground: Cultural Routines for Children's Development. New York: Guilford Press. [19]

———. 2009. The Anthropology of Childhood: Cherubs, Chattel and Changlings. Cambridge Cambridge Univ. Press. [19]

———. 2010. Learning from nobody: The limited role of teaching in folk models of children's development. *Childhood in the Past* **3**:79–106. [19]

Langer, E. J. 1975. The illusion of control. *J. Pers. Soc. Psychol.* **32**:311–328. [10]

Langergraber, K., G. Schubert, C. Rowney, et al. 2011. Genetic differentiation and the evolution of cooperation in chimpanzees and humans. *Proc. Roy. Soc. B* doi:10.1098/rspb.2010.2592. [18]

Langford, D. J., S. E. Crager, Z. Shehzad, et al. 2006. Social modulation of pain as evidence for empathy in mice. *Science* **312**:1967–1970. [18]

Lansing, J. S. 2006. Perfect Order: Recognizing Complexity in Bali. Princeton: Princeton Univ. Press. [8]

Lashley, K. 1929. Brain Mechanisms and Intelligence: A Quantitative Study of Injuries to the Brain. Chicago: Univ. of Chicago Press. [9]

Laubichler, M., E. H. Hagen, and P. Hammerstein. 2005. The strategy concept and John Maynard Smith's influence on theoretical biology. *Biol. Phil.* **20**:1041–1050. [7]

Laubichler, M., and J. Maienschein, eds. 2009. From Embryology to Evo-Devo: A History of Developmental Evolution. Cambridge, MA: MIT Press. [8]

Layard, R. 2005. Happiness: Lessons from a New Science. New York: Penguin Press. [14]

Ledoux, J. 1998. The Emotional Brain. New York: Simon and Schuster. [19]

Lee, D. Y. 2006. Neural basis of quasi-rational decision making. *Curr. Opin. Neurobiol.* **16**:191–198. [6]

Lehmann, L., M. Feldman, and R. Kaeuffer. 2010. Cumulative cultural dynamics and the coevolution of cultural innovation and transmission: An ESS model for panmictic and structured populations. *J. Evol. Biol.* **23**:2356–2369. [19]

Lengyel, M., and P. Dayan. 2008. Hippocampal contributions to control: The third way. In: Advances in Neural Information Processing Systems 20, ed. J. Platt et al., pp. 889–896. Cambridge, MA: MIT Press. [9]

Lerner, J. S., and L. Z. Tiedens. 2006. Portrait of the angry decision maker: How appraisal tendencies shape anger's influence on cognition. *J. Behav. Decis. Mak.* **19**:115–137. [19]

Lesne, A. 2008. Robustness: Confronting lessons from physics and biology. *Biol. Rev.* **83**:509–532. [8]

Lettvin, J. Y., H. Maturana, W. McCulloch, and W. Pitts. 1959. What the frog's eye tells the frog's brain. *Proc. Inst. Radio Eng.* **47**:1940–1951. [4]

Lévi-Strauss, C. 1969. Elementary Structures of Kinship (translated by J. H. Bell, J. R. von Sturmer, and Rodney Needham. Boston: Beacon Press. [8]

Levin, S. A., B. Grenfell, A. Hastings, and A. S. Perelson. 1997. Mathematical and computational challenges in population biology and ecosystems science. *Science* **275**:334–343. [8]

Levine, D. K. 1998. Modeling altruism and spitefulness in experiments. *Review of Economic Dynamics* **1**:593–622. [17]

Levins, R. 1966. The strategy of model-building in population biology. *Am. Sci.* **54**:421–431. [12]

Levinson, M. 1987. Using science to bid for business. *Business Month* **129**:50–51. [19]

Lichtenstein, S., and P. Slovic, eds. 2006. The Construction of Preference. Cambridge: Cambridge Univ. Press. [14]

Lieberman, E., C. Hauert, and M. Nowak. 2005. Evolutionary dynamics on graphs. *Nature* **433**:312–316. [15]

Lindauer, M. 1957. Sonnenorientierung der Bienen unter der Aequatorsonne und zur Nachtzeit. *Naturwiss.* **44**:1–6. [3]

———. 1959. Angeborene und erlente Komponenten in der Sonnesorientierung der Bienen. *Z. Vgl. Physiol.* **42**:43–63. [3]

Linville, P. W., P. Salovey, and G. W. Fischer. 1989. Perceived distributions of the characteristics of in-group and out-group members: Empirical evidence and a computer simulation. *J. Pers. Soc. Psychol.* **57**:165–188. [10]

Liszkowski, U. 2006. Infant pointing at twelve months: Communicative goals, motives, and social-cognitive abilities. In: Roots of Human Sociality: Culture, Cognition, and Interaction, ed. N. J. Engfield and S. C. Levinson, pp. 153–178. Oxford: Berg. [18]

Littman, M., J. Goldsmith, and M. Mundhenk. 1998. The computational complexity of probabilistic planning. *J. Artif. Intell. Res.* **9**:36. [9]

Livnat, A., and N. Pippenger. 2006. An optimal brain can be composed of conflicting agents. *PNAS* **103**:3198–3202. [7]

Ljungberg, T., P. Apicella, and W. Schultz. 1992. Responses of monkey dopamine neurons during learning of behavioral reactions. *J. Neurophysiol.* **67**:145–163. [9]

Loewenstein, G. 1992. The fall and rise of psychological explanations in the economics of intertemporal choice. In: Choice over Time, ed. G. Loewenstein and J. Elster, pp. 3–35. New York: Russell Sage Foundation. [7]

———. 2003. The role of affect in decision making. In: Handbook of Affective Sciences, ed. R. J. Davidson, pp. 619–642. Oxford: Oxford Univ. Press. [8]

Loewenstein, G., and J. Elster, eds. 1992. Choice over Time. New York: Russell Sage Foundation. [4]

Logue, A. W. 1988. Research on self–control: An integrating framework. *Behav. Brain Sci.* **11**:665–709. [7]

Loomes, G. 2005. Modelling the stochastic component of behaviour in experiments: Some issues for the interpretation of data. *Exp. Econ.* **8**:301–323. [14]

———. 2010. Modelling choice and valuation in decision experiments. *Psychol. Rev.* **117**:902–924. [4]

Loomes, G., J. L. Pinto-Prades, J. M. Abellan-Perpinan, and E. Rodriguez-Miguez. 2012. Modelling noise and imprecision in individual decisions. *Psychol. Rev.*, in press. [7]

Loomes, G., C. Starmer, and R. Sugden. 1991. Observing violations of transitivity by experimental methods. *Econometrica* **59**:425–439. [7]

Loomes, G., and R. Sugden. 1995. Incorporating a stochastic element into decision theories. *Eur. Econ. Rev.* **39**:641–648. [14]

Lòpez-Moliner, J., and J. Ma Sopena. 1993. Variable binding using serial order in recurrent neural networks. In: Lecture Notes in Computer Science, vol. 686, pp. 90–95. Berlin: Springer. [3]

Lorincz, E. N., C. I. Baker, and D. I. Perrett. 1999. Visual cues for attention following in rhesus monkeys. *Cahiers Psychol. Cogn.* **18**:973–1003. [6]

Lotem, A., M. A. Fishman, and L. Stone. 1999. Evolution of cooperation between individuals. *Nature* **400**:226–227. [14]

Lucas, C., T. L. Griffiths, F. Xu, and C. Fawcett. 2009. A rational model of preference learning and choice prediction by children. *Adv. Neural Inf. Proc. Syst.* 21 http://cocosci.berkeley.edu/tom/papers/preferences1.pdf (accessed 20 June 2012). [20]

Lukaszewski, A. W., and J. R. Roney. 2011. The origins of extraversion: Joint effects of facultative calibration and genetic polymorphism. *Pers. Soc. Psychol. Bull.* **37**:409–421. [14, 15]

Lyons, D. E., A. G. Young, and F. C. Keil. 2007. The hidden structure of overimitation. *PNAS* **104**:19,751–19,756. [19, 20]

MacDonald, K. 2007. Cross-cultural comparison of learning in human hunting: Implications for life history evolution. *Human Nature* **18**:386–402. [19]

MacFarlane, G. R., S. P. Blomberg, and P. L. Vasey. 2010. Homosexual behaviour in birds: Frequency of expression is related to parental care disparity between the sexes. *Anim. Behav.* **80**:375–390. [5]

Mackintosh, N. J. 1983. Conditioning and Associative Learning. Oxford: Oxford Univ. Press. [9]

Macrae, C. N., and G. V. Bodenhausen. 2000. Social cognition: Thinking categorically about others. *Ann. Rev. Psychol.* **51**:93–120. [16]

Magnhagen, C., and F. Staffan. 2005. Is boldness affected by group composition in young-of-the-year perch (*Perca fluviatilis*)? *Behav. Ecol. Sociobiol.* **57**:295–303. [15]

Magurran, A. E. 1998. Population differentiation without speciation. *Phil. Trans. R. Soc. B* **353**:275–286. [15]

Mahajan, N., M. A. Martinez, N. L. Gutierrez, et al. 2011. The evolution of intergroup bias: Perceptions and attitudes in rhesus macaques. *J. Pers. Soc. Psychol.* **100**:387–405. [6]

Maier, S. F., J. Amat, M. V. Baratta, E. Paul, and L. R. Watkins. 2006. Behavioral control, the medial prefrontal cortex, and resilience. *Dialogues Clin. Neurosci.* **8**:397–406. [9]

Mandelbrot, B. 1983. The Fractal Geometry of Nature. New York: W. H. Freeman. [9]

Mangel, M., and C. W. Clark. 1986. Towards a Unified Foraging Theory. *Ecology* **67**:1127–1138. [7]

Manski, C. F. 1993. Identification of endogenous social effects: The reflection problem. *Rev. Econ. Stud.* **60**:531–542. [15]

Manyika, J., and H. Durrant-Whyte. 1995. Data Fusion and Sensor Management: A Decentralized Information-Theoretic Approach. Upper Saddle River, NJ: Prentice Hall. [9]

Marewski, J. N., and L. J. Schooler. 2011. Cognitive niches: An ecological model of strategy selection. *Psychol. Rev.* **118**:393–437. [1, 11]

Marlowe, F. W., and J. C. Berbesque. 2008. More "altruistic" punishment in larger societies. *Proc. Roy. Soc. B* **275**:587–590. [18]

Marr, D. 1982. Vision: A Computational Investigation into the Human Representation and Processing of Visual Information. San Francisco: W. H. Freeman. [4, 6, 7, 9, 15, 20]

Marsh, B., C. Schuck-Paim, and A. Kacelnik. 2004. Energetic state during learning affects foraging choices in starlings. *Behav. Ecol.* **15**:396–399. [7]

Martin, R. 2008. The St. Petersburg paradox. In: The Stanford Encyclopedia of Philosophy, Fall 2008 edition, ed. E. N. Zalta. Stanford: Stanford University. [7]

Martin, S. J., and R. G. M. Morris. 2000. Synaptic plasticity and memory: An evaluation of the hypothesis. *Ann. Rev. Neurosci.* **23**:649–711. [3]

Massen, J. J. M., L. M. v. d. Berg, B. M. Spruijt, and E. H. M. Sterck. 2010. Generous leaders and selfish underdogs: Pro-sociality in despotic macaques. *PLoS ONE* **5**:e9734. [20]

Massey, C., and G. Wu. 2005. Detecting regime shifts: The causes of under- and over-reaction. *Manag. Sci.* **51**:932–947. [10]

Mata, R., L. J. Schooler, and J. Rieskamp. 2007. The aging decision maker: Cognitive aging and the adaptive selection of decision strategies. *Psychol. Aging* **22**:796–810. [14]

Mathew, S., and R. Boyd. 2011. Punishment sustains large-scale cooperation in prestate warfare. *PNAS* **108**:11,375–11,380. [18, 19]

Matsumoto, K., T. Nakayama, and H. Sakai. 1999. Neuronal apoptosis inhibitory protein (NAIP) may enhance the survival of granulose cells thus indirectly affecting oocyte survival. *Mol. Reprod. Dev.* **54**:103–111. [8]

Maynard Smith, J. 1982. Evolution and the Theory of Games. Cambridge: Cambridge Univ. Press. [7, 14]

Maynard Smith, J., and G. R. Price. 1973. Logic of animal conflict. *Nature* **246**:15–18. [7, 14, 20]

Mayr, E., and W. B. Provine, eds. 1998. The Evolutionary Synthesis: Perspectives on the Unification of Biology, 2nd edition. Boston: Harvard Univ. Press. [8]

Mazar, M., B. Koszegi, and D. Ariely. 2009. Price-sensitive preferences. Unpublished manuscript available at http://elsa.berkeley.edu/~botond/PriceSensitivePreferences.pdf. (accessed 15 June 2012). [14]

Mazur, J. E. 1984. Tests of an equivalence rule for fixed and variable reinforcer delays. *J. Exp. Psychol. Anim. Behav.* **10**:426–436. [7]

———. 1988. Estimation of indifference points with an adjusting-delay procedure. *J. Exp. Psychol. Anim. Behav.* **49**:37–47. [7]

McClelland, J. 1990. Parallel Distributed Processing: Implications for Cognition and Development. Oxford: Oxford Univ. Press. [8]

McClure, S. M., K. M. Ericson, D. I. Laibson, G. Loewenstein, and J. D. Cohen. 2007. Time discounting for primary rewards. *J. Neurosci.* **27**:5796–5804. [7]

McClure, S. M., D. I. Laibson, G. Loewenstein, and J. D. Cohen. 2004. Separate neural systems value immediate and delayed monetary rewards. *Science* **306**:503–507. [7]

McCrae, R. R., and J. Allik. 2002. The Five-Factor Model of Personality across Cultures. New York: Kluwer Academic. [15]

McCrae, R. R., and P. T. Costa. 1987. Validation of the 5-factor model of personality across instruments and observers. *J. Pers. Soc. Psychol.* **52**:81–90. [14]

McCulloch, W. S., and W. Pitts. 1943. A logical calculus of the ideas immanent in nervous activity. *Bull. Math. Biophys.* **5**:115–133. [3]

McElreath, R., T. H. Clutton-Brock, E. Fehr, et al. 2003. The role of cognition and emotion in cooperation. In: Genetic and Cultural Evolution of Cooperation, ed. P. Hammerstein, pp. 125–152, Dahlem Workshop Reports, J. Lupp, series ed. Cambridge, MA: MIT Press. [18]

McFadden, D., and K. Train. 2000. Mixed MNL models for discrete response. *J. Appl. Econ.* **15**:447–470. [20]

McGregor, I., and B. R. Little. 1998. Personal projects, happiness, and meaning: On doing well and being yourself. *J. Pers. Soc. Psychol.* **74**:494–512. [14]

McGuire, M. T. 1991. Moralistic aggression and the sense of justice. *Am. Behav. Sci.* **34**:371–385. [18]

McKay, R., and C. Efferson. 2010. The subtleties of error management. *Evol. Human Behav.* **31**:309–319. [5]

McKelvey, R., and T. Palfrey. 1995. Quantal response equilibria for normal form games. *Games Econ. Behav.* **10**:6–38. [20]

———. 1998. Quantal response equilibria for extensive form games. *Exp. Econ.* **1**:9–41. [20]

McNamara, J. M. 1996. Risk-prone behaviour under rules which have evolved in a changing environment. *Am. Zool.* **36**:484–495. [12]

McNamara, J. M., B. Barta, L. Fromhage, and A. I. Houston. 2008. The coevolution of choosiness and cooperation. *Nature* **451**:189–192. [15, 17]

McNamara, J. M., Z. Barta, and A. I. Houston. 2004. Variation in behaviour promotes cooperation in the Prisoner's Dilemma game. *Nature* **428**:745–748. [14, 15]

McNamara, J. M., C. E. Gasson, and A. I. Houston. 1999. Incorporating rules for responding into evolutionary games. *Nature* **401**:368–371. [2]

McNamara, J. M., and A. I. Houston. 1980. The application of statistical decision theory to animal behaviour. *J. Theor. Biol.* **85**:673–690. [7]

———. 1985a. Optimal foraging and learning. *J. Theor. Biol.* **117**:231–249. [10]

———. 1985b. A simple model of information use in the exploitation of patchily distributed food. *Anim. Behav.* **33**:553–560. [12]

———. 1986. The common currency for behavioral decisions. *Am. Natural.* **127**:358–378. [7]

———. 1987. Memory and the efficient use of information. *J. Theor. Biol.* **125**:385–395. [10]

———. 1992. Risk-sensitive foraging: A review of the theory. *Bull. Math. Biol.* **54**:355–378. [7]

———. 2009. Integrating function and mechanism. *Trends Ecol. Evol.* **24**:670–675. [1, 2, 5, 12]

McNamara, J. M., and O. Leimar. 2010 Variation and the response to variation as a basis for successful cooperation. *Phil. Trans. R. Soc. B* **365**:2627–2633. [15]

McNamara, J. M., P. A. Stephens, S. R. X. Dall, and A. I. Houston. 2009. Evolution of trust and trustworthiness: social awareness favours personality differences. *Proc. Roy. Soc. B* **276**:605–613. [14, 15]

McNamara, J. M., P. C. Trimmer, A. Eriksson, J. A. R. Marshall, and A. I. Houston. 2011. Environmental variability can select for optimism or pessimism. *Ecol. Lett.* **14**:58–62. [5, 12]

McNamara, J. M., P. C. Trimmer, and A. I. Houston. 2012. The ecological rationality of state-dependent valuation. *Psychol. Rev.* **119**:114–119. [7]

McNaughton, B. L., F. P. Battaglia, O. Jensen, E. I. Moser, and M. B. Moser. 2006. Path integration and the neural basis of the cognitive map. *Nat. Neurosci.* **7**:663–678. [7]

Meier, B. P., M. D. Robinson, M. S. Carter, and V. B. Hinsz. 2010. Are sociable people more beautiful? A zero-acquaintance analysis of agreeableness, extraversion, and attractiveness. *J. Res. Personality* **44**:293–296. [15]

Melis, A. P., B. Hare, and M. Tomasello. 2006a. Chimpanzees recruit the best collaborators. *Science* **311**:1297–1300. [17]

———. 2006b. Engineering cooperation in chimpanzees: Tolerance constraints in chimpanzees. *Anim. Behav.* **72**:275–286. [17]

———. 2008. Do chimpanzees reciprocate received favours? *Anim. Behav.* **76**:951–962. [17]

Melis, A. P., F. Warneken, K. Jensen, et al. 2010. Chimpanzees help conspecifics obtain food and non-food items. *Proc. Roy. Soc. B* **278**:1405–1413. [17, 18]

Mellers, B. A. 1982. Equity judgment: A revision of Aristotelian views. *J. Exp. Psychol. Gen.* **111**:242–270. [14]

Meltzoff, A. N. 1995. Understanding the intentions of others: Re-enactment of intended acts by 18-month-old children. *Dev. Psychol.* **31**:1–16. [17]

Menzel, R., U. Greggers, A. Smith, et al. 2005. Honey bees navigate according to a map-like spatial memory. *PNAS* **102**:3040–3045. [7]

Menzel, R., A. Kirbach, W.-D. Haass, et al. 2011. A common frame of reference for learned and communicated vectors in honeybee navigation. *Curr. Biol.* **21**:645–650. [3]

Metz, J. A. J., R. M. Nisbet, and S. A. H. Geritz. 1992. How should we define fitness for general ecological scenarios? *Trends Ecol. Evol.* **7**:198–202. [5]

Miller, G. A. 1956. The magical number seven, plus or minus two: Some limits on our capacity to process information. *Psychol. Rev.* **63**:81–97. [9, 10]

Miller, G. F., and P. M. Todd. 1998. Mate choice turns cognitive. *Trends Cogn. Sci.* **2**:190–198. [15]

Millikan, R. G. 1984. Naturalistic reflections on knowledge. *Pac. Phil. Q.* **65**:315–334. [5]

Mischel, W. 1968. Personality and Assessment. Mahwah, NJ: Lawrence Erlbaum. [14]

Mizell, L. 1997. Aggressive driving. In: Aggressive Driving Three Studies. Washington, D.C.: AAA Foundation for Traffic Safety. [19]

Moore, C. 2009. Fairness in children's resource allocation depends on the recipient. *Psychol. Sci.* **20**:944–948. [17]

Moore, T., K. M. Armstrong, and M. Fallah. 2003. Visuomotor origins of covert spatial attention. *Neuron* **40**:671–683. [6]

Morales, A. J. 2002. Absolutely expedient imitative behavior. *Intl. J. Game Theory* **31**:475–492. [10]

Moses, P., and J. Stiles. 2002. The lesion methodology: Contrasting views from adult and child studies. *Dev. Psychobiol.* **40**:266–277. [9]

Mottley, K., and L. A. Giraldeau. 2000. Experimental evidence that group foragers can converge on predicted producer-scrounger equilibria. *Anim. Behav.* **60**:341–350. [15]

Muller, M., and J. C. Mitani. 2005. Conflict and cooperation in wild chimpanzees. *Adv. Study Behav.* **35**:275–331. [17]

Munafò, M. R., B. Yalcin, S. A. Willis-Owen, and J. Flint. 2008. Association of the dopamine D4 receptor (DRD4) gene and approach-related personality traits: Meta-analysis and new data. *Biol. Psych.* **63**:197–206. [13]

Murray, J., and E. Stanley. 1986. On the spatial spread of rabies among foxes. *Proc. R. Soc. B* **229**:111–150. [8]

Mussweiler, T. 2003. Comparison processes in social judgment: Mechanisms and consequences. *Psychol. Rev.* **110**:472–489. [16]

Mussweiler, T., and K. Epstude. 2009. Relatively fast! Efficiency advantages of comparative thinking. *J. Exp. Psychol. Gen.* **138**:1–21. [16]

Mussweiler, T., and A.-C. Posten. 2012. Relatively certain! Comparative thinking reduces uncertainty. *Cognition* **122**:236–240. [16]

Mussweiler, T., and K. Rüter. 2003. What friends are for! The use of routine standards in social comparison. *J. Pers. Soc. Psychol.* **85**:467–481. [16]

Myers, J. L., and E. Sadler. 1960. Effects of range of payoffs as a variable in risk taking. *J. Exp. Psychol.* **60**:306–309. [15]

Myerson, R. B. 1978. Refinement of the Nash Equilibrium Concept. *Intl. J. Game Theory* **7**:73–80. [10]

Nagel, R. 1995. Unraveling in guessing games: An experimental study. *Am. Econ. Rev.* **85**:1313–1326. [15]

Navarick, D. J., and E. Fantino. 1972. Transitivity as a property of choice. *J. Exp. Anal. Behav.* **18**:389–401. [7]

———. 1974. Stochastic transitivity and unidimensional behavior theories. *Psychol. Rev.* **81**:426–441. [7]

———. 1975. Stochastic transitivity and the unidimensional control of choice. *Learn. Motiv.* **6**:179–201. [7]

Nelissen, R. A., and M. Zeelenberg. 2009. Moral emotions as determinants of third-party punishment: Anger, guilt, and the functions of altruistic sanctions. *Judgm. Decis. Mak.* **4**:543–553. [1, 19]

Nesse, R. M. 2001a. Natural selection and the capacity for subjective commitment. In: Evolution and the Capacity for Commitment, ed. R. M. Nesse. New York: Russell Sage Foundation. [19]

———. 2001b. The smoke detector principle: Natural selection and the regulation of defenses. *Ann. NY Acad. Sci.* **935**:75–85. [1, 5]

———. 2005. Natural selection and the regulation of defenses: A signal detection analysis of the smoke detector principle. *Evol. Human Behav.* **26**:283–286. [5]

Nesse, R. M., and G. C. Williams. 1996. Why We Get Sick: The New Science of Darwinian Medicine. New York: Vintage. [7]

Nettle, D. 2004. Adaptive illusions: Optimism, control, and human rationality. In: Evolution, Emotion and Rationality, ed. D. Evans and P. Cruse. Oxford: Oxford Univ. Press. [5]

———. 2005. An evolutionary approach to the extraversion continuum. *Evol. Human Behav.* **26**:363–373. [14]

———. 2006. The evolution of personality variation in humans and other animals. *Am. Psychol.* **61**:622–631. [14]

———. 2009. An evolutionary model of low mood states. *J. Theor. Biol.* **257**:100–103. [14]

Nettle, D., and L. Penke. 2010. Personality: bridging the literatures from human psychology and behavioural ecology. *Phil. Trans. R. Soc. B* **365**:4043–4050. [14]

Neuhoff, J. G. 2001. An adaptive bias in the perception of looming auditory motion. *Ecol. Psychol.* **13**:87–110. [5]

Neves, G., S. F. Cooke, and T. V. Bliss. 2008. Synaptic plasticity, memory and the hippocampus: A neural network approach to causality. *Nat. Rev. Neurosci.* **9**:65–75. [3]

Newell, A., and H. A. Simon. 1972. Human Problem Solving. Englewood Cliffs, NJ: Prentice-Hall. [4]

Newport, E. L. 1988. Constraints on learning and their role in language acquisition: Studies of the acquisition of American Sign Language. *Lang. Sci.* **10**:147–172. [10]

———. 1990. Maturational constraints on language learning. *Cogn. Sci.* **14**:11–28. [10]

Ng, A. Y., D. Harada, and S. Russell. 1999. Policy invariance under reward transformations: Theory and application to reward shaping. In: Proc. 16th Intl. Conf. on Machine Learning, pp. 278–287. San Francisco: Morgan Kaufmann. [9]

Nickerson, R. 1998. Confirmation bias: A ubiquitous phenomenon in many guises. *Rev. Gen. Psych.* **2**:175–220. [10]

Niedrich, R. W., S. Sharma, and D. H. Wedell. 2001. Reference price and price perceptions: A comparison of alternative models. *J. Consum. Res.* **28**:339–354. [14]

Niedrich, R. W., D. Weathers, R. C. Hill, and D. R. Bell. 2009. Specifying price judgments with range-frequency theory in models of brand choice. *J. Mark. Res.* **46**:693–702. [14]

Nielsen, M. 2009. 12-month-olds produce others' intended but unfulfilled acts. *Infancy* **14**:377–389. [17]

Nielsen, M., and K. Tomaselli. 2010. Overimitation in Kalahari Bushman children and the origins of human cultural cognition. *Psychol. Sci.* **21**:729–736. [19]

Nisbett, R., and D. Cohen. 1996. The Culture of Honor: The Psychology of Violence in the South. Boulder Westview Press. [19]

Noel, J. A., W. Teizer, and W. Hwang. 2009. Surface manipulation of microtubules using self-assembled monolayers and electrophoresis. *ACS Nano* **3**:1938–1946. [8]

Nowak, M., and R. Highfield. 2011. Supercooperators. Edinburgh: Canongate. [14]

Nowak, M., and R. M. May. 1992. Evolutionary games and spatial chaos. *Nature* **359**:826–829. [10]

Nowak, M., and K. Sigmund. 1993. A strategy of win-stay, lose-shift that outperforms tit-for-tat in the Prisoner's Dilemma game. *Nature* **364**:56–58. [9, 10]

———. 1998. Evolution of indirect reciprocity by image scoring. *Nature* **393**:573–577. [14]

———. 2005. Evolution of indirect reciprocity. *Nature* **437**:1291–1298. [11]

Nowak, M., C. E. Tarnita, and E. O. Wilson. 2010. The evolution of eusociality. *Nature* **466**:1057–1062. [8]

Nussey, D. H., A. J. Wilson, and J. E. Brommer. 2007. The evolutionary ecology of individual phenotypic plasticity in wild populations. *J. Evol. Biol.* **20**:831–844. [14, 15]

Nygren, T. E., A. M. Isen, P. J. Taylor, and J. Dulin. 1996. The influence of positive affect on the decision rule in risk situations: Focus on outcome (and especially avoidance of loss) rather than probability. *Org. Behav. Hum. Decis. Process.* **66**:59–72. [1, 14]

Oaksford, M., and N. Chater. 2007. Bayesian Rationality. Oxford: Oxford Univ. Press. [4]

Oaksford, M., F. Morris, B. Grainger, and J. M. G. Williams. 1996. Mood, reasoning, and central executive processes. *J. Exp. Psychol. Learn. Mem. Cogn.* **22**:476–492. [14]

Odling-Smee, F. J., K. N. Laland, and M. W. Feldman. 2003. Niche Construction: The Neglected Process in Evolution. vol. 37. Princeton Princeton Univ. Press. [8]

Offerman, T. 2002. Hurting hurts more than helping helps. *Eur. Econ. Rev.* **46**:1423–1437. [18]

Öhman, A., and S. Mineka. 2001. Fears, phobias, and preparedness: Toward an evolved module of fear and fear learning. *Psychol. Rev.* **108**:483–522. [16]

Olivola, C. Y., and N. Sagara. 2009. Distributions of observed death tolls govern sensitivity to human fatalities. *PNAS* **106**:22,151–22,156. [1, 14]

Olshausen, B. A., and D. J. Field. 1996. Emergence of simple-cell receptive field properties by learning a sparse code for natural images. *Nature* **381**:607–609. [9]

———. Vision and the coding of natural images. *Am. Sci.* **88**:238–245. [7]

Olson, K. R., and E. S. Spelke. 2008. Foundations of cooperation in preschool children. *Cognition* **108**:222–231. [17]

Oppenheimer, D. M. 2008. The secret life of fluency. *Trends Cogn. Sci.* **12**:237–241. [7]

Oram, M., and D. Perrett. 1994. Responses of anterior superior temporal polysensory (STPa) neurons to "biological motion" stimuli. *J. Cogn. Neurosci.* **6**:99–116. [6]

Orians, G. H. 1969. On evolution of mating systems in birds and mammals. *Am. Natural.* **103**:589–603. [6]

Ortony, A., G. L. Clore, and A. Collins. 1988. The Cognitive Structure of Emotions. Cambridge: Cambridge Univ. Press. [18]

Ossher, H., and P. Tarr. 2001. Using multidimensional separation of concerns to (re) shape evolving software. *Commun. ACM* **44**:43–50. [11]

Oswald, A. J., and S. Wu. 2010. Objective confirmation of subjective measures of human well-being: Evidence from the USA. *Science* **327**:576–579. [14]

Ozer, D. J., and V. Benet-Martinez. 2006. Personality and the prediction of consequential outcomes. *Ann. Rev. Psychol.* **57**:401–421. [13, 15]

Oztop, E., M. Kawato, and M. Arbib. 2006. Mirror neurons and imitation: A computationally guided review. *Neural Netw.* **19**:254–271. [9, 12]

Pachur, T., R. Hertwig, and J. Rieskamp. 2011. The mind as an intuitive pollster: Frugal search in social spaces. In: Simple Heuristics in a Social World, ed. R. Hertwig et al. New York: Oxford Univ. Press. [15]

Packer, C. 1986. The ecology of sociality in felids. In: Ecological Aspects of Social Evolution, ed. D. I. Rubenstein and W. Wrangham, pp. 429–451. Princeton: Princeton Univ. Press. [15]

Padoa-Schioppa, C. 2007. Orbitofrontal cortex and the computation of economic value. *Ann. NY Acad. Sci.* **1121**:232–253. [6]

———. 2009. Range-adapting representation of economic value in the orbitofrontal cortex. *J. Neurosci.* **29**:14,004–14,014. [9]

Parducci, A., S. Knobel, and C. Thomas. 1976. Independent contexts for category ratings: Range-frequency analysis. *Percept. Psychophys.* **20**:360–366. [14]

Parducci, A., and L. F. Perrett. 1971. Category rating scales: Effects of relative spacing and frequency of stimulus values. *J. Exp. Psychol.* **89**:427–452. [14]

Parker, G. A., and R. A. Stuart. 1976. Animal behavior as a strategy optimizer: Evolution of resource assessment strategies and optimal emigration thresholds. *Am. Natural.* **110**:1055–1076. [2]

Parr, L. A., J. T. Winslow, W. D. Hopkins, and F. B. M. de Waal. 2000. Recognizing facial cues: Individual discrimination by chimpanzees (*Pan troglodytes*)and rhesus monkeys (*Macaca mulatta*). *J. Comp. Psychol.* **114**:47. [6]

Pavlov, I. 1928. The Inhibitory Type of Nervous Systems in the Dog. Lectures on Conditioned Reflexes:Twenty-Five Years of Objective Study of the Higher Nervous Activity (Behavior) of Animals. New York: Liverwright Publ. [13]

Payne, J. W., J. R. Bettman, and E. J. Johnson. 1993. The Adaptive Decision Maker. Cambridge: Cambridge Univ. Press. [4, 8, 15]

Pearce, J. M., and G. Hall. 1980. A model for Pavlovian learning: Variations in the effectiveness of conditioned but not of unconditioned stimuli. *Psychol. Rev.* **87**:532–552. [9]

Pearl, J. 2000. Causality: Models, Reasoning and Inference. Cambridge: Cambridge Univ. Press. [4]

Pearl, J. 2010. Causality, 2nd edition. Cambridge: Cambridge Univ. Press. [8]

Pellegrini, A. D., and J. D. Long. 2002. A longitudinal study of bullying, dominance and victimization during the transition from primary through secondary school. *Br. J. Dev. Psychol.* **20**:259–280. [15]

Penke, L., J. J. A. Denissen, and G. F. Miller. 2007a. Evolution, genes, and inter-disciplinary personality research. *Eur. J. Pers.* **21**:639–665. [14]

———. 2007b. The evolutionary genetics of personality. *Eur. J. Pers.* **21**:549–587. [14]

Perkins, H. W., ed. 2003. The social norms approach to preventing school and college age substance abuse. New York: Jossey-Bass. [14]

Perreault, C., C. Moya, and R. Boyd. 2012. A Bayesean approach to the evolution of social learning. *Evol. Human Behav.*, in press. [19]

Perrett, D., P. Smith, D. Potter, et al. 1984. Neurones responsive to faces in the temporal cortex: Studies of functional organization, sensitivity to identity and relation to perception. *Human Neurobiol.* **3**:197. [6]

Perrett, D. I., P. A. Smith, D. D. Potter, et al. 1985. Visual cells in the temporal cortex sensitive to face view and gaze direction. *Proc. Roy. Soc. B* **223**:293–317. [6]

Pervin, L., and O. P. John. 1999. Handbook of Personality: Theory and Research. New York: Guilford Press. [13]

Petty, R. E., R. H. Fazio, and P. Briñol, eds. 2008. Attitudes: Insights from the new implicit measures. New York: Psychology Press. [16]

Phillips, W., J. L. Barnes, N. Mahajan, M. Yamaguchi, and L. R. Santos. 2009. "Unwilling" versus "unable": Capuchin monkeys' (*Cebus apella*) understanding of human intentional action. *Dev. Sci.* **12**:938–945. [17]

Pillutla, M. M., and J. K. Murnighan. 1996. Unfairness, anger, and spite: Emotional rejections of ultimatum offers. *Org. Behav. Hum. Decis. Process.* **68**:208–224. [18]

Pinker, S. 1995. The Language Instinct: How the Mind Creates Language. New York: Harper Collins. [18]

———. 1997. How the Mind Works. New York: W. W. Norton. [11]

———. 2010. The cognitive niche: Coevolution of intelligence, sociality, and language. *PNAS* **107**:8993–8999. [19]

Pinsker, H. M., W. A. Hening, T. J. Carew, and E. R. Kandel. 1973. Long-term sensitization of a defensive withdrawal reflex in aplysia. *Science* **182**:1039–1042. [9]

Piper, W. H., and H. Wiley. 1989. Correlates of dominance in wintering white-throated sparrows: Age, sex and location. *Anim. Behav.* **37**:298–310. [15]

Platt, M. L., and P. W. Glimcher. 1999. Neural correlates of decision variables in parietal cortex. *Nature* **400**:233–238. [6]

Plott, C. R. 1996. Rational individual behavior in markets and social choiceprocesses: The discovered preference hypothesis. In: Rational Foundations of Economic Behavior, ed. K. Arrow et al., pp. 225–250. London: Macmillan. [14]

Pompilio, L., and A. Kacelnik. 2005. State-dependent learning and suboptimal choice: When starlings prefer long over short delays to food. *Anim. Behav.* **70**:571–578. [7]

Pompilio, L., A. Kacelnik, and S. T. Behmer. 2006. State-dependent learned valuation drives choice in an invertebrate. *Science* **311**:1613–1615. [2, 7]

Ponte, E., E. Bracco, J. Faix, and S. Bozzaro. 1998. Detection of subtle phenotypes: the case of the cell adhesion molecule csA in Dictyostelium. *PNAS* **95**:9360–9365. [8]

Posch, M. 1999. Win-stay, lose-shift strategies for repeated games: Memory length, aspiration levels and noise. *J. Theor. Biol.* **198**:183–195. [10]

Posch, M., A. Pichler, and K. Sigmund. 1999. The efficiency of adapting aspiration levels. *Proc. Roy. Soc. B* **266**:1427–1435. [10]

Posner, M. I., C. R. Snyder, and B. J. Davidson. 1980. Attention and the detection of signals. *J. Exp. Psychol.* **109**:160–174. [6]

Pouget, A., P. Dayan, and R. Zemel. 2000. Information processing with population codes. *Nat. Rev. Neurosci.* **1**:125–132. [9]

———. 2003. Inference and computation with population codes. *Ann. Rev. Neurosci.* **26**:381–410. [9]

Prelec, D., B. Wernerfelt, and F. Zettelmeyer. 1997. The role of inference in context effects: Inferring what you want from what is available. *J. Consum. Res.* **24**:118–125. [14]

Preston, S. D., and F. B. M. de Waal. 2002. Empathy: Its ultimate and proximate bases. *Behav. Brain Sci.* **25**:1–20. [18]

Price, M. E., L. Cosmides, and J. Tooby. 2002. Punitive sentiment as an anti-free rider psychological device. *Evol. Human Behav.* **23**:203–231. [18]

Prins, H. H. T. 1996. Ecology and Behaviour of the African Buffalo: Social Inequality and Decision Making. London: Chapman and Hall. [6]

Prinz, J. J. 2007. The Emotional Construction of Morals. Oxford: Oxford Univ. Press. [18]

Prior, I. 2001. Compartmentalization of Ras proteins. *J. Cell Sci.* **114**:1603–1608. [8]

Pruitt, D. G. 1971. Choice shifts in group discussion: An introductory review. *J. Pers. Soc. Psychol.* **20**:339–360. [15]

Puterman, M. L. 2005. Markov Decision Processes: Discrete Stochastic Dynamic Programming. Chichester: Wiley. [9]

Pylyshyn, Z. W. 1984. Computation and Cognition. Cambridge, MA: MIT Press. [4]

Queitsch, C., T. A. Sangster, and S. Lindquist. 2002. Hsp90 as a capacitor of phenotypic variation. *Nature* **417**:618–624. [8]

Quine, W. V. O. 1951. Two dogmas of empiricism. *Philos. Rev.* **60**:20–43. [4]

Rabin, M. 1993. Incorporating fairness into game theory and economics. *Am. Econ. Rev.* **83**:1281–1302. [17]

Rachlin, H., and L. Green. 1972. Commitment, choice and self control. *J. Exp. Anal. Behav.* **17**:15–22. [7]

Raff, E. C., and R. A. Raff. 2000. Dissociability, modularity, evolvability. *Evol. Devel.* **2**:235–237. [8]

Raff, R. A., and B. J. Sly. 2000. Modularity and dissociation in the evolution of gene expression territories in development. *Evol. Devel.* **2**:102–113. [8]

Raihani, N. J., A. S. Grutter, and R. Bshary. 2010. Punishers benefit from third-party punishment in fish. *Science* **327**:171. [18]

Raihani, N. J., A. I. Pinto, A. S. Grutter, S. Wismer, and R. Bshary. 2012a. Male cleaner wrasses adjust punishment of female partners according to the stakes. *Proc. R. Soc. B* **279**: [18]

Raihani, N. J., A. Thornton, and R. Bshary. 2012b. Punishment and cooperation in nature. *Trends Ecol. Evol.* **27**:288–295. [18]

Rakoczy, H., F. Warneken, and M. Tomasello. 2008. The sources of normativity: Young children's awareness of the normative structure of games. *Dev. Psychol.* **44**:875–881. [18]

Rakow, T., and K. Miler. 2009. Doomed to repeat the successes of the past: History is best forgotten for repeated choices with nonstationary payoffs. *Mem. Cogn.* **37**:985–1000. [10]

Ramist, L., C. Lewis, and L. McCamley-Jenkins. 1993. Student group differences in predicting college grades: Sex, language, and ethnic groups. College Board Report No. 93–1. New York: The College Board. [15]

Ramsey, F. 1926/1980. Truth and probability. In: Studies in Subjective Probability, 2nd edition, ed. H. E. Kyburg and H. E. Smokler, pp. 25–52. New York: Robert Krieger. [15]

Range, F., Z. Virányi, and L. Huber. 2007. Selective imitation in dogs. *Curr. Biol.* **17**:868–872. [20]

Rasmussen, C. E., and C. K. I. Williams. 2006. Gaussian Processes for Machine Learning. London: MIT Press. [9]

Ratcliff, R. 1978. A theory of memory retrieval. *Psychol. Rev.* **85**:59. [6]

Ratcliff, R., and G. McKoon. 2008. The diffusion decision model: Theory and data for two-choice decision tasks. *Neural Comput.* **20**:873–922. [6]

Ratcliffe, J. M., M. B. Fenton, and B. G. Galef. 2003. An exception to the rule: Common vampire bats do not learn taste aversions. *Anim. Behav.* **65**:385–389. [7, 19]

Ratneshwar, S., A. D. Shocker, and D. W. Stewart. 1987. Toward understanding the attraction effect: The implications of product stimulus meaningfulness and familiarity. *J. Consumer Res.* **13**:520–533. [7]

Ray, D., B. King-Casas, P. R. Montague, and P. Dayan. 2009. Bayesian model of behaviour in economic games. In: Advances in Neural Information Processing Systems 21, ed. D. Koller et al., pp. 1345–1352. San Mateo, CA: Morgan Kaufmann. [9]

Read, D., and N. L. Read. 2004. Time discounting over the lifespan. *Org. Behav. Hum. Decis. Process.* **94**:22–32. [14]

Réale, D., S. M. Reader, D. Sol, P. McDougall, and N. J. Dingemanse. 2007. Integrating temperament in ecology and evolutionary biology. *Biol. Rev.* **82**:291–318. [13]

Reber, R., and C. Unkelbach. 2010. The epistemic status of processing fluency as source for judgments of truth. *Rev. Phil. Psych.* **1**:563–583. [7]

Reimers, S., E. A. Maylor, N. Stewart, and N. Chater. 2009. Associations between a one-shot delay discounting measure and age, income, education and real-world impulsive behavior. *Pers. Ind. Diff.* **47**:973–978. [14]

Rendell, L., R. Boyd, D. Cownden, et al. 2010. Why copy others? Insights from the social learning strategies tournament. *Science* **328**:208–213. [19]

Resnick, R., and H. R. Varian. 1997. Recommender systems: Guest editors' introduction. *Commun. ACM* **40**:56–58. [15]

Reynolds, C. W. 1987. Flocks, herds, and schools: A distributed behavioral model. *Comp. Graph.* **21**:25–34. [15]

Reynolds, S. M., and K. C. Berridge. 2001. Fear and feeding in the nucleus accumbens shell: Rostrocaudal segregation of gaba-elicited defensive behavior versus eating behavior. *J. Neurosci.* **21**:3261–3270. [9]

———. 2008. Emotional environments retune the valence of appetitive versus fearful functions in nucleus accumbens. *Nat. Neurosci.* **11**:423–425. [9]

Richerson, P. J., and R. Boyd. 2005. Not by Genes Alone: How Culture Transformed Human Evolution. Chicago: Univ. of Chicago Press. [18, 20]

Richeson, J. A., and S. Trawalter. 2008. The threat of appearing prejudiced and race-based attentional biases. *Psychol. Sci.* **19**:98–102. [16]

Rieke, F., D. Warland, R. de Ruyter van Steveninck, and W. Bialek. 1997. Spikes: Exploring the Neural Code. Cambridge, MA: MIT Press. [7]

Rieskamp, J., J. R. Busemeyer, and B. A. Mellers. 2006. Extending the bounds of rationality: Evidence and theories of preferential choice. *J. Econ. Lit.* **44**:631–661. [7]

Rieskamp, J., and P. E. Otto. 2006. SSL: A theory of how people learn to select strategies. *J. Exp. Psychol. Gen.* **135**:207–236. [1]

Rigotti, M., D. B. D. Rubin, X.-J. Wang, and S. Fusi. 2010. Internal representation of task rules by recurrent dynamics: The importance of the diversity of neural responses. Front Comput Neurosci, 4:24. . *Front. Comput. Neurosci.* **4**: [9]

Rilling, J. K., and A. G. Sanfey. 2011. The neuroscience of social decision-making. *Annu. Rev. Psychol.* **62**:23–48. [8]

Rips, L. J. 1994. The Psychology of Proof. Cambridge, MA: MIT Press. [4]

Riskey, D. R., A. Parducci, and G. K. Beauchamp. 1979. Effects of context in judgments of sweetness and pleasantness. *Percept. Psychophys.* **26**:171–176. [14]

Roberts, B. W., and E. M. Donahue. 1994. One personality, multiple selves: Integrating personality and social roles. *J. Pers.* **62**:199–218. [14]

Roberts, B. W., N. R. Kuncel, R. Shiner, A. Caspi, and L. R. Goldberg. 2007. The power of personality: The comparative validity of personality traits, socioeconomic status, and cognitive ability for predicting important life outcomes. *Perspect. Psychol. Sci.* **2**:313–345. [13–15]

Roberts, B. W., K. E. Walton, and W. Viechtbauer. 2006. Patterns of mean-level change in personality traits across the life course: A metaanalysis of longitudinal studies. *Psychol. Bull.* **132**:1–25. [15]

Roberts, B. W., and D. Wood. 2006. Personality development in the context of the neo-socioanalytic model of personality. In: Handbook of Personality Development, ed. D. K. Mroczek and T. D. Little, pp. 11–39. Mahwah, NJ: Erlbaum. [15]

Robinson, G. E. 1992. Regulation of division of labour in insect societies. *Annual Rev. Entomol.* **37**:637–665. [12]

Robinson, S., S. M. Sandstrom, V. H. Denenberg, and R. D. Palmiter. 2005. Distinguishing whether dopamine regulates liking, wanting, and/or learning about rewards. *Behav. Neurosci.* **119**:5–15. [9]

Rochat, P., M. D. G. Dias, L. Guo, et al. 2009. Fairness in distributive justice by 3- and 5-year-olds across seven cultures. *J. Cross Cult. Psychol.* **40**:416–442. [18]

Roesch, M. R., and C. R. Olson. 2005. Neuronal activity in primate orbitofrontal cortex reflects the value of time. *J. Neurophysiol.* **94**:2457–2471. [9]

Roese, N. J., and J. W. Sherman. 2007. Expectancies. In: Social Psychology: Handbook of Basic Principles, 2nd edition, ed. E. T. Higgins and A. W. Kruglanski, pp. 91–115. New York: Guilford Press. [16]

Rogers, A. 1988. Does biology constrain culture? *Am. Anthropol.* **90**:819–831. [19, 20]

Rohde, K. I. M. 2005. The hyperbolic factor: A measure of decreasing impatience, Research Memoranda 044. Maastricht METEOR, Maastricht Research School of Economics of Technology and Organization. [7]

Roma, P. G., A. Silberberg, A. M. Ruggiero, and S. J. Suomi. 2006. Capuchin monkeys, inequity aversion, and the frustration effect. *J. Comp. Psychol.* **120**:67–73. [18]

Romer, P. 1990. Endogenous technological change. *J. Pol. Econ.* **98**:71–102. [19]

Rosati, A. G., L. R. Santos, and B. Hare. 2010. Primate social cognition: Thirty years after Premack and Woodruff. In: Primate Neuroethology, ed. M. L. Platt and A. A. Ghazanfar, pp. 117–143. Oxford: Oxford Univ. Press. [17]

Rosati, A. G., and J. R. Stevens. 2009. Rational decisions: The adaptive nature of context-dependent choice. In: Rational Animals, Irrational Humans, ed. S. Watanabe et al., pp. 101–117. Tokyo: Keio Univ. Press. [1]

Rosati, A. G., J. R. Stevens, B. Hare, and M. Hauser. 2007. The evolutionary origins of human patience: Temporal preferences in chimpanzees, bonobos, and adult humans. *Curr. Biol.* **17**:1663–1668. [1]

Rosenhead, J., M. Elton, and S. K. Gupta. 1972. Robustness and optimality as criteria for strategic decisions. *J. Oper. Res. Soc.* **23**:413–431. [12]

Roskos-Ewoldsen, D. R., and R. H. Fazio. 1992. On the orienting value of attitudes: Attitude accessibility as a determinant of an object's attraction of visual attention. *J. Pers. Soc. Psychol.* **63**:198–211. [16]

Roth, A. E., and I. Erev. 1995. Learning in extensive-form games: Experimental data and simple dynamic models in the intermediate term. *Games Econ. Behav.* **8**:164–212. [19]

Rozin, P., and J. W. Kalat. 1971. Specific hungers and poison avoidance as adaptive specializations of learning. *Psychol. Rev.* **78**:459–486. [7]

Rubin, K. H., R. Caplan, X. Chen, A. A. Buskirk, and J. C. Wojslawowicz. 2005. Peer relationships in childhood. In: Developmental Science: An Advanced Textbook, ed. M. H. Bornstein and M. E. Lamb, pp. 469–512. Mahwah, NJ: Lawrence Erlbaum. [17]

Ruckstuhl, K. E. 2007. Sexual segregation in vertebrates: Proximate and ultimate causes. *Integr. Comp. Biol.* **47**:245–257. [15]

Russo, J. E., and B. A. Dosher. 1983. Strategies for multiattribute binary choice. *J. Exp. Psychol. Learn. Mem. Cogn.* **9**:676–696. [7]

Sachs, J., B. Bard, and M. L. Johnson. 1981. Language learning with restricted input: Case studies of two hearing children of deaf parents. *Appl. Psycholing.* **2**:33–54. [15]

Sackett, G. P. 1966. Monkeys reared in isolation with pictures as visual input: Evidence for an innate releasing mechanism. *Science* **154**:1468. [6]

Sahlins, M. D. 1976. The Use and Abuse of Biology: An Anthropological Critique of Sociobiology. Ann Arbor: Univ. of Michigan Press. [1]

Sales, S. M., and J. House. 1971. Job dissatisfaction as a possible risk factor in coronary heart disease. *J. Chron. Dis.* **23**:861–873. [14]

Sally, D., and E. Hill. 2006. The development of interpersonal strategy: Autism, theory-of-mind, cooperation and fairness. *J. Econ. Psychol.* **27**:73–97. [18]

Samejima, K., and K. Doya. 2007. Multiple representations of belief states and action values in corticobasal ganglia loops. *Ann. NY Acad. Sci.* **1104**:213–228. [9]

Samuel, A. 1959. Some studies in machine learning using the game of checkers. *IBM J. Res. Dev.* **3**:210–229. [9]

Samuelson, P. 1937. A note on measurement of utility. *Rev. Econ. Stud.* **4**:155–161. [7]

———. 1938. A note on the pure theory of consumers' behaviour. *Economica* **5**:61–71. [7, 15]

Sandholm, W. H. 1998. Simple and clever decision rules for a model of evolution. *Econ. Lett.* **61**:165–170. [10]

Sanfey, A. G., and R. Hastie. 1999. Judgment and decision making across the adult life span: A tutorial review of psychological research. In: Aging and Cognition: A Primer, ed. D. Park and N. Schwarz, pp. 253–273. Philadelphia: Psychology Press. [14]

Sanfey, A. G., J. K. Rilling, J. A. Aronson, L. E. Nystrom, and J. D. Cohen. 2003. The neural basis of economic decision-making in the ultimatum game. *Science* **300**:1755–1758. [17]

Sapolsky, R. M. 1990. Adrenocortical functions, social rank, and personality among wild baboons. *Biol. Psych.* **28**:862–878. [15]

Sarafidis, Y. 2007. What have you done for me lately? Release of information and strategic manipulation of memories. *Econ. J.* **117**:307–326. [14]

Savage, L. J. 1954. The Foundations of Statistics. New York: Wiley. [4]

Scaife, M., and J. S. Bruner. 1975. The capacity for joint visual attention in the infant. *Nature* **253**:265–266 [6]

Schaal, S., A. Ijspeert, and A. Billard. 2003. Computational approaches to motor learning by imitation. *Phil. Trans. R. Soc. B* **358**:537–547. [9]

Schaller, G. B. 1972. The Serengeti Lion. Chicago: Univ. of Chicago Press. [15]

Scheibehenne, B., A. Wilke, and P. M. Todd. 2011. Expectations of clumpy resources influence predictions of sequential events. *Evol. Human Behav.* **32**:326–333. [11]

Schelling, T. 1960. The Strategy of Conflict. Cambridge, MA: Harvard Univ. Press. [19]

———. 1978. Micromotives and Macrobehavior. New York: W. W. Norton. [15]

Schino, G., and F. Aureli. 2010. Primate reciprocity and its cognitive requirements. *Evol. Anthropol.* **19**:130–135. [17]

Schlag, K. H. 1999. Which one should I imitate? *J. Math. Econ.* **31**:493–522. [10]

Schlosser, G., and G. P. Wagner, eds. 2004. Modularity in Development and Evolution. Chicago: Univ. of Chicago Press. [8]

Schneider, A. S., D. M. Oppenheimer, and G. Detre. 2007. Application of voting geometry to multialternative choice. Paper presented at the 29th Annual Conf. of the Cognitive Science Society. http://csjarchive.cogsci.rpi.edu/Proceedings/2007/docs/p635.pdf. (accessed 3 July 2012). [7]

Scholten, M., and D. Read. 2010. The psychology of intertemporal tradeoffs. *Psychol. Rev.* **117**:925–944. [4]

Schooler, L. J., and R. Hertwig. 2005. How forgetting aids heuristic inference. *Psychol. Rev.* **112**:610–628. [7]

Schuck-Paim, C., L. Pompilio, and A. Kacelnik. 2004. State-dependent decisions cause apparent violations of rationality in animal choice. *PLoS Biol.* **2**:2305–2315. [7]

Schuett, W., S. R. X. Dall, and N. J. Royle. 2011. Pairs of zebra finches with similar "personalities" make better parents. *Anim. Behav.* **81**:609–618. [15]

Schulke, O., J. Bhagavatula, L. Vigilant, and J. Ostner. 2010. Social bonds enhance reproductive success in male macaques. *Curr. Biol.* **20**:2207–2210. [6]

Schultz, W. 1998. Predictive reward signal of dopamine neurons. *J. Neurophysiol.* **80**:1–27. [9]

Schulz, D. J., Z. Y. Huang, and G. E. Robinson. 1998. Effects of colony food shortage on behavioral development in honey bees. *Behav. Ecol. Sociobiol.* **42**:295–303. [12]

Schuster, P. 1999. Chance and necessity in evolution: Lessons from RNA. *Physica D* **133**:427–452. [8]

Schwartz, O., T. J. Sejnowski, and P. Dayan. 2006. Soft mixer assignment in a hierarchical generative model of natural scene statistics. *Neural Comput.* **18**:2680–2718. [9]

Seeley, T. D., and P. K. Visscher. 2004. Group decision making in nest-site selection by honey bees. *Apidologie* **35**:101–116. [6]

Seligman, M. E. 1970. On generality of the laws of learning. *Psychol. Rev.* **77**:406–418. [19]

———. 1975. Helplessness: On depression, development, and death. San Francisco: W. H. Freeman. [10]

Seligman, M. E., and S. F. Maier. 1967. Failure to escape traumatic shock. *J. Exp. Psychol.* **74**:1–9. [9]

Selten, R. 1975. A reexamination of the perfectness concept for equilibrium points in extensive games. *Intl. J. Game Theory* **4**:25–55. [8, 10, 14, 15]

———. 1998. Features of experimentally observed bounded rationality. *Eur. Econ. Rev.* **42**:413-436. [19]

Selten, R. 2001. What is bounded rationality? In: Bounded Rationality: The Adaptive Toolbox, ed. G. Gigerenzer and R. Selten, pp. 13–36. Dahlem Workshop Reports, vol. 84, J. Lupp, series ed. Cambridge, MA: MIT Press. [10]

Selten, R., K. Abbink, and R. Cox. 2005. Learning direction theory and the winner's curse. *Exp. Econ.* **8**:5–20. [19]

Selten, R., and T. Chmura. 2008. Stationary concepts for experimental 2x2 games. *Am. Econ. Rev.* **98**:938–966. [10]

Selten, R., and P. Hammerstein. 1984. Gaps in Harley's argument on evolutionarily stable learning rules and in the logic of tit for tat. *Behav. Brain Sci* **7**:115–116. [12]

Selten, R., S. Pittnauer, and M. Hohnisch. 2011. Experimental results on the process of goal formation and aspiration adaptation. Discussion Paper, Bonn: Econ Lab. [20]

Selten, R., and R. Stoecker. 1986. End behavior in sequences of finite prisoners-dilemma supergames: A learning-theory approach. *J. Econ. Behav. Org.* **7**:47–70. [14, 19]

Shadlen, M. N., K. H. Britten, W. T. Newsome, and J. A. Movshon. 1996. A computational analysis of the relationship between neuronal and behavioral responses to visual motion. *J. Neurosci.* **16**:1486–1510. [9]

Shafir, S. 1994. Intransitivity of preferences in honey bees: Support for "comparative" evaluation of foraging options. *Anim. Behav.* **48**:55–67. [7, 20]

Shafir, S., T. Reich, E. Tsur, I. Erev, and A. Lotem. 2008. Perceptual accuracy and conflicting effects of certainty on risk-taking behaviour. *Nature* **453**:917–920. [15]

Shafir, S., and J. Roughgarden. 1994. The effect of memory length on the foraging behavior of a lizard. In: From Animals to Animats: Proc. of the 3rd Intl. Conf. on the Simulation of Adaptive Behavior, ed. D. Cliff et al., pp. 221–225. Cambridge, MA: MIT Press. [10]

Shafir, S., T. A. Waite, and B. H. Smith. 2002. Context-dependent violations of rational choice in honeybees (*Apis mellifera*) and gray jays (*Perisoreus canadensis*). *Behav. Ecol. Sociobiol.* **51**:180–187. [7]

Shah, A. K., and D. M. Oppenheimer. 2008. Heuristics made easy: An effort-reduction framework. *Psychol. Bull.* **134**:207–222. [7]

Shallice, T. 1988. From Neuropsychology to Mental Structure. Cambridge: Cambridge Univ. Press. [9]

Shanks, D. R. 1995. Is human learning rational? *Q. J. Exp. Psychol. A* **48**:257–279. [10]

Shannon, C. 1948. A mathematical theory of communication. *Bell System Tech. J.* **27**:623. [9]

Shariff, A. F., and A. Norenzayan. 2007. God is watching you: Priming god concepts increases prosocial behavior in an anonymous economic game. *Psychol. Sci.* **18**:803–809. [20]

Sharpe, K. M., R. Staelin, and J. Huber. 2008. Using extremeness aversion to fight obesity: Policy implications of context dependent demand. *J. Consum. Res.* **35**:406–422. [14]

Sheldon, K. M., R. M. Ryan, L. J. Rawsthorne, and B. Ilardi. 1997. Trait self and true self: Cross-role variation in the big-five personality traits and its relations with psychological authenticity and subjective well-being. *J. Pers. Soc. Psychol.* **73**:1380–1393. [14]

Sheldon, K. M., M. S. Sheldon, and C. P. Nichols. 2007. Traits and trade-offs are insufficient for evolutionary personality psychology. *Am. Psychol.* **62**:1073–1074. [14]

Shepard, R. N. 1987. Toward a universal law of generalization for psychological science. *Science* **237**:1317–1323. [4]

Sherry, D. F., and D. L. Schacter. 1987. The evolution of multiple memory systems. *Psychol. Rev.* **94**:439–454. [11]

Shettleworth, S. J. 1998. Cognition, Evolution, and Behavior. Oxford: Oxford Univ. Press. [15]

———. 2000. Modularity and the evolution of cognition. In: The Evolution of Cognition, ed. C. M. Heyes and L. Huber, pp. 43–60. Cambridge, MA: MIT Press. [1]

Shettleworth, S. J., J. R. Krebs, D. W. Stephens, and J. Gibbon. 1988. Tracking a fluctuating environment: A study of sampling. *Anim. Behav.* **36**:87–105. [10]

Shizgal, P. 1997. Neural basis of utility estimation. *Curr. Opin. Neurobiol.* **7**:198–208. [7]

Siegal, M., and A. Bergman. 2002. Waddington's canalization revisited: Developmental stability and evolution. *PNAS* **99**:10,528–10,532. [8]

Sigmund, K., and M. Nowak. 1993. A strategy of win-stay, lose-shift that outperforms tit-for-tat in the prisoner's dilemma game. *Nature* **364**:56–58. [8]

Sigmund, K., H. D. Silva, A. Traulsen, and C. Hauert. 2010. Social learning promotes institutions for governing the commons. *Nature* **466**:861–863. [8]

Sih, A., A. M. Bell, J. C. Johnson, and R. E. Ziemba. 2004. Behavioral syndromes: An integrative overview. *Q. Rev. Biol.* **79**:241–277. [13]

Sih, A., J. Cote, M. Evans, S. Fogarty, and J. Pruitt. 2012. Ecological implications of behavioural syndromes. *Ecol. Lett.* **15**:278–289. [13]

Silk, J. B., S. C. Alberts, and J. Altmann. 2003. Social bonds of female baboons enhance infant survival. *Science* **302**:1231–1234. [6]

Silk, J. B., J. C. Beehner, T. J. Bergman, et al. 2010. Strong and consistent social bonds enhance the longevity of female baboons. *Curr. Biol.* **20**:1359–1361. [6]

Silk, J. B., S. F. Brosnan, J. Vonk, et al. 2005. Chimpanzees are indifferent to the welfare of unrelated group members. *Nature* **437**:1357–1359. [18, 20]

Silk, J. B., and B. R. House. 2011. Evolutionary foundations of human prosocial sentiments. *PNAS* **108**:10,910–10,917. [18]

Simmons, R., D. Apfelbaum, W. Burgard, et al. 2000. Coordination for multi-robot exploration and mapping. In: Proc. Natl. Conf. on Artificial Intelligence, pp. 852–858. Cambridge, MA: MIT Press. [9]

Simon, H. A. 1955. A behavioral model of rational choice. *Q. J. Econ.* **69**:99–118. [4]

———. 1956. Rational choice and the structure of the environment. *Psychol. Rev.* **63**:129–138. [1, 7, 20]

———. 1962. The architecture of complexity. *Proc. Amer. Philos. Soc.* **106**:467–482. [8]

———. 1990. Invariants of human behavior. *Ann. Rev. Psychol.* **41**:1–19. [7, 10]

———. 1996. The Sciences of the Artificial, 3rd edition. Cambridge, MA: MIT Press. [12]

Simons, A. M. 2004. Many wrongs: The advantage of group navigation. *Trends Ecol. Evol.* **19**:453–455. [12]

Simonson, I. 2008. Will I like a "medium" pillow? Another look at constructed and inherent preferences. *J. Consum. Psychol.* **18**:155–169. [14]

Singer, T., B. Seymour, J. P. O'Doherty, et al. 2004. Empathy for pain involves the affective but not sensory components of pain. *Science* **303**:1157–1162. [19]

———. 2006. Empathic neural responses are modulated by the perceived fairness of others. [Letter]. *Nature* **439**:466–469. [18, 19]

Slotine, J. J., and W. Lohmiller. 2001. Modularity, evolution, and the binding problem: A view from stability theory. *Neural Netw.* **14**:137–145. [8]

Smith, A. 1759/2000. The Theory of Moral Sentiments. http://www.econlib.org/library/Smith/smMSCover.html. [18]

———. 1776/2005. An Inquiry into the Nature and Causes of the Wealth of Nations, Pennsylvania State Univ. http://www2.hn.psu.edu/faculty/jmanis/adam-smith/Wealth-Nations.pdf (accessed May 8, 2012). [11, 18]

Smith, J. 2009. Imperfect memory and the preference for increasing payments. *J. Inst. Theor. Econ.* **165**:684–700. [14]

Smith, M. A., and A. Kohn. 2008. Spatial and temporal scales of neuronal correlation in primary visual cortex. *J. Neurosci.* **48**:12,591–12,603. [9]

Smith, P. L., and R. Ratcliff. 2004. Psychology and neurobiology of simple decisions. *Trends Neurosci.* **27**:161–168. [9]

Smith, R. H., E. Diener, and D. H. Wedell. 1989. Intrapersonal and social comparison determinants of happiness: A range-frequency analysis. *J. Pers. Soc. Psychol.* **56**:317–325. [14]

Smith, V. L. 1998. The two faces of Adam Smith. *Southern Econ. J.* **65**:2–19. [11]

Smolensky, P. 1986. Information processing in dynamical systems: Foundations of harmony theory. In: Parallel Distributed Processing: Foundations, ed. D. E. Rumelhart and J. L. McClelland, vol. 1, pp. 194–281. Cambridge, MA: MIT Press. [3]

———. 1990. Tensor product variable binding and the representation of symbolic structures in connectionist systems. *Art. Intell.* **46**:159–216. [3]

Soffer, S., and Y. Kareev. 2011. The effects of problem content and scientific background on information search and the assessment and valuation of correlations. *Mem. Cogn.* **39**:107–116. [10]

Soler, L., E. Trizio, T. Nickles, and W. Wimsatt, eds. 2011. Characterizing the Robustness of Science after the Practice Turn in Philosophy of Science. New York: Springer. [12]

Soto, C. J., O. P. John, S. Gosling, and J. Potter. 2008. The developmental psychometrics of Big Five self-reports: Acquiescence, factor structure, coherence, and differentiation from ages 10 to 20. *J. Pers. Soc. Psychol.* **94**:718–737. [15]

———. 2011. Age differences in personality traits from 10 to 65: Big-five domains and facets in a large cross-sectional sample. *J. Pers. Soc. Psychol.* **100**:330–348. [15]

Sougné, J. 1998. Connectionism and the problem of multiple instantiation. *Trends Cogn. Sci.* **2**:183–189. [3]

Soyer, O. S., and T. Pfeifer. 2010. Evolution under fluctuating environments explains observed robustness in metabolic networks. *PLoS Comp. Biol.* **6**:e1000907. [8]

Speekenbrink, M., and D. R. Shanks. 2010. Learning in a changing environment. *J. Exp. Psychol. Gen.* **139**:266–298. [10]

Sperber, D. 1994. The modularity of thought and the epidemiology of representations. In: Mapping the Mind: Domain Specificity in Cognition and Culture, ed. L. A. Hirschfeld and S. A. Gelman, pp. 39–67. New York: Cambridge Univ. Press. [11]

———. 2005. Modularity and relevance: How can a massively modular mind be flexible and context-sensitive? In: The Innate Mind: Structure and Content ed. P. Carruthers et al., pp. 53–68. New York: Oxford Univ. Press. [11]

Sprenger, A., and M. R. Dougherty. 2006. Differences between probability and frequency judgments: The role of individual differences in working memory capacity. *Org. Behav. Hum. Decis. Process.* **99**:202–211. [14, 15]

Srivastava, S., O. P. John, S. Gosling, and J. Potter. 2003. Development of personality in early and middle adulthood: Set like plaster or persistent change? *J. Pers. Soc. Psychol.* **84**:1041–1053. [15]

Stamps, J. 2003. Behavioural processes affecting development: Tinbergen's fourth question comes of age. *Anim. Behav.* **66**:1–13. [8]

Stanovich, K. E., and R. F. West. 2000. Individual differences in reasoning: Implications for the rationality debate? *Behav. Brain Sci.* **23**:645–665. [14]

Stephens, D. W. 2002. Discrimination, discounting and impulsivity: a role for an informational constraint. *Phil. Trans. R. Soc. B* **357**:1527–1537. [7]

———. 2007. Models of information use. In: Foraging: Behavior and Ecology, ed. D. W. Stephens et al., pp. 31–58. Chicago: Univ. of Chicago Press. [15]

———. 2008. Decision ecology: Foraging and the ecology of animal decision making. *Cogn. Affect. Behav. Neurosci.* **8**:475–484. [7]

Stephens, D. W., and D. Anderson. 2001. The adaptive value of preferences for immediacy: When shortsighted rules have far sighted consequences. *Behav. Ecol.* **12**:330–339. [7, 15]

Stephens, D. W., J. S. Brown, and R. C. Ydenberg. 2007. Foraging: Behavior and Ecology. Chicago: Univ. of Chicago Press. [6]

Stephens, D. W., B. Kerr, and E. Fernandez-Juricic. 2004. Impulsiveness without discounting: the ecological rationality hypothesis. *Proc. R. Soc. B* **271**:2459–2465. [7]

Stephens, D. W., and J. R. Krebs. 1986. Foraging Theory. Princeton: Princeton Univ. Press. [2, 6, 7, 15]

Stevens, J. R. 2008. The evolutionary biology of decision making. In: Better than Conscious? Decision Making, the Human Mind, and Implications For Institutions, ed. C. Engel and W. Singer, vol. 1, pp. 285–304, J. Lupp, series ed. Cambridge, MA: MIT Press. [1]

———. 2010. Donor payoffs and other-regarding preferences in cotton-top tamarins (*Saguinus oedipus*). *Anim. Cogn.* **13**:663–670. [18]

Stevens, J. R., and M. Hauser. 2004. Why be nice? Psychological constraints on the evolution of cooperation. *Trends Cogn. Sci.* **8**:60–65. [18]

Stevens, J. R., J. Volstorf, L. J. Schooler, and J. Rieskamp. 2011. Forgetting constrains the emergence of cooperative decision strategies. *Front. Psychol.* **1**:1–12. [9, 14]

Stewart, N. 2009. Decision by sampling: The role of the decision environment in risky choice. *Q. J. Exp. Psychol.* **62**:1041–1062. [14]

Stewart, N., G. D. A. Brown, and N. Chater. 2005. Absolute identification by relative judgment. *Psychol. Rev.* **112**:881–911. [4]

Stewart, N., N. Chater, and G. D. A. Brown. 2006. Decision by sampling. *Cogn. Psychol.* **53**:1–26. [1, 4, 10, 14, 15]

Stich, S. 1990. The Fragmentation of Reason. Cambridge, MA: MIT Press. [5]

Stigler, G., and G. Becker. 1977. *De gustibus non est disputandum. Am. Econ. Rev.* **67**:76–90. [15]

Stoner, J. A. F. 1961. A comparison of individual and group decisions involving risk. Master's Thesis, Massachusetts Institute of Technology, Cambridge, MA. [15]

Strack, F., and R. Deutsch. 2004. Reflective and impulsive determinants of social behavior. *Pers. Soc. Psychol. Rev.* **8**:220–247. [16]

Stutzer, A., and B. S. Frey. 2006. Does marriage make people happy, or do happy people get married? *J. Socio-Econ.* **35**:326–347. [14]

Sugrue, L. P., G. S. Corrado, and W. T. Newsome. 2005. Choosing the greater of two goods: Neural currencies for valuation and decision making. *Nat. Rev. Neurosci.* **6**:363–375. [6]

Sumpter, D. J. T. 2006. The principles of collective animal behaviour. *Phil. Trans. R. Soc. B* **361**:5–22. [8]

Sun, R. 1992. On variable binding in connectionist networks. *Conn. Sci.* **4**:93–124. [3]

Suomi, S. J. 2006. Risk, resilience, and gene×environment interactions in rhesus monkeys. *Ann. NY Acad. Sci.* **1094**:52–62. [9]

Surowiecki, J. 2004. The Wisdom of Crowds. Garden City: Doubleday. [8]

Sutter, M. 2007. Outcomes versus intentions: On the nature of fair behavior and its development with age. *J. Econ. Psychol.* **28**:69–78. [18]

Sutter, M., and M. Kocher. 2007. Trust and trustworthiness across different age groups. *Games Econ. Behav.* **59**:364–382. [15]

Sutton, R. 1988. Learning to predict by the methods of temporal differences. *Mach. Learn.* **3**:9–44. [9]

———. 1990. Integrated architectures for learning, planning, and reacting based on approximating dynamic programming. In: Proc. of the 7th Intl. Conf. on Machine Learning, pp. 216–224. Waltham, MA: Morgan Kaufmann. [9]

Sutton, R., and A. G. Barto. 1998. Reinforcement Learning: An Introduction. Cambridge, MA: MIT Press. [9]

Sutton, R., A. Koop, and D. Silver. 2007. On the role of tracking in stationary environments. http://webdocs.cs.ualberta.ca/~sutton/papers/SKS-07.pdf. [9]

Tajfel, H., M. G. Billig, R. P. Bundy, and C. Flament. 1971. Social categorization and intergroup behavior. *Eur. J. Soc. Psychol.* **1**:149–178. [20]

Takagishi, H., S. Kameshima, J. Schug, M. Koizumi, and T. Yamagishi. 2010. Theory of mind enhances preference for fairness. *J. Exp. Child Psychol.* **105**:130–137. [18]

Taleb, N. 2007. Black swans and the domains of statistics. *Am. Statist.* **61**:198–200. [9]

Tautz, D. 1992. Redundancies, development and the flow of information. *BioEssays* **14**:263–266. [8]

Taylor, S. E. 1981. The interface of cognitive and social psychology. In: Cognition, Social Behavior, and the Environment, ed. J. Harvey, pp. 182–211. Hillsdale, NJ: Erlbaum. [16]

Tenenbaum, J., C. Kemp, T. Griffiths, and N. Goodman. 2011. How to grow a mind: statistics, structure, and abstraction. *Science* **331**:1279. [15]

Thaler, R. H. 1988. Anomalies: The winner's curse. *J. Econ. Persp.* **2**:191–202. [19]

Thaler, R. H., and H. M. Shefrin. 1981. An economic theory of self-control. *J. Polit. Econ.* **89**:392–406. [7]

Thaler, R. H., and C. R. Sunstein. 2008. Nudge: Improving Decisions about Health, Wealth, and Happiness. New Haven: Yale Univ. Press. [14]

Theraulaz, G., J. Gautrais, S. Camazine, and J.-L. Deneubourg. 2003. The formation of spatial patterns in social insects: From simple behaviours to complex structures. *Phil. Trans. R. Soc. A* **361**:1263–1282. [8]

Thompson, C., J. Barresi, and C. Moore. 1997. The development of future-oriented prudence and altruism in preschoolers. *Cogn. Dev.* **12**:199–212. [18]

Thorndike, E. L. 1911. Animal Intelligence. New York: Macmillan. [19]

Thornton, A., and K. McAuliffe. 2006. Teaching in wild meerkats. *Science* **313**:227–229. [20]

Thorpe, S., D. Fize, and C. Marlot. 1996. Speed of processing in the human visual system. *Nature* **381**:520–522. [9]

Thrun, S. 1992. Efficient exploration in reinforcement learning. In: Technical Report CS-92-102. Pittsburgh: CMU. [9]

Tinbergen, N. 1963. On aims and methods of ethology. *Z. Tierpsychol.* **20**:410–433. [2, 15, 17]

Tinbergen, N., G. J. Broekhuysef, N. Feekes, et al. 1963. Egg shell removal by the black-headed gull, *Larus ridibundus* L.: A behaviour component of camouflage. *Behaviour* **19**:74–117. [2]

Tinker, M. T., M. Mangel, and J. A. Estes. 2009. Learning to be different: acquired skills, social learning, frequency dependence, and environmental variation can cause behaviourally mediated foraging specializations. *Evol. Ecol. Res.* **11**:841–869. [15]

Todd, A. R., K. Hanko, A. D. Galinsky, and T. Mussweiler. 2011a. When focusing on differences leads to similar perspectives. *Psychol. Sci.* **22**:134–141. [16]

Todd, P. M., G. Gigerenzer, and the ABC Research Group. 2000. Précis of *Simple Heuristics that Make Us Smart. Behav. Brain Sci.* **23**:727–741. [2]

———. 2011b. Ecological Rationality: Intelligence in the World. New York: Oxford Univ. Press. [11, 15]

Tolman, E. 1948. Cognitive maps in rats and men. *Psychol. Rev.* **55**:189–208. [7]

Tomasello, M., and J. Call. 1997. Primate Cognition. New York: Oxford Univ. Press. [20]

Tomasello, M., M. Carpenter, J. Call, T. Behne, and H. Moll. 2005. Understanding and sharing intentions: The origins of cultural cognition. *Behav. Brain Sci.* **28**:675–691. [1, 17, 19]

Tooby, J., and L. Cosmides. 1992. The psychological foundations of culture. In: The Adapted Mind, ed. J. H. Barkow et al., pp. 19–136. New York: Oxford Univ. Press. [11]

———. 2008. The evolutionary psychology of the emotions and their relationship to internal regulatory variables. In: Handbook of Emotions, 3rd edition, ed. M. Lewis et al., pp. 114–137. New York: Guilford Press. [11]

Trawalter, S., A. R. Todd, A. A. Baird, and J. A. Richeson. 2008. Attending to threat: Race-based patterns of selective attention. *J. Exp. Soc. Psychol.* **44**:1322–1327. [16]

Trewavas, A. 2005. Green plants as intelligent organisms. *Trends Plant Sci.* **10**:413–419. [7]

Trivers, R. L. 1971. The evolution of reciprocal altruism. *Q. Rev. Biol.* **46**:35–57. [14, 18]

Tsang, J. A. 2006. Gratitude and prosocial behaviour: An experimental test of gratitude. *Cogn. Emot.* **20**:138–148. [14]

Tsao, D. Y., W. A. Freiwald, T. A. Knutsen, J. B. Mandeville, and R. B. Tootell. 2003. Faces and objects in macaque cerebral cortex. *Nat. Neurosci.* **6**:989–995. [6]

Tsodyks, M., and T. Sejnowski. 1995. Rapid state switching in balanced cortical network models. *Network* **6**:111–124. [9]

Turing, A. M. 1936. On computable numbers, with an application to the *Entscheidungsproblem. Proc. Lond. Math. Soc.* **42**:230–265. [3]

Turkewitz, G., and P. A. Kenny. 1982. Limitations on input as a basis for neural organization and perceptual development: A preliminary theoretical statement. *Dev. Psychobiol.* **15**:357–368. [10]

Tversky, A. 1969. Intransitivity of preferences. *Psychol. Rev.* **76**:31–48. [7]

Tversky, A., and D. Kahneman. 1974. Judgment under uncertainty: Heuristics and biases. *Science* **185**:1124–1131. [10, 12]

Tversky, A., and I. Simonson. 1993. Context-dependent preferences. *Manag. Sci.* **39**:1179–1189. [7, 20]

Vaish, A., M. Carpenter, and M. Tomasello. 2009. Sympathy through affective perspective taking and its relation to prosocial behavior in toddlers. *Dev. Psychol.* **45**:534–543. [18, 20]

Vaish, A., M. Carpenter, and M. Tomasello. 2010a. Young children selectively avoid helping people with harmful intentions. *Child Devel.* **81**:1661–1669. [17]

Vaish, A., M. Missana, and M. Tomasello. 2010b. Three-year-old children intervene in third-party moral transgressions. *Br. J. Dev. Psychol.* **29**:124–130. [18]

van't Wout, M., R. Kahn, A. Sanfey, and A. Aleman. 2006. Affective state and decision-making in the Ultimatum Game. *Exp. Brain Res.* **169**:564–568. [18]

van Alphen, J. J. M., C. Bernstein, and G. Driessen. 2003. Information acquisition and time allocation in insect parasitoids. *Trends Cogn. Sci.* **18**:81–87. [2]

van der Meer, M. A., and A. D. Redish. 2010. Expectancies in decision making, reinforcement learning, and ventral striatum. *Front. Neurosci.* **4**:6. [9]

van Doorn, G. S., G. M. Hengeveld, and F. J. Weissing. 2003. The evolution of social dominance. II. Multi-player models. *Behaviour* **140**:1333–1358. [15]

van Hooff, J. A. R. A. M. 1967. The facial displays of the Catarrhine monkey and apes. In: Primate Ethology, ed. D. Morris. Chicago: Aldine. [6]

van Oers, K., G. de Jong, A. J. van Noordwijk, B. Kempenaers, and P. J. Drent. 2005. Contribution of genetics to the study of animal personalities: A review of case studies. *Behaviour* **142**:1191–1212. [13]

Vaughan, W. 1981. Melioration, matching, and maximizing. *J. Exper. Anal. Behav.* **36**:141–149. [3]

Verbeek, M. E. M., P. J. Drent, and P. R. Wiepkema. 1994. Consistent individual differences in early exploratory behavior of male great tits. *Anim. Behav.* **48**:1113–1121. [13]

Verner, J., and M. F. Willson. 1966. Influence of habitats on mating systems of North American passerine birds. *Ecology* **47**:143–147. [6]

Vlaev, I., N. Chater, N. Stewart, and G. D. A. Brown. 2011. Does the brain calculate value? *Trends Cogn. Sci.* **15**:546–554. [20]

Vlaev, I., B. Seymour, R. J. Dolan, and N. Chater. 2009. The price of pain and the value of suffering. *Psychol. Sci.* **20**:309–317. [4]

Volk, S., C. Thöni, and W. Ruigrok. 2012. Temporal stability and psychological foundations of cooperation preferences. *J. Econ. Behav. Org.* **81**:664–676. [15]

Vonfrisch, K., and M. Lindauer. 1956. The language and orientation of the honey bee. *Ann. Rev. Entomol.* **1**:45–58. [6]

Vonk, J., S. F. Brosnan, J. B. Silk, et al. 2008. Chimpanzees do not take advantage of very low cost opportunities to deliver food to unrelated group members. *Anim. Behav.* **75**:1757–1770. [18, 20]

von Neumann, J. 1956. Probabilistic logics and the synthesis of reliable organisms from unreliable components. In: Automata Studies, ed. C. Shannon and J. McCarthy, pp. 43–98. Princeton: Princeton Univ. Press. [8]

von Neumann, J., and O. Morgenstern. 1947. Theory of Games and Economic Behavior, 2nd ed. Princeton: Princeton Univ. Press. [7]

Vul, E., D. Hanus, and N. Kanwisher. 2009. Attention as inference: Selection is probabilistic; responses are all-or-none samples. *Journal of Experimental Psychology: General* **138**:546–560. [15]

Vul, E., and H. Pashler. 2008. Measuring the crowd within: Probabilistic representations within individuals. *Psychol. Sci.* **19**:645–647. [14, 15]

Waddington, C. 1942. Canalization of development and the inheritance of acquired characters *Nature* **3811**:563–565. [8]

Wade, M. J. 1978. A critical review of the models of group selection. *Q. Rev. Biol.* **53**:101–114. [12]

Wagner, A. 1996. Genetic redundancy caused by gene duplications and its evolution in networks of transcriptional regulators. *Biol. Cybern.* **74**:557–567. [8]

———. 2005. Robustness and Evolvability in Living Systems. Princeton: Princeton Univ. Press. [12]

———. 2007. Robustness and Evolvability in Living Systems. Princeton: Princeton Univ. Press. [8]

———. 2008. Robustness and evolvability: A paradox resolved. *Proc. R. Soc. B* **276**:91–100. [8]

Waite, T. A. 2001. Intransitive preferences in hoarding gray jays (*Perisoreus canadensis*). *Behav. Ecol. Sociobiol.* **50**:116–121. [7]

Wald, A. 1945. Statistical decision functions which minimize the maximum risk. *Ann. Math.* **46**:265–280. [9, 12]

Wald, A., and J. Wolfowitz. 1948. Optimum character of the sequential probability ratio test. *Ann. Math. Stat.* **19**:326–339. [6]

Wallsten, T. S., R. H. Bender, and Y. L. Li. 1999. Dissociating judgment from response processes in statement verification: The effects of experience on each component. *J. Exp. Psychol. Learn. Mem. Cogn.* **25**:96–115. [10]

Wallsten, T. S., and C. González-Vallejo. 1994. Statement verification: A stochastic model of judgment and response. *Psychol. Rev.* **101**:490–504. [10]

Wang, X. T., and R. D. Dvorak. 2010. Sweet future: Fluctuating blood glucose levels affect future discounting. *Psychol. Sci.* **21**:183–188. [14]

Wansink, B., D. R. Just, and C. R. Payne. 2009. Mindless eating and healthy heuristics for the irrational. *Am. Econ. Rev.* **99**:165–169. [14]

Ward, A. J. W., D. J. T. Sumpter, I. D. Couzin, P. J. B. Hart, and J. Krause. 2008. Quorum decision making facilitates information transfer in fish shoals. *PNAS* **105**:6948–6953. [12]

Warneken, F., F. Chen, and M. Tomasello. 2006. Cooperative activities in young children and chimpanzees. *Child Devel.* **77**:640–663. [17, 18]

Warneken, F., M. Gräfenhain, and M. Tomasello. 2012. Collaborative partner or social tool? New evidence for young children's understanding of shared intentions in collaborative activities. *Dev. Sci.* **15**:54–61. [17]

Warneken, F., B. Hare, A. P. Melis, D. Hanus, and M. Tomasello. 2007. Spontaneous altruism by chimpanzees and young children. *PLoS Biol.* **5**:1414–1420. [17, 20]

Warneken, F., and M. Tomasello. 2006. Altruistic helping in human infants and young chimpanzees. *Science* **311**:1301–1303. [17, 18, 20]

———. 2007. Helping and cooperation at 14 months of age. *Infancy* **11**:271–294. [17, 20]

———. 2008. Extrinsic rewards undermine altruistic tendencies in 20-month-olds. *Dev. Psychol.* **44**:1785–1788. [18]

———. 2009. Varieties of altruism in children and chimpanzees. *Trends Cogn. Sci.* **13**:397–402. [17, 18, 20]

Wascher, C. A. F., I. B. R. Scheiber, and K. Kotrschal. 2008. Heart rate modulation in bystanding geese watching social and non-social events. *Proc. Roy. Soc. B* **275**:1653–1659. [18]

Wason, P. C. 1983. Realism and rationality in the selection task. In: Thinking and Reasoning: Psychological Approaches, ed. J. S. B. T. Evans. London: Routledge & Kegan Paul. [11]

Watkins, C. J. C. H. 1989. Learning from Delayed Rewards. Ph.D. thesis, Univ. of Cambridge. [4, 9]

Watson, K. K., J. H. Ghodasra, and M. L. Platt. 2009. Serotonin transporter genotype modulates social reward and punishment in rhesus macaques. *PLoS ONE* **4**:e4156. [6]

Weber, E. U., A.-R. Blais, and N. E. Betz. 2002. A domain-specific risk-attitude scale: Measuring risk perceptions and risk behaviors. *J. Behav. Decis. Mak.* **15**:263–290. [15]

Weber, E. U., S. Shafir, and A.-R. Blais. 2004. Predicting risk sensitivity in humans and lower animals: Risk as variance or coefficient of variation. *Psychol. Rev.* **111**:430–445. [9]

Webster, D. M., L. Richter, and A. W. Kruglanski. 1996. On leaping to conclusions when feeling tired: Mental fatigue effects on impressional primacy. *J. Exp. Soc. Psychol.* **32**:181–195. [14]

Wechsler, D. 1949. Manual for the Wechsler Intelligence Scale for Children. Oxford: The Psychological Corporation. [10]

Wedell, D. H., E. M. Santoyo, and J. C. Pettibone. 2005. The thick and the thin of it: Contextual effects in body perception. *Basic Appl Soc. Psychol.* **27**:213–227. [14]

Wehner, R., and S. Rossel. 1985. The bee's celestial compass: A case study in behavioural neurobiology. *Fort. Zool.* **31**:11–53. [7]

Wehner, R., and M. V. Srinivasan. 1981. Searching behavior of desert ants, genus *Cataglyphis* (*Formicidae*, *Hymenoptera*). *J. Comp. Physiol.* **142**:315–338. [3, 7]

Weinstein, T. A. R., J. P. Capitanio, and S. Gosling. 2008. Personality in animals. In: Handbook of Personality Theory and Research, ed. O. P. John et al., pp. 328–348. New York: Guilford Press. [15]

Weiss, A., J. E. King, and A. J. Figueredo. 2000. The heritability of personality factors in chimpanzees (*Pan troglodytes*). *Behav. Genet.* **30**:213–221. [13]

Werner, N. S., S. Duschek, and R. Schandry. 2009. Relationships between affective states and decision-making. *Intl. J. Psychophys.* **74**:259–265. [14]

West, G. B., J. H. Brown, and B. J. Enquist. 1997. A general model for the origin of allometric scaling laws in biology. *Science* **276**:122–126. [8]

West-Eberhard, M. J. 2003. Developmental plasticity and evolution. New York: Oxford Univ. Press. [15]

Whitehead, H. 2007. Learning, climate and the evolution of cultural capacity. *J. Theor. Biol.* **245**:341. [15]

Whitehead, H., and P. J. Richerson. 2009. The evolution of conformist social learning can cause population collapse in realistically variable environments. *Evol. Human Behav.* **30**:261–273. [19]

Whiten, A., N. McGuigan, S. Marshall-Pescini, and L. M. Hopper. 2009. Emulation, imitation, over-imitation and the scope of culture for child and chimpanzee. *Phil. Trans. R. Soc. B* **364**:2417–2428. [19]

Whitfield, J. T., W. H. Pako, J. Collinge, and M. P. Alper. 2008. Mortuary rites of the South Fore and kuru. *Phil. Trans. R. Soc. B* **363**:3721–3724. [1, 19]

Widdig, A. 2007. Paternal kin discrimination: The evidence and likely mechanisms. *Biol. Rev.* **82**:319–334. [17]

Wiegmann, D. D., L. A. Real, T. A. Capone, and S. Ellner. 1996. Some distinguishing features of models of search behavior and mate choice. *Am. Natural.* **147**:188–204. [2]

Wilcoxon, H. C., W. B. Dragoin, and P. A. Kral. 1971. Illness-induced aversions in rat and quail: Relative salience of visual and gustatory cues. *Science* **171**:826–828. [19]

Wilke, A., and H. C. Barrett. 2009. The hot hand phenomenon as a cognitive adaptation to clumped resources. *Evol. Human Behav.* **30**:161–169. [1, 11]

Wilke, A., J. M. C. Hutchinson, P. M. Todd, and U. Czienskowski. 2009. Fishing for the right words: Decision rules for human foraging behavior in internal search tasks. *Cogn. Sci.* **33**:497–529. [1]

Wilke, A., and P. M. Todd. 2010. Past and present environments: The evolution of decision making *Psicothema* **22**:4–8. [5]

Williams, D. R., and H. Williams. 1969. Auto-maintenance in the pigeon: Sustained pecking despite contingent non-reinforcement. *J. Exp. Anal. Behav.* **12**:511–520. [2, 9]

Williams, J., and E. Taylor. 2006. The evolution of hyperactivity, impulsivity and cognitive diversity. *J. R. Soc. Interface* **3**:399–413. [9]

Wilson, D. S. 1998a. Adaptive individual differences within single populations. *Phil. Trans. R. Soc. B* **353**:199–205. [13]

Wilson, E. O. 1998b. Consilience: The Unity of Knowledge. New York: Alfred A. Knopf. [1]

Wilson, R. 2002. Four Colors Suffice. London: Penguin Press. [3]

Wimmer, H., and J. Perner. 1983. Beliefs about beliefs: Representation and constraining function of wrong beliefs in young children's understanding of deception. *Cognition* **13**:103–128. [4]

Wimsatt, W. C. 1980. Reductionistic research strategies and their biases in the units of selection controversy. In: Scientific Discovery: Case Studies, ed. T. Nickles, pp. 213–259. Dordrecht: Reidel. [12]

———. 1981. Robustness, reliability, and multiple-determination. In: Scientific Inquiry and the Social Sciences, ed. M. Brewer and B. Collins, pp. 124–163. San Francisco: Jossey-Bass. [12]

———. 2001. Generative entrenchment and the developmental systems approach to evolutionary processes. In: Cycles of Contingency: Developmental Systems and Evolution, ed. S. Oyama and R. Gray, pp. 219–237. Cambridge, MA: MIT Press. [8]

———. 2007. On building reliable pictures with unreliable data: An evolutionary and developmental coda for the new systems biology. In: Systems Biology: Philosophical Foundations, ed. F. C. Boogerd et al., pp. 103–120. Amsterdam: Reed-Elsevier. [8]

Winther, R. G. 2001. Varieties of modules: Kinds, levels, origins, and behaviors. *J. Exp. Zool.* **291**:116–129. [8]

Wittig, R. M., and C. Boesch. 2010. Receiving post-conflict affiliation from the enemy's friend reconciles former opponents. *PLoS One* **5**:e13995. [17]

Wolf, M., G. S. van Doorn, O. Leimar, and F. J. Weissing. 2007. Life-history trade-offs favour the evolution of animal personalities. *Nature* **447**:581–584. [15]

Wolf, M., G. S. van Doorn, and F. J. Weissing. 2008. Evolutionary emergence of responsive and unresponsive personalities. *PNAS* **105**:15,825–15,830. [15]

———. 2011. On the coevolution of social responsiveness and behavioural consistency. *Proc. Roy. Soc. B* **278**:440–448. [15]

Wolf, M., and F. J. Weissing. 2010. An explanatory framework for adaptive personality differences. *Phil. Trans. R. Soc. B* **365**:3959–3968. [13–15]

Wolf, M., and F. J. Weissing. 2012. Animal personalities: Consequences for ecology and evolution. *Trends Ecol. Evol.*, in press. [13]

Wolpert, D. M., K. Doya, and M. Kawato. 2003. A unifying computational framework for motor control and social interaction. *Phil. Trans. R. Soc. B* **358**:593–602. [9]

Wood, A. M., G. D. A. Brown, and J. Maltby. 2012. Social norm influences on evaluations of the risks associated with alcohol consumption: Applying the rank-based decision by sampling model to health judgments. *Alcohol and Alcoholism* **47**:57–62. [14]

Wood, A. M., S. Joseph, and J. Maltby. 2008. Gratitude uniquely predicts satisfaction with life: Incremental validity above the domains and facets of the five factor model. *Pers. Ind. Diff.* **45**:49–54. [14]

———. 2009. Gratitude predicts psychological well-being above the Big Five facets. *Pers. Ind. Diff.* **46**:443–447. [14]

Woodward, A. L. 1998. Infants selectively encode the goal object of an actor's reach. *Cognition* **69**:1–34. [17]

Wright, H. 1935. The evolution of civilizations. In: American Archaeology Past and Future, ed. D. Meltzer et al., pp. 323–365. Washington, D. C.: Smithsonian Institution Press. [8]

Wynn, K. 2008. Some innate foundations of social and moral cognition. In: The Innate Mind: Foundations and the Future, ed. S. Carruthers et al. Oxford: Oxford Univ. Press. [18]

Yamamoto, S., T. Humle, and M. Tanaka. 2009. Chimpanzees help each other upon request. *PloS One* **4**:e7416. [20]

Yamamoto, S., and M. Tanaka. 2009. How did altruism and reciprocity evolve in humans? Perspectives from experiments on chimpanzees (*Pan troglodytes*). *Interaction Stud.* **10**:150–182. [18]

Yerkes, R. M. 1939. The life history and personality of the chimpanzee. *Am. Natural.* **73**:97–112. [13]

Yerkes, R. M., and A. W. Yerkes. 1929. The Great Apes: A Study of Anthropoid Life. New Haven: Yale Univ. Press. http://www.econlib.org/library/Smith/smMSCover.html. [18]

Yoshida, W., R. J. Dolan, and K. J. Friston. 2008. Game theory of mind. *PLoS Comp. Biol.* **4**:e1000254. [9]

Yoshida, W., B. Seymour, K. J. Friston, and R. J. Dolan. 2010. Neural mechanisms of belief inference during cooperative games. *J. Neurosci.* **30**:10,744–10,751. [9]

Yu, A. J., and P. Dayan. 2005. Uncertainty, neuromodulation, and attention. *Neuron* **46**:681–692. [9]

Yuille, A., and D. Kersten. 2006. Vision as Bayesian inference: Analysis by synthesis? *Trends Cogn. Sci.* **10**:301–308. [4, 15]

Zhang, W., and S. J. Luck. 2008. Discrete fixed-resolution representations in visual working memory. *Nature* **453**:233–235. [9]

Zhaoping, L. 2006. Theoretical understanding of the early visual processes by data compression and data selection. *Network* **17**:301–334. [9]

Zizzo, D. J., and A. J. Oswald. 2001. Are people willing to pay to reduce others' incomes? *Ann. Econ. Stat.* **63–64**:39–65. [18]

Subject Index

www.ingramcontent.com/pod-product-compliance
Lightning Source LLC
LaVergne TN
LVHW050926210926
848482LV00013B/546
9780262551502